Lecture Notes of the Institute for Computer Sciences, Social Informatics and Telecommunications Engineering   565

The LNICST series publishes ICST's conferences, symposia and workshops.

LNICST reports state-of-the-art results in areas related to the scope of the Institute. The type of material published includes

- Proceedings (published in time for the respective event)
- Other edited monographs (such as project reports or invited volumes)

LNICST topics span the following areas:

- General Computer Science
- E-Economy
- E-Medicine
- Knowledge Management
- Multimedia
- Operations, Management and Policy
- Social Informatics
- Systems

Anthony L. Brooks

Editor

ArtsIT, Interactivity and Game Creation

12th EAI International Conference, ArtsIT 2023
São Paulo, Brazil, November 27–29, 2023
Proceedings, Part II

 Springer

Editor
Anthony L. Brooks ⓘ
Aalborg University
Aalborg, Denmark

ISSN 1867-8211　　　　　　　　ISSN 1867-822X (electronic)
Lecture Notes of the Institute for Computer Sciences, Social Informatics
and Telecommunications Engineering
ISBN 978-3-031-55311-0　　　　ISBN 978-3-031-55312-7 (eBook)
https://doi.org/10.1007/978-3-031-55312-7

This Springer imprint is published by the registered company Springer Nature Switzerland AG
The registered company address is: Gewerbestrasse 11, 6330 Cham, Switzerland

Paper in this product is recyclable.

Preface

I am delighted to introduce the proceedings of the twelfth edition of the European Alliance for Innovation (EAI) International Conference on Arts and Technologies, Interactivity, and Game Creation (ArtsIT 2023). This conference brought together researchers, developers and practitioners around the world who are leveraging and developing topics associated with IT in the Arts.

The technical program of ArtsIT 2023 consisted of 40 full papers presented in face-to-face or remote online sessions. Special track sessions at the event matched the groupings in these proceedings, namely 'Exploring new frontiers in Music Therapy' – Special Track (4 papers); 'Network Dance & Technology' – Special Track (4 papers); and 'Computational Art and the Creative Process' – Special Track (6 papers). Other papers are from the main proceedings tracks, which are grouped herein under 'Alternative realities, immersion experiences, and arts-based research' (7 papers); 'Games' (4 papers); 'Interactive technologies, multimedia, and musical art' (8 papers); and 'Human at Centre' (7 papers).

Coordination with the conference managers was essential for the success of the conference and the whole team acknowledge and appreciate their constant support and guidance. From my side, it was also a great pleasure to work with such an excellent organizing committee team and I acknowledge their hard work in organizing and supporting the conference.

Anthony L. Brooks

Preface

Editorial

The 12th EAI International Conference: ArtsIT, Interactivity & Game Creation (ArtsIT 2023) was hosted November 27th – 29th at the Institute of Philosophy and Human Sciences – IFCH – Marielle Franco and Fausto Castilho Auditoriums, UNICAMP, the State University of Campinas, São Paulo, in Brazil.

The ArtsIT conference once again brought together a wide array of cross-/inter-/trans-disciplinary researchers, practitioners, artists, and academia to present and discuss the symbiosis between art (in its widest definition) and information technology applied in a variety of distinct fields. Since 2009 ArtsIT has been steered[1] towards becoming acknowledged as a high quality scholarly leading scientific forum for the dissemination of cutting-edge research results in the intersection between art, science, culture, performing arts, media, and technology. The role of artistic practice using digital media also serves as a tool for analysis and critical reflection on how technologies influence our lives, culture, and society. ArtsIT is therefore considered not only as a place to discuss technological progress but also a place to reflect on the impact of art and technology on sustainability, responsibility, and human dignity.

Campinas is a Brazilian city located in the southeast region of São Paulo State. It is situated a few kilometres from the Tropic of Capricorn, which, at about 23°27' S, passes through the region north of São Paulo, roughly marking the boundary between the tropical and temperate areas of South America. São Paulo (SP) is the first Brazilian city to become a metropolis without being a capital, exerting significant national influence, and, because of its elevation, it experiences a temperate climate. The city is home to several universities, including Unicamp University, which ranks in the top 15% of the best universities globally, and among the top 3 in Latin America, which as the State University of Campinas, in its 57 years of existence has achieved international recognition through its building and disseminating of knowledge and innovation.

ArtsIT 2023 was hosted under Unicamp by the Interdisciplinary Nucleus of Sound Studies (NICS), as part of its 40th anniversary. Formed in 1983, NICS' primary objective has been to research different manifestations of sound, in all its aspects and applications, as a source of informational, cognitive, and creative content. The nucleus was founded with a vocation for interdisciplinary research, addressing contemporary innovative themes that ongoingly are updated and include Analysis of Musical Practices and Sound Art Studies; Interactive Environments and Interfaces in Extended Music; Analysis, Synthesis, and Perception of Sound Scenes; Sound pollution; Analysis of sounds in Brazilian forests, phonetics of their fauna and native populations; electronic music; music therapy, and ethnomusicological records (amongst others). Such themes associated well with the hosting of ArtsIT as is evident by the scope of submissions call for texts – see at the web site https://artsit.eai-conferences.org/2023/ under URL menu item "For Authors/Calls".

---

[1] Since 2009 the (near) annual international conference ArtsIT has been steered by Anthony Lewis Brooks.

Further evidence of alignment between ArtsIT and NICS extended beyond the main track call with the offering of special tracks on eight specialist subjects. These were, namely, special track on:- *'Affective and Emphatic Physical and/or Digital Experiences to Human Body'* - chaired by Dr. Rachel Zuanon (Arts Institute and School of Civil Engineering, Architecture and Urban Design, State University of Campinas [UNICAMP]); *'Amplifying Creativity: Exploring Musical Interfaces – An Artistic and Historical Approach'* - chaired by Elena Partesotti (NICS Lab, Unicamp University, Brazil) and Marcelo Wanderley (Music Research – IDMIL – McGill University, Canada); *'Artificially Generated Content: Identification and Ownership Protection in the Age of AI'* – chaired by Paula Dornhofer Paro Costa (Dept. of Computer Engineering and Automation, School of Electrical and Computer Engineering, University of Campinas, Brazil); *'Computational Art and the Creative Process: Exploring Human-Computer Co-creation'* – co-chaired by Artemis Moroni (DISCF – CTI Renato Archer, Brazil), Jônatas Manzolli (IA/NICS – UNICAMP University, Brazil), Elena Partesotti (NICS – UNICAMP University, Brazil), and Manuel Falleiros (NICS – UNICAMP University, Brazil); *'Music Therapy, Technology, Embodied Cognition and Rehabilitation: Exploring New Frontiers'* – co-chaired by Elena Partesotti (NICS Lab, Unicamp University, Brazil) and Wendy Magee (Music Therapy Program, Temple University, USA); *'Expression in Human-Robot Interaction'* – co-chaired by Murillo Rehder Batista and Artemis Moroni (both from CTI Renato Archer, Brazil); *'Network Dance & Technology'* – chaired by Daniela Gatti (Corporal arts Department – Arts Institute, Unicamp University, Brazil); and a multi-language (English, Portuguese, Spanish or Italian text submission) offering of a *'Special Track on Musical Meanings in Ubimus/Significados Musicais em Ubimus'* – co-chaired by Damián Keller and Luzilei Aliel (both from NAP, Universidade Federal do Acre, Universidade Federal da Paraíba, Brazil) and Micael Antunes (Universidade de Campinas, Brazil). Each of these eight special tracks offered authors opportunities for distinct specialized submissions that were handled by luminary chairpersons alongside selected panels of experts in the subject field as track committee members. Each special track had its own web presence under ArtsIT 2023 with each URL site (page) detailing with abstracts/calls on the distinct speciality on offer.

Additionally, each chair and aligned special track technical program committee (TPC) were offered to conduct their own specialized physical session at the actual event where the chair and TPC could lead and manage the session as a self-contained 'sub-event'. Thus, offering new opportunities for young career academics in the specialized fields to meet pioneering leaders, luminaries in their field, and peers for networking and insightful debate and discussions at greater depth than in a more general session. Groupings of the special tracks is also reflected in these proceedings from the ArtsIT 2023 event. A posters session supported the main and special tracks.

The following text offers insight to the actual physical 'happenings' that were incorporated under ArtsIT 2023 to enhance the delegates' experiences.

Having a commitment to diversity and inclusivity, the local organizing team gave space at the venue to the native indigenous community whose representative Brazilian artisans realized handmade nametags for delegates. Such representatives, who have connections with UNICAMP University, were present during the first two days of the conference offering typical indigenous body painting, thus providing delegates a unique opportunity to engage with and appreciate their rich cultural heritage. This artistic choice aimed to contribute to a deeper and more inclusive intercultural dialogue, promoting an understanding and appreciation of the cultural roots that shape the richness and diversity of Brazilian society, as well as the political voice of indigenous communities in the pursuit of a more just and equitable society. An outcome from this understanding and appreciation was ever present each day in the form of the indigenous logo of ArtsIT 2023 – realized as a manifestation inspired by the intersection of art and technology, as shown in image two of the web page automatic 'slider' (see above link) and herein shared below as Fig. 1.

Fig. 1. Brazilian indigenous logo of ArtsIT 2023

On a special note, on the first day after the Technical Sessions, a virtual reality experience was shared at the CPV Hotel by Professor Ivani Santana, Artemis Moroni, Felipe Mammoli, and their team who presented 'GaiaSenses', showcasing audio-visual compositions created using data from planetary platforms (see https://www1.cti.gov.br/sites/default/files/jicc-2021-paper-7.pdf).

ArtsIT 2023 featured four luminary plenary keynote speakers who were each offered the opportunity to publish individually in this proceedings, however, none were forthcoming.

The first keynote talk, on the 27th of November opening morning, was by Wendy Magee who is professor under the Music Therapy Program at Temple University, USA. Magee is also on the Editorial Board of the Journal of Music Therapy, and board member of the International Association of Music and Medicine. The title of the keynote was *'Access, agency and aesthetics: developments music therapy and technology'*.

The second ArtsIT 2023 keynote talk was given on the afternoon of 27th of November by Marcelo Wanderley who is Professor of Music Technology at McGill University, in Montreal, Canada. Wanderley's research interests include the design and evaluation of digital musical instruments and the analysis of performer movements. He co-edited the electronic book "Trends in Gestural Control of Music" in 2000, co-authored the textbook "New Digital Musical Instruments: Control and Interaction Beyond the Keyboard" in 2006, and chaired the 2003 International Conference on New Interfaces for Musical Expression (NIME03). He is a member of the Computer Music Journal's Editorial Advisory Board and a senior member of the ACM and the IEEE. The title of the keynote was *'Five Decades of Computer Music Interfaces: from ICMC & CMJ to NIME'*.

The third ArtsIT 2023 keynote talk, given on the morning of 28th of November, was by Dr Frédéric Bevilacqua, Head of the Sound Music Movement Interaction team at IRCAM in Paris. Bevilacqua is co-founder of the International Conference on Movement and Computing and is part of the joint research lab on Science & Technology for Music and Sound (between IRCAM, CNRS, and Sorbonne Université), a collaboration focusing upon the modelling and design of interaction between human movement, sound, and the development of gesture-based digital musical instruments. The title of the keynote was *'Sound-Music-Movement Interaction: from listening to performing, from general public to musician, from solo to collective experiences'*.

The final ArtsIT 2023 keynote was given by Professor Anderson Rocha on the event's closing morning on the 29th of November. Rocha is full professor of Artificial Intelligence and Digital Forensics at the Institute of Computing, University of Campinas (Unicamp), Brazil. He is the Director of the Artificial Intelligence Lab (Recod.ai) and Institute Director for the 2019–2023 period. The title of Rocha's keynote was *'How to Live with Synthetic Realities: ChatGPT, Midjourney, Dall-E2, Stable Diffusion, and others'*.

More details on the keynotes and the speakers' profiles are available at the conference web site. Also available at the web site is the full program of delegates' presentations aligned to the texts within this proceedings.

Extra to the proceedings was an optional excursion for delegates to Sirius, the new Brazilian synchrotron light source hosted at the Brazilian Synchrotron Light Laboratory (LNLS) on 30th November – see https://lnls.cnpem.br/sirius-en/.

There were two best papers awarded at ArtsIT 2023 and heartfelt congratulations are shared in acknowledging the authors who each received a gratis registration to ArtsIT 2024 (UAE). These prize-winning papers are herein listed and are available for reading within this volume:

(1) *Art as an expanded field: the case of the r/place social experiment*. Authors: Marcela Jatene Cavalcante Botelho and Hosana Celeste Oliveira (both from The Federal University of Pará [UFPA])

(2) *Fostering Collaboration in Science: Designing an Exploratory Time Travel Visualization*. Authors: Bruno Azevedo (Centro ALGORITMI, EngageLab – University of Minho); Francisco Cunha (University of Minho); Pedro Branco (University of Minho, Dep. of Information Systems [DSI])

Overviewing the ArtsIT event is to once again emphasize that it has a rich history and conjoined is its growing community of global scholars who are acknowledged as leading within their respective fields.

Previous editions of ArtsIT have been hosted at the following locations: ArtsIT 2022 was hosted by The Centro de Investigação em Artes e Comunicação (CIAC) under The University of Algarve, in Faro, Portugal. In 2021, ArtsIT was targeted to be hosted as a hybrid event at the UNESCO Creative City of Media Arts Karlsruhe, Germany, and in Cyberspace, however, it had to be changed to a fully-fledged online conference due to the COVID-19/Coronavirus pandemic. ArtsIT 2020 was also hosted as a fully-fledged online conference due to the pandemic. In 2019, ArtsIT was hosted physically at CREATE (The Institute for Architecture, Design and Media Technology) under Aalborg University main campus, in Aalborg Denmark. In 2018 the University of Minho in Braga, Portugal hosted ArtsIT, and in 2017 The Technological Educational Institute of Crete in its capital city of Heraklion, Greece hosted the ArtsIT event. ArtsIT 2016 was again hosted at CREATE (The Institute for Architecture, Design and Media Technology) but this time at Aalborg University's Esbjerg campus, Denmark, where CREATE's foundational Medialogy education was originated and established. ArtsIT was not run in 2015. In 2014 the ArtsIT international conference was hosted in Turkey, at the historic Minerva Han, the Communications Centre of Sabancı University, located in Karaköy in the very heart of Istanbul city. ArtsIT 2013 was hosted under the Department of Informatics, Systems and Communication (DISCo), University of Milano-Bicocca in Milan, Italy and in 2011, ArtsIT was again hosted under CREATE at Aalborg University's Esbjerg city campus in Denmark. The inaugural ArtsIT was held in 2009 at Yi-Lan, Taiwan. Hyperlinks to all prior ArtsIT event web sites with details of hosts, committees, etc., are found on the latest ArtsIT front web page (scroll down to view on lower right side).

Acknowledgements

In closing this editorial text of the ArtsIT 2023 proceedings, acknowledgement is stated of the associated EAI team and especially the conference managers Veronika Kissova and Radka Vasileiadis. Distinctive thanks are of course sent to the local Brazilian team who worked so hard to organize the ArtsIT 2023 hosting, these being: local organizing chair Elena Partesotti (NICS Lab, Unicamp University, Brazil); general co-chairs, Manuel Falleiros (NICS, Unicamp University, Brazil) and Artemis Moroni (ICT Renato Archer, Brazil); technical program committee chairs, Marcelo Wanderley (Music Research – IDMIL – McGill University, Canada) and Wendy Magee (Music Therapy Program, Temple University, USA); and the many others as listed at the conference web site under committees. Recognition also to Dante Pezzin, Edelson Costantino and Professor Dr. José Fornari for their commitment to their roles in supporting and realizing the event. Additionally, Rafael Brandão and the COCEN team are acknowledged for their efforts in advertising the event. Gratitude to Dr. Josué Ramos from CTI Renato Archer for supporting this event, as well as to Murillo Batista and Cleide Elizeu da Silva for the extensive promotion on social media. Also, thanks to Bruno Azevedo for his support in the physical running of the event in sessions alongside the local organizing chair. Special gratitude is also extended to the Brazilian Synchrotron Light Laboratory (LNLS), especially Gustavo Moreno, who made the excursion to Sirius possible.

Behind the scenes are the 'technical crew' who are often hidden to attendees; thus, it is important to recognise their contribution and support. Heartfelt thanks therefore are extended to Dante Pezzin, Letícia Zima, Edelson Costantino, Ingryd Sousa, Jeovane Lima, Edson Pfützenreuter, Eduardo Goldenberg, Kimberly Oliveira, Caroline Okuyama, Vinícius Lima, Aden Moreira, Franco Simões, Ingryd Sousa, Guilherme Zanchetta, Ricardo Vieira Cioldin, Márcio Massamitsu Ota, Valério Freire Paiva, José Maria Otávio, - these the crew who worked tirelessly behind the scenes to ensure the seamless execution of the conference. Their expertise and dedication were instrumental in overcoming any technical challenges, contributing to the overall success of the event and hopefully no name is missed from this valuable involvement.

The invaluable and important contributions of ArtsIT 2023 student volunteers deserve special recognition due to their enthusiasm, commitment, and proactive involvement that significantly enhanced the conference atmosphere… ArtsIT ongoingly posits its mission statement to support students in their gaining positive experiences from attending such high quality international conferences as offered by the ArtsIT series of editions, where volunteering can influence, inspire and motivate becoming themselves authors and presenters that can be life-changing vocationally and otherwise…. whilst names are not herein listed, 'you know who you are': Thanks and good luck in your future endeavours, we hope to see you again at a future ArtsIT event.

For those whose poster, very short paper, and papers from the Ubimus Special Track couldn't be published due to EAI/Springer guidelines but were showcased during the event, the efforts are recognised herein alongside a mention of the importance of contributions to the conference's vibrant atmosphere: These were namely: Douglas Bazo de Castro for his Poster presentation; Camila Gonçalves, Rafael S. Oliveira, Audrey T. Tsunoda, Percy Nohamafî for their very short paper presentation; Ivan Simurra, Damián Keller, Celio Marcos and Marcello Messina for their presentation of the papers related to Ubimus' Special Track.

Finally, it is important to state that everyone involved with the growing ArtsIT community of scholars and its (near) annual hosting look forward to welcoming, if possible, all readers at one of its upcoming editions. Attendance as either an audience member, as a passive delegate attendee, or even (ideally) as an active future presenter at an ArtsIT event to share your own research within the scope of fields covered by the title *Arts and Technologies, Interactivity, and Game Creation*, which between November 13th to 15th 2024, in its 13th edition, will be hosted at New York University in Abu Dhabi, United Arab Emirates: see https://artsit.eai-conferences.org/2024/ - we all hope to meet and greet you there and meanwhile wish you good health, happiness, and well-being – until we meet again….

The ArtsIT 2023 team.

(Editorial text composed and edited by Anthony Brooks, with important contributions from: Elena Partesotti, Artemis Moroni, Manuel Falleiros, Bruno Azevedo, and Leticia Zima).

Organization

Steering Committee

Anthony Brooks   Aalborg University, Denmark
Imrich Chlamtac   Bruno Kessler Professor, University of Trento, Italy

Organizing Committee

General Chair

Elena Partesotti   Unicamp, Brazil

General Co-chairs

Manuel Falleiros   Unicamp, Brazil
Artemis Moroni   CTI Renato Archer, Brazil

TPC Chairs

Marcelo Wanderley   McGill University, Canada
Wendy Magee   Temple University, USA

TPC Co-chairs

Jin Hyun Kim   University of the Arts, Helsinki, Finland
Luca Turchet   Trento University, Italy

Sponsorship and Exhibits Chair

Dante Pezzin   Unicamp, Brazil

Local Chairs

Dante Pezzin   Unicamp, Brazil
Cleide Elizeu da Silva   CTI Renato Archer, Brazil

Publicity and Social Media Chair

Guilherme Zanchetta   Unicamp, Brazil

Publications Chair

José Eduardo Fornari Novo Junior   Unicamp, Brazil

Web Chairs

Rodolfo Luis Tonoli   Unicamp, Brazil
Gustavo Araújo Morais   Unicamp, Brazil

Posters and PhD Track Chair

Bruno Azevedo Minho   Universidade do Minho, Portugal

Panels Chairs

Elena Partesotti   Unicamp, Brazil
Manuel Falleiros   Unicamp, Brazil
Artemis Moroni   CTI Renato Archer, Brazil
Bruno Azevedo   Universidade do Minho, Portugal

Technical Program Committee

Adriano Claro Monteiro   Federal University of Goiás, Brazil
Alex Street Anglia   Ruskin University, UK
Alexandre Zamith Almeida   Unicamp, Brazil
Alfredo Raglio   University of Pavia, Italy
Anésio Azevedo Costa Neto   Federal Institute of Education, Science and
                                Technology of São Paulo, Brazil

Antonio Rodà   University of Padova, Italy
Artemis Moroni   CTI Renato Archer, Brazil
Atau Tanaka   Goldsmiths, University of London, UK
Bruno Azevedo   Universidade do Minho, Portugal
Camila Acosta Gonçalves   Certified Music Therapist, Brazil
Carlos Augusto Nóbrega   Federal University of Rio de Janeiro, Brazil
Carlos Mario Gómez Mejía   NAP, Federal University of Paraiba, Brazil
Claudia Núñez Pacheco   KTH Royal Institute of Technology, Sweden

Claudia Zanini	Universidade Federal de Goiás, Brazil
Damián Keller	Federal University of Acre, Brazil
Daniela Gatti	Unicamp, Brazil
David Gamella	International University of La Rioja, Spain
Eduardo Hebling	Unicamp, Brazil
Elena Partesotti	Unicamp, Brazil
Esteban Walter Gonzalez	Universidade Federal Fluminense, Brazil
Felipe Mammoli	CTI Renato Archer, Brazil
Flavio Soares Correa da Silva	University of São Paulo, Brazil
Francisco Z. de Oliveira	Unicamp, Brazil
Gabriele Trovato	Shibaura Institute of Technology, Japan
Gilberto Prado	University of São Paulo, Brazil
Giovanni De Poli	University of Padua, Italy
Hélio Azevedo	CTI Renato Archer, Brazil
Ivan Simurra	NAP, Federal University of Acre, Brazil
Ivani Santana	Federal University of Río de Janeiro, Brazil
Jan Schacher	University of the Arts Helsinki, Finland
José Eduardo Fornari Novo Junior	Unicamp, Brazil
Josué Ramos	CTI Renato Archer, Brazil
Luca Truchet	Trento University, Italy
Manuel Falleiros	Unicamp, Brazil
Marcelo Caetano	PRISM Laboratory, France
Marcello Lussana	Humboldt University, Germany
Marcello Messina	NAP, Southern Federal University, Russia
María Teresa Del Moral Marcos	Universidad Pontificia de Salamanca, AEMP, Spain
Marijke Groothuis	ArtEZ University of the Arts, Enschede, The Netherlands
Melissa Mercadal-Brotons	Escola Superior de Música de Catalunya, Spain
Murillo Batista	CTI Renato Archer, Brazil
Paula Costa	Unicamp, Brazil
Rachel Zuanon Dias	Unicamp, Brazil
Regis Rossi Alves Faria	University of São Paulo, Brazil
Ricardo Del Farra	Concordia University, Canada
Rodrigo Bonacin	CTI Renato Archer, Brazil
Rodolfo Luis Tonoli	Unicamp, Brazil
Tadeu Moraes Taffarello	Unicamp, Brazil
Tereza Raquel De Melo Alcantara Silva	Federal University of Goiás, Brazil
Tiago Fernandez Tavares	Insper, Brazil
Uwe Seifert	Universität zu Köln, Germany

Contents – Part II

Interactive Technologies, Multimedia, and Musical Art

Human at Centre

Contents – Part I

Computational Art and the Creative Process

Alternative Realities, Immersion Experiences, and Arts-Based Research

Marker-Based and Area-Target-Based User Tracking for Virtual Reconstruction of Cultural Heritage in Mixed Reality

Sophie Schauer[✉] and Jürgen Sieck

HTW University of Applied Sciences, 12459 Berlin, Germany
{sophie.schauer,juergen.sieck}@htw-berlin.de

Abstract. The focus of this paper is the potential of mixed reality (MR) technology for the creation of virtual reconstructions of tangible and intangible cultural heritage. It delves into the key components of immersive and engaging MR experiences. The paper provides insights into the development process of an MR application prototype that virtualises cultural heritage pieces. It covers concept design and implementation, offering a comprehensive understanding of the steps involved. Additionally, it discusses the advantages of tracking approaches for user localisation, specifically marker-based and area-target-based tracking, in the use case of cultural heritage reconstruction. In summary, this paper offers insights into the current state-of-the-art by analysing exemplary projects and the potential of MR technology for virtual reconstruction of cultural heritage sites. With the presentation of an MR prototype and the key development steps involved, the paper contributes to the understanding and utilisation of MR in preserving and presenting our shared cultural heritage.

Keywords: Digital Cultural Heritage · Extended Reality · Digitalisation

1 Introduction

The preservation of cultural heritage and historical information has always been reliant on archival documentation. However, traditional storage and access methods have limitations, like the risk of loss of or damage to physical documents and difficulties in interpreting their content accurately, that increase over time. Fortunately, digitalisation has emerged as a practical solution to these challenges, enabling easier storage, access, and preservation of important archival documents.

As technology advances, new opportunities arise to enhance our engagement with archival documentation. One such technology is mixed reality, which has the potential to revolutionise the way we interpret and interact with historical artefacts long-term. Through MR, we can transform archival documents into immersive experiences that offer deeper insights into historical events and cultural heritage.

A. L. Brooks (Ed.): ArtsIT 2023, LNICST 565, pp. 3–13, 2024.
https://doi.org/10.1007/978-3-031-55312-7_1

A successor project of the recent EU-funded project AURA[1] explored the digitalisation and MR visualisation of archival documentation from the Archivio Storico[2] of the Teatro del Maggio Musicale Fiorentino[3] in Italy. The project aimed to create an immersive and interactive experience of the historical scenography of Shakespeare's „Othello" tragedy, performed at the theatre in 1937.

This paper delves into the development of a digitalised 3D model and MR visualisation of archival documentation, emphasising the significance of reinterpretation in MR as a means to improve access and preservation of the information contained in these documents. It discusses the technology required to display the augmented scenography correctly on a theatre stage, considering the unique conditions of theatre halls and their suitability for presenting virtual objects in MR. Additionally, the paper will discuss a marker-based and area-target-based tracking approach and find advantages and disadvantages for this use case.

Finally, the paper concludes with a future outlook on the digitalisation of archives and the visualisation of archival documents in MR. By leveraging MR technology, we can engage with historical artefacts in immersive and interactive ways, fostering greater appreciation and understanding of our cultural heritage. The potential for MR to revolutionise access and interaction with archival documentation is immense, and further exploration and development of its applications in this field are crucial.

2 State of the Art

Digitalising and virtualising tangible as well as intangible cultural heritage has been of interest for several years. Different forms of applications that fall on the spectrum of reality and virtuality have explored the potentials of modern technology for the reinterpretation of culture and history and have opened up new ways of engaging users in creative and innovative ways.

On the spectrum of reality-virtuality, environments are categorised based on their level of immersion and interaction. This can range from the real world to virtual environments and everything in between [1]. In augmented reality (AR), just like mixed reality, digital information or models are inserted in the user's field of view to replace or add on top of the existing objects in the real world. This technology has been proven to be of high value for the handling and conservation of Cultural Heritage. In [2] eight trending topics for AR application in the context of Cultural Heritage were disclosed: 3D reconstruction of cultural artefacts, digital heritage, virtual museums, user experience, education, tourism, intangible cultural heritage, and gamification.

Numerous projects have specifically explored these topics in AR and MR. A significant benefit of the technologies is the ability to recreate and visualise ancient structures and artefacts that may no longer exist or are in remote locations. AR and MR technologies enable users to virtually walk through historical sites, interact with virtual exhibits, and experience cultural heritage in previously impossible ways. It enhances the learning

---

[1] https://aura-project.eu/en/.

[2] https://www.maggiofiorentino.com/archivio-storico/.

[3] https://www.maggiofiorentino.com.

and educational experiences for both students and tourists. Through interactive story-telling and virtual guides, users can delve into the rich history and context of cultural heritage, fostering a deeper understanding and appreciation of our past [3, 4].

To correctly position the augmented content, a form of marker tracking has to be implemented. Generally, it can be distinguished between marker-based and marker-less tracking. Marker-based applications rely on visual markers, also known as fiducial markers, like icons, images, QR codes, etc. combined with computer vision recognition algorithms that ensure correct identification. These markers can be recognised by the device's camera stream, allowing for position and orientation calculation relative to the camera or vice versa. In MR, the virtual elements are then rendered in front of or in relation to the marker [5].

Marker-less tracking, on the other hand, involves the recognition and tracking of key feature points within a predefined environment, also called an area target. Unlike marker-based systems, markerless tracking does not require the placement of physical markers. The virtual content can be rendered seamlessly in the environment. However, this form of tracking can require more computational power [5].

One example was the marker-less AR application, developed by Indrawan [6], where a gyroscope was utilised to demonstrate the position of Dewata Nawa Sanga, one of the Hindu gods. Users could learn and understand the properties of Dewata Nawa Sanga by using the gyroscope sensor to identify the deities' coordinates. It also provides informative 3D animations about Dewata Nawa Sanga. After evaluating the application's usefulness, functionality, and impact on users' motivation. 84.8% of the participants found the application very useful and were highly satisfied with its use [6].

In the following, a concept design for an MR application will be described featuring the 3D digitalisation of an archival document, user interface conventions in MR and interaction systems for immersive applications.

3   Concept Design

Upon evaluating the prevailing trends in digitalisation and virtual reconstruction of cultural heritage, the prerequisites for an innovative application were found, and a design concept for an MR application of an augmented scenography piece was formulated.

The application concept will be deployed on a HoloLens. The objective is to create an immersive and interactive experience that allows users to explore and gain knowledge about a specific cultural artefact, primarily focusing on historical scenography documents from the Archivio Storico of the Teatro del Maggio Musicale Fiorentino. In particular, the scenography of Shakespeare's "Othello" tragedy was to be modelled and virtualised in MR on the stage of an opera hall (Fig. 1).

To ensure precise projection of the 3D model on the stage, a form of user tracking is essential. As such, the application should incorporate advanced tracking technology to accurately map the position and orientation of the artefact in real-time. This enables users to observe it from various angles and perspectives, enhancing the overall experience. Hereby two approaches are to be tested, the projection of content through the recognition of an area target and a marker. For area-target-based tracking, the whole room, including the stage, needs to be scanned and can be tracked by the application, whereas marker-based only facilitates one image, QR code or similar figures.

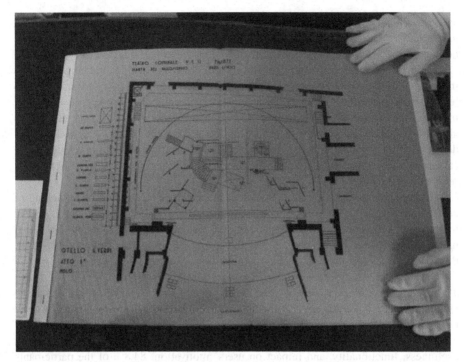

Fig. 1. Scenography document of a scene from "Othello"

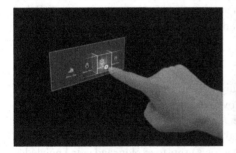

Fig. 2. UI and interaction examples for HoloLens [6]

Optimised for the HoloLens 2, the application should take advantage of its implemented hand gestures and offer basic interaction capabilities. Gesture recognition empowers users to engage with the artefact by manipulating it through intuitive hand gestures.

To maximise accessibility, the design should follow a user-centric approach, emphasising usability. The user interface (UI) should be intuitive, facilitating easy navigation and providing clear instructions, adhering to best-practice examples and UI conventions for MR (Fig. 2).

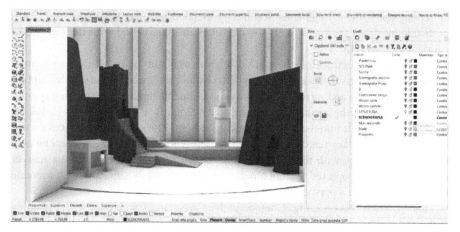

Fig. 3. 3D model of scenography in Rhinoceros

In summary, the application concept aims to deliver a distinct and captivating way for users to immerse themselves in and comprehend cultural heritage pieces. It is intended to offer a unique opportunity to learn about and engage with these cultural artefacts uniquely and memorably.

4   Implementation

Firstly, to implement the concept, the scenography drawings from the archive were detailedly documented. With the 3D modelling software Rhinoceros, a reproduction of the scenography was created (Fig. 3). The original documentation lacked specified information on used materials, colours, and textures. Through observation of photographs, these details were estimated. From the modelling software, the 3D object was exported as a.fbx file and easily imported into the project in the Unity engine, which would be used for further MR development.

During the digitalisation process, the development of the MR application was already started. Unity was chosen not only because of its broad documentation and compatibility with several frameworks but also for the fast and simple deployment process onto the HoloLens device and other MR and VR head-mounted devices, where only slight adjustments have to be made.

The MR functionality was added through the Mixed Reality Toolkit (MRTK), which holds essential features and resources for MR development. After setting the build platform to Universal Windows Platform, seamless deployment to HoloLens was ensured. The scene was adjusted by integrating configuration and components like MixedRealityToolkit, MixedRealityPlayspace, including the main camera, and MixedRealitySceneContent, which serve as the foundation for MR applications.

Following the initial concept design, special care was taken to ensure the UI elements adhered to established conventions and best practices for MR applications. A floating menu was introduced to facilitate intuitive navigation and offer additional interaction options. Employing a consistent visual language, including appropriate colours, fonts,

and iconography, the UI elements seamlessly blended into the MR experience without distracting from the main focus—exploration and interaction with the virtualised
historical scenography.

To implement user tracking, the tool Vuforia was used, through which the application
can utilise the built-in sensors and cameras of the HoloLens device. These sensors capture
the user's movements and surroundings, providing valuable input for the tracking system.

During runtime, the application continuously captures and analyses the HoloLens
sensor data, feeding it into the Vuforia tracking system. The system then processes this
data in real-time. The key feature points from the device's camera stream are continuously
analysed for matches with marker data from the Vuforia database. In case a marker
target is correctly recognised, virtual objects can be rendered dynamically with accurate
alignment to the user's perspective.

This approach was once implemented with an area target, where the hall, including
the theatre's stage, was captured. With the help of the Vuforia Area Target Creator app,
the hall of the Teatro della Pergola was scanned through a LiDAR scanner. The Teatro
della Pergola[4] was chosen because of its accessibility and fitting stage dimension, which
adhere to the original stage the "Othello" tragedy was performed on (Fig. 7).

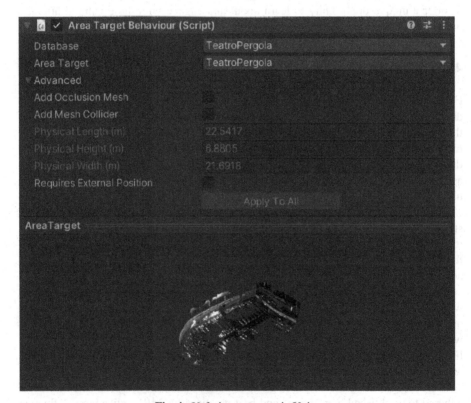

Fig. 4. Vuforia area target in Unity

---

[4] https://www.teatrodellapergola.com.

The area target could then be selected in Unity (Fig. 4), and an event handler implemented the functionality for displaying content once the target was found (Fig. 5). In case the target is no longer within the camera stream, the content will be disabled.

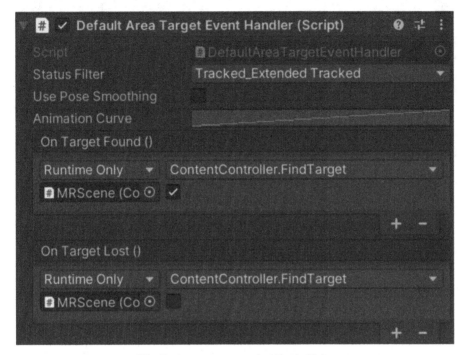

Fig. 5. Area target event handler in Unity

A second approach was implemented using an image marker which can be added by simply configuring an ImageTarget game object from the Vuforia library and inserting the desired image as a.jpg or.png file (Fig. 6).

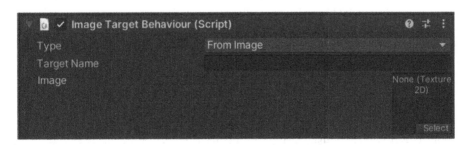

Fig. 6. Vuforia image target in Unity

The original scenography document was chosen as the image marker and uploaded to the Vuforia database. The augmentability was rated five stars out of five by Vuforia, making it very suitable for tracking. The key feature points of the image can be seen in the figure below.

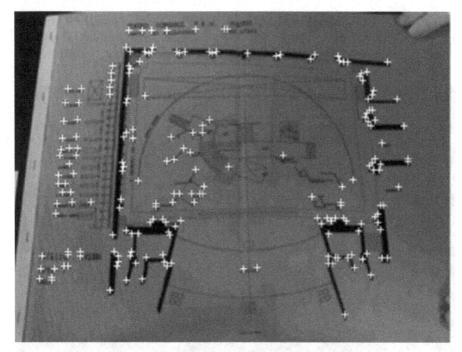

Fig. 7. Feature points of image target in Vuforia database

The resulting MR application visualised a reconstructed scenography from the Archivio Storico in Florence, Italy, enabling users to view and interact with it on a theatre stage (Fig. 8).

5 Discussion

Comparing the two approaches revealed advantages and disadvantages for either version. The advantages of marker-based tracking, through an image marker, were, among others, the reliable and robust tracking performances. Well-chosen markers with many key feature points serve as distinct visual references, allowing for precise alignment of the virtual content with the real world.

Furthermore, high accuracy in tracking the user's position and orientation can be achieved since the markers provide a fixed reference point, enabling accurate spatial registration of virtual objects within the physical space.

Moreover, marker-based tracking is scalable and implementable in various environments and scenarios, making it easy to embed in cultural heritage sites for both small-scale and large-scale applications. In this use case, the marker could simply be placed on the stage or moved to a different space if needed.

However, a disadvantage that became apparent was the overall dependency and presence of markers in the environment. This can limit the freedom of movement for users and may not be suitable for scenarios where markers are not feasible or desirable, such as in outdoor or dynamic environments. This also includes occlusion errors that

Fig. 8. Scenography model in MR application on the stage

occur when markers are partially or completely hidden from the camera's view. Users or objects obstructing the markers can cause temporary tracking disruptions, impacting the user experience.

Markerless tracking, through the use of area targets, provides users more freedom of movement and comes with a different range of benefits but weaknesses as well. Area-target-based methods can perform real-time mapping of the environment, allowing for dynamic interactions with the virtual content. This is particularly advantageous when exploring cultural heritage sites with complex architectural structures.

The tracking and integration of virtual content within the physical space work seamlessly, as it uses the environment itself as a reference, enhancing the sense of realism and immersion of the mixed reality application.

Some disadvantages of the area-target-based approach, however, are the environmental limitations, such as lighting changes or textureless surfaces. In challenging lighting or featureless environments, tracking accuracy may be compromised, which affects the alignment of virtual content. In very unsuitable conditions, the area might not be recognised at all. This can especially be the case in theatre halls that might only be partially lit or do not cover many unique feature points.

Additionally, area targets often involve computationally intensive processes, such as depth sensing, mapping, or point cloud registration. This complexity, also depending on the size of the area, may require more powerful hardware or computational resources, which can impact the overall performance and usability of the mixed reality application on devices like HoloLens. Since Vuforia has limitations on size and scanning time, large theatre halls might be too big to be scanned, making it impossible to use the entire halls as a marker.

Also, inaccurate mapping of the environment can lead to misalignments between virtual and physical objects. These errors can occur when capturing the environment or when handling dynamic scenes and can potentially impact the fidelity and accuracy of the virtual cultural heritage representation.

To sum up, the choice between marker-based and area-target-based tracking depends on the specific requirements of the use case, constraints within the environment, and characteristics of the cultural heritage application. Both approaches have their strengths and limitations, and the suitability of each method should be carefully considered based on the context and goals of the virtual reconstruction application.

6 Conclusion

In conclusion, digitalising and visualising cultural heritage in MR can bring exciting opportunities to explore and experience history. An important part of an MR application is the chosen form of user tracking to ensure accurate alignment and seamless integration of virtual content in the cultural space.

Marker-based tracking provides robustness and accuracy, relying on predefined markers as visual references. Correctly designed makers can lead to precise spatial registration, making it suitable for scenarios where high precision and detailed virtual reconstructions are necessary. On the other hand, markerless tracking utilises the natural features and structure of the environment and uses the whole area as a marker. This approach offers real-time mapping, enhancing dynamic interactions with virtual content and creating more authentic mixed reality experiences.

Both marker-based and area-target-based tracking approaches have their strengths and limitations. Ultimately, the use case and the formulated requirements will determine the better suitable technology.

In the future, improvements through further progress in sensor technologies and machine learning can be expected, leading to enhancements in tracking accuracy, robustness, and overall performance of cultural heritage MR applications and thus making the user experience more immersive.

As the field continues to evolve, researchers and stakeholders need to collaborate and explore innovative solutions for reconstructing and preserving cultural heritage. By leveraging the potential of MR, new dimensions of knowledge, engagement, and appreciation for the shared cultural heritage can be achieved.

References

1. Milgram, P., Kishino, F.: A taxonomy of mixed reality visual displays. IEICE Trans. Inf. Syst. E77-D(12), 1321–1329 (1994)
2. Boboc, R.G., Băutu, E., Gîrbacia, F., Popovici, N., Popovici, D.M.: Augmented reality in cultural heritage: an overview of the last decade of applications. Appl. Sci. 12(19), 9859 (2022)

3. Peteva, I., Denchev, S., Trenchev, I.: Mixed reality as new opportunity in the preservation and promotion of the cultural and historical heritage. In: EDULEARN19 Proceedings, pp. 2451–2456. IATED (2019)
4. Kolivand, H., El Rhalibi, A., Tajdini, M., Abdulazeez, S., Praiwattana, P.: Cultural heritage in marker-less augmented reality: a survey. IntechOpen(2019). https://doi.org/10.5772/intech open.80975
5. Costanza, E., Kunz, A., Fjeld, M.: Mixed Reality: A Survey, pp. 47–68. Springer, Berlin Heidelberg (2009)
6. Brondi, R., Carrozzino, M., Lorenzini, C., Tecchia, F.: Using mixed reality and natural interaction in cultural heritage applications. Informatica **40**(3) (2016)
7. Cre8ivepark, January 8. Mixed reality UX elements - Mixed Reality. Microsoft Learn (2021). https://learn.microsoft.com/en-us/windows/mixed-reality/design/app-patterns-landin gpage. Accessed 14 July 2023

Point-Based Stylization: An Interactive Rendering Approach for Non-uniform Artistic Representation

Yun-Chen Lee$^{(\boxtimes)}$ ⬤ and June-Hao Hou ⬤

Graduate Institute of Architecture, National Yang Ming Chiao Tung University,
Hsinchu, Taiwan
{yclee,jhou}@arch.nycu.edu.tw

Abstract. Inspired by real-world point-based art forms such as Impressionism, embroidery, and pebble mosaics, we have discovered the creative potential of arranging points with distinct appearances within a unified form of expression. We propose an approach that facilitates the creation of non-uniform, art-directed representations for point-based geometry, allowing for interactive rendering in 3D space. Rather than replacing all points with a set of static image textures as input, a dynamic representation of points must be designed in a controllable way that enhances visual richness and creative freedom to the artist. We met this challenge by defining each point as multi-dimensional spatial data and replacing it with parameterized textured proxy geometry, resulting in varied painterly strokes. Notably, the textured proxy geometry remains interactive, empowering artists to dynamically modify the strokes after placement. To aid the creative process, we introduce a sketch-based authoring tool that allows artists to compose various styles by sketching a few curves in 3D space. We provide examples generated by our prototype in an oil pastel painting style to demonstrate its ability to create non-uniform stylizations for point-based geometry.

Keywords: Non-Photorealistic Rendering (NPR) · Point Cloud · Sketch-Based Interaction · Parametric Proxy Geometry

1 Introduction

Points are the basic geometric elements in three-dimensional space. Well-known for their simplicity, flexibility, and compactness, point-based geometry employs discrete points to represent geometric models. Point-based geometry can be obtained using 3D scanning like LiDAR sensors and photogrammetry or by sampling from polygonal mesh models. This has become an affordable and feasible source for design and art creation in recent years. The flexibility and efficiency of points enhance their potential applicability across various creative fields such as animation, data visualization, VR, architecture, and art. As evidenced by works such as "Ghost Cell" by Antoine Delach (2015) [8], "Virtual Depictions: San Francisco" by Refik Anadol with Kilroy Realty Corporation (2015) [26], "In The

A. L. Brooks (Ed.): ArtsIT 2023, LNICST 565, pp. 14–28, 2024.
https://doi.org/10.1007/978-3-031-55312-7_2

Eyes of The Animal" by Marshmallow Laser Feast collective (MLF) (2016) [12], a colored pencil painting technique by Hin Sun Lee (2022) [6], "What Homes Are Made Of" by Lucija Ivsic (2021) [2], etc., points can be transformed into different representations, ranging from realistic scenes to science fiction or non-photorealistic styles.

2   Research Context and Goal

Observing various forms of expression in the real world, we can identify numerous cases that can be described from a point-based perspective. This type of point-based expression is prevalent across different art and design fields, including painting, crafts, and architecture. We studied examples from styles, such as Impressionism, Pointillism, metal texturing, embroidery, torn paper art, mosaic tiles, and pebble floors (See Fig. 1). We found that arranging diverse elements in specific patterns on the canvas or space can yield a wide spectrum of visual variations. For instance, in Impressionism, the character of the works lies in the composition of brushstrokes. Artists manipulate various factors, such as color, position, size, and shape, to arrange the strokes, resulting in an image that conveys rich information through the points. How artists manage these elements directly influences the representation of points, reflecting the artist's personality and painting style.

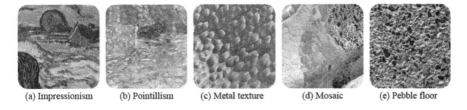

(a) Impressionism   (b) Pointillism   (c) Metal texture   (d) Mosaic   (e) Pebble floor

Fig. 1. Point-based expressions in real world. Detail from (a) Two Peasant Women, Vincent van Gogh, 1889 (b) Bridge in London, Jan Toorop, 1889 (d) Park Guell, Antoni Gaudi. photoed by tiburi.

Despite the vast potential of point-based representation, exploration of art-directed rendering for points in the digital environment remains relatively under-researched. Existing studies on point-based geometry rendering mainly focus on geometric representations of point clouds, like point set surface representation [1,16] and point cloud visualization [21,22]. However, these studies pay less attention to the non-uniform stylization and artistic expression achievable through art-directed parameterization and interactive data control. This study aims to fill the research gap by devising a novel method for non-photorealistic rendering of point-based geometry.

In this paper, we put forward a point-based stylization approach grounded on the concept of extended spatial data [17,25]. Spatial data is defined to store multi-dimensional information related to coordinate. We extend spatial data to temporarily store a set of dynamic parameters that produce parametric textured geometry, enabling the creation of images in an oil painting style with diverse stroke types. We concentrate on the process of creation rather than traditional rendering techniques or automated rendering processes. A prototype of sketch-based interactive technique is presented to continuously modify spatial data, leading to dynamic rendering results that allow the combination of different painting techniques. The proposed non-photorealistic rendering approach for point-based geometry is not confined to the oil pastel style, but can be utilized to define other artistic representations.

The following section discusses the research background of stroke-based non-photorealistic rendering and explores methods closely related to the concept of parametric proxy geometry. Section 4 describes the design concept and implementation of a point-based stylization system, presenting the dynamic creation process and discussing the limitations and next steps for the current system prototype. Finally, our work is concluded in Sect. 5.

3   Background and Related Work

In non-photorealistic rendering (NPR), points have often been depicted using strokes that encompass a wide variety of subjects, including but not limited to oil painting [10,15], sketching [19], colored pencils [28], mosaics [7], and plants [5]. Stroke-based NPR has been extensively explored across different rendering domains, such as images, models, and physically-based digital painting systems. Varied strokes can be achieved through image processing and texture-based rendering techniques [10,11], with the distribution of strokes being determined by properties of the input [29,30], including the colors and contours of the image, and the normal vectors and curvature of the model. These features enable the computer to generate effects that resemble human-made art.

Interactive NPR emphasizes on empowering users to modify rendering results via parameter adjustments, modifications to the input stroke texture, and alterations to the source images. The design of Interactive NPR should accommodate both the algorithm and user interaction, focusing on flexible and controllable parameter modifications that leverage the capabilities of both the computer and user creativity [13]. An image abstraction filter by Haeberli [10] allows for user-guided stroke direction using a second input image with a black and white gradient. Schwarz et al. [23] provide users control over stroke unit properties through an interactive canvas, diverse tools, and color palettes. Semmo et al. [24] facilitate manipulation of the generated stroke flow field using intuitive finger-based interactions. Chiew et al. [3], while not exclusively focused on strokes, afford users the ability to add effects such as smudging and painting through shading techniques applied to the original model. Compared to a fully automated process, an adjustable parameter field opens the door for modification or replacement of parameters, enabling artists to engage in an ongoing cycle of experimentation and refinement of visual effects. An essential consideration is the design of a specific toolkit that allows artists to access parameters in an intuitive and meaningful manner.

The image and texture-based methods mentioned above cannot be directly applied to point-based geometry for artistic representation. The former calculates strokes on the entire canvas, while the latter operates on individual strokes, resulting in fundamentally different approaches. However, NPR offers a viable solution through the utilization of proxy geometry, a technique generates new geometric strokes based on the original point coordinates. Proxy geometry can be created automatically from the model [4,14] or drawn by the artist in the digital painting environment [20,31]. The rendered proxy geometry may be controlled through algorithms and input images, and further modified based on the neighborhood relationships and attributes of the model [9,27]. This utilization of proxy geometry serves as the core of technology in our proposed system.

Traditional methods of altering appearance typically fall into two extremes: global parameters, e.g. filter-like adjustments, and pixel-level modifications, e.g. physical particle simulations. Our proposed method is situated between these extremes, recognizing that each point in the rendering carries unique information that influences the representation of points. This concept presents a challenge. First, we aim to preserve the ability to create unpredictability and complexity in the artwork through pixel-based techniques. Second, our system should allow users without artistic training to easily control and stylize points using a limited set of parameters. This approach seeks a balance between visual richness and a concise parameter set for our non-photorealistic rendered strokes, fusing the strengths of both methodologies. A potential solution to this problem involves utilizing spatial data as a parameter field, continuously modifying the proxy geometry parameters to guide the point representation. The parameter field must be interactively adjustable during the rendering process, a feature that may introduce additional complexity and unforeseen outcomes during iterative design and interaction.

4   Point-Based Stylization Approach

4.1   Concept and Design

The viewer's perception of texture in painting or drawing extends beyond the mere presence or absence of individual strokes. When multiple strokes combine, they collectively create an organized representation of texture [18]. These grouped strokes can effectively portray surface texture, tactile qualities, or other visual attributes of an object. Even if individual strokes are altered or removed, the group maintains a relatively consistent representation of texture. This is because the viewer's perception mainly depends on the collection and arrangement of the overall strokes. Such grouping of strokes in painting can yield richer visual effects and a sense of texture. Strokes that share similar appearances or are organized in a particular pattern are visually coherent, implying that they belong to the same ensemble.

In traditional NPR, most of the stroke placement is automated, lacking an design of the interaction that enables users to manipulate visual representation through point grouping. Taking inspiration from the real-world practice of sketching before painting, artists often simplify and deconstruct a scene's composition with preliminary lines on paper. Sketching assists artists in considering the fundamental elements of a composition and creatively recombining various colors and strokes. It serves as a tool to achieve stylized effects. Our proposed system includes a sketch-based authoring tool for point-based stylization with adjustment layers, allowing artists to utilize sketches to steer the stylization process (Fig. 2).

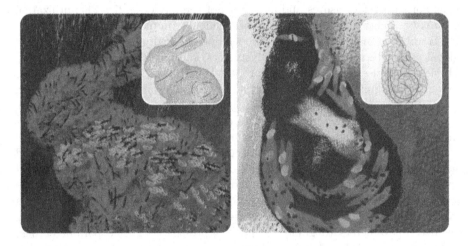

Fig. 2. Our technique's examples illustrate the concept of sketching before painting.

In an adjustment layer, points in three-dimensional space can be clustered by one or multiple sketched curves. The adjustment layer not only serves as a parameter buffer to modify the representation of the spatial data group, but also as a reference for the placement of textured proxy geometry in space, creating continuous variations within the cluster. The rendering pipeline begins with the input points model, where our system replaces each point with a textured proxy geometry, initially rendered in an oil pastel painting style. Adjustment layers follow the initial attribute setting and can be sequentially applied to any local point. Once all adjustment layers are in place, each spatial data receives a set of modifications from the original data, which is then used to generate textured proxy geometry.

Art-Directed Representation for Spatial Data The proxy geometry, which replaces the original point, can be modeled as different types of instances. Both the geometric properties and rendered texture of the proxy geometry can be customized dynamically at any time. Thus, art-directed controls are condensed into a concise set of key factors that can alter the appearance of proxy geometry rendering. Section 4.2 illustrates the generation of textured proxy geometry, created from a line or an arbitrary curve, and automatically produces a vector coordinate for forming the basic rendered stroke. By utilizing the vector coordinate, various effects, such as pressure, taper, smudging, edge effects, etc., can be achieved. All parameters that affect the rendering and geometric properties of the proxy geometry are stored individually in each spatial data, allowing automatic or manual modification.

Stylization by Sketch-Based Authoring Tool Each point-based geometry is defined as multi-dimensional spatial data. A collection of spatial data can be considered as a parameter field that stores information about the rendering process. This field allows parameters to be adjusted interactively at any stage of the rendering to create textured proxy geometry. The point-based stylization approach is achieved through the dynamic modification of spatial data, enabling variations in stroke shape, color, orientation, and painting effect in the generation and rendering of proxy geometry. Each adjustment layer comprises one or more curves sketched in three-dimensional space, clustering the points based on spatial relationships. The proposed sketch-based authoring tool promotes a nonlinear workflow, allowing interactive adjustments between sketching and painting, including the editing of curves and modifications to rendering parameters of textured proxy geometry, respectively (Fig. 3).

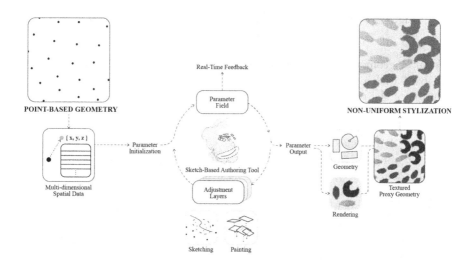

Fig. 3. Process of point-based stylization.

4.2   Create Textured Proxy Geometry from Spatial Data

Each spatial data holds the essential information required to create a stroke. Based on the information of each spatial data, we replace each point with proxy geometry that is rendered as a parametric stroke texture. Compared to a consistent representation for point-based geometry, this approach enables a more diverse and non-uniform appearance through textured proxy geometry. Such geometry must be adaptable to various sizes and shapes to produce coherent strokes. Our method automatically generates standardized two-dimensional vector coordinates, suitable for rendering proxy geometry with different source curves and geometric attributes. Parametric stroke is rendered using an alpha mask based on vector coordinates. We use the oil pastel painting style as an example to illustrate the process of proxy geometry generation and rendering. This fundamental method can be extended to other parameterized strokes, not limited to the oil pastel style (Fig. 2).

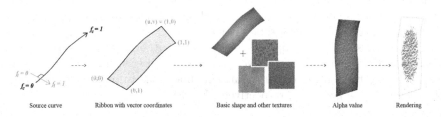

Fig. 4. Method to generate and render a textured proxy geometry.

A proxy geometry is derived from a source curve C and a perpendicular line primitive L with a length l. When L follows C, a ribbon of width l is created. The resolution of the ribbon can be adjusted based on the number of vertices in C, considering both the visual effect and computing performance. The parameters f_c and f_l are factors related to C and L respectively, with $0 \leq f_c, f_l \leq 1$. These factors are determined by the proportion of the curve's total length at each vertex. By applying these factors, we can derive vector coordinates $uv = (u, v)$, where u is defined by f_c and v by f_l. Generating standardized vector coordinates for proxy geometry from any given source curve C becomes straightforward. Moreover, the rectangular geometry, defined by texture coordinates uv with u,v values from 0 to 1, can be expanded to fit diverse stroke shapes. This flexibility allows artists to experiment with their preferred stroke styles, such as curly lines, short strokes, dots, circular strokes, and other types of curves.

Moving into the rendering phase, an alpha mask channel is necessary to define the stroke's shape, while a color channel sets its color. By remapping normalized coordinates to $uv + (-0.5, -0.5)$, we can generate circular gradient images. In addition, by adjusting the sampling position of the noise texture using random seed values assigned to each spatial data, we can overlay this noise texture onto the original circular image. This process generates diverse stroke patterns through alpha blending. Afterwards, the material's texture can be either layered on top of this alpha mask or blended with the color output.

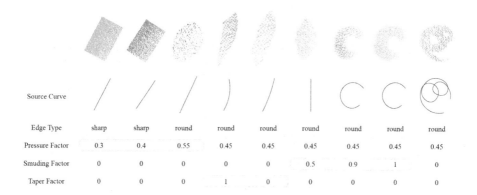

Source Curve									
Edge Type	sharp	sharp	round	round	round	round	round	round	round
Pressure Factor	0.3	0.4	0.55	0.45	0.45	0.45	0.45	0.45	0.45
Smuding Factor	0	0	0	0	0	0.5	0.9	1	0
Taper Factor	0	0	0	1	0	0	0	0	0

Fig. 5. Stroke variations.

The variability of painterly strokes is affected by a series of high-level and complex decisions made by the artists, such as brush pressure, painting motion, canvas texture, and more. We believe that several factors that contribute to the form of strokes can be condensed into a few key parameters. Each spatial data, capable of storing different data, can then be rendered with a distinct appearance. Using oil pastels as an illustrative case, Fig. 5 lists the parameters of the generated strokes using this method, showcasing the different styles that can be manipulated through these parameters. Accessible effects include pressure, smudge, and taper effects, with additional stroke textures generated by inputting paper texture.

4.3   Sketch-Based Authoring Tool for Point-Based Geometry

Based on the previous section, we realize that by transforming each source point into spatial data, we can create diverse representations. By individually modifying this data, we generate various textured proxy geometries, enhancing flexibility and control in the composition of different styles for point-based stylization.

Adjustment layers store a set of modifications written into a parameter buffer, which controls the style of the selected point groups. Multiple layers of adjustments can be applied sequentially, enabling changes to the parameters of the proxy geometry. This affords artists the flexibility to revisit and modify any adjustment layer at their convenience, altering both the parameters and the selection of point groups.

The adjustment layers use a sketched curve drawn in space, which allows batch modifications on selected neighboring point groups. The workflow involving adjustment layers is non-linear and consists of two parts: Sketching and Painting, allowing artists to freely switch between the two.

Sketching. This step involves drawing arbitrary 3D curves within space. Spatial data near these curves is selected based on geometric proximity, resulting in a subset of spatial data. This subset might contain points from one or several groups distributed throughout the model.

Painting. During the painting step, a collection of modified parameters is applied to the spatial data subset. The system offers four adjustment modes:

1. Stylization: Modifies the rendering results of textured proxy geometry.
2. Transformation: Alters the geometric properties, including rotations, translations, and scaling.
3. Duplication: Enables spatial data to be replicated, resulting in multiple proxy geometries representing a single point.
4. Density: Controls the probability of generating proxy geometry from the spatial data. By adjusting this parameter, the density of proxy geometries within the spatial data subset can be manipulated (Fig. 6).

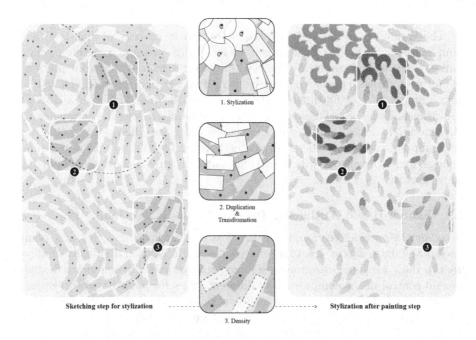

Fig. 6. Interactive workflow - sketching and painting.

The sketched curves and the rendering state of the proxy geometry are dynamic. This means that the curves can be edited during the sketching part, and the parameter buffer of the proxy geometry can be modified during the painting part. This interactive workflow empowers artists to create non-photorealistic styles for point-based geometry, encouraging constant experimentation and creative exploration with different textured proxy geometries (Fig. 7).

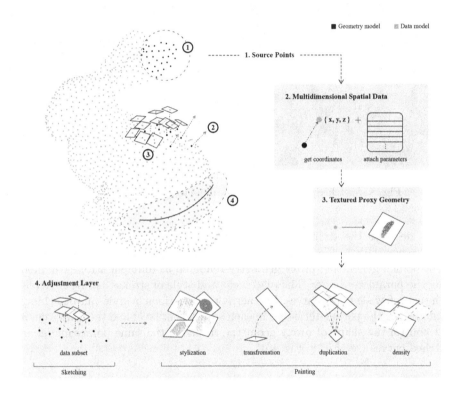

Fig. 7. System diagram of sketch-based authoring tool.

4.4   Dynamic Stylization Process

This section demonstrates the ability of our prototype in point-based stylization to create diverse non-realistic rendering results for point-based geometry, emphasizing the process of stylization and illustrating how different choices made during the transition between sketching and painting parts can significantly affect the final rendering. The stylization of the model is modified using different sketched curves and parameters, allowing for unique stylizations. For example in Fig. 8 reveals how managing the density of two elementary stroke types, horizontal and vertical, can lead to distinct rendering outcomes (Fig. 8).

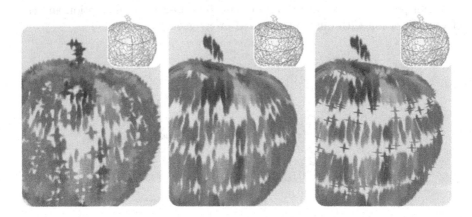

Fig. 8. Sketched curves determine the placement of the strokes.

We describe the stylization process through a series of adjustment layers added sequentially in Fig. 9. Each layer can employ sketched curves to select different points, generating proxy geometry rendered as different strokes by modifying the parameter buffer. The placement and style of strokes can be controlled with just a few sketched curves. Furthermore, variations in style can be achieved by modifying the parameter buffer in each adjustment layer and the source curves that control the shape of proxy geometry, all using the same points model and sketched curves (see Fig. 9, Fig. 10).

4.5   Discussion

This prototype was developed using Blender version 3.5, a functional and free open-source 3D software. Many designers and artists in animation, gaming, digital art, and design fields actively explore Blender plugins, which provide the possibility of integrating this technique into existing creative workflows. Additionally, the proposed point-based stylization approach utilizes multi-dimensional spatial data as parameter fields for textured proxy geometry. This approach proposes a fundamental concept and is not dependent on the techniques provided by Blender. It can be implemented in other development environments. Furthermore, by referring to the studies of spatial data structures and algorithms, and designing a specialized data structure, more efficient point selection and processing of larger, more complex points can be achieved.

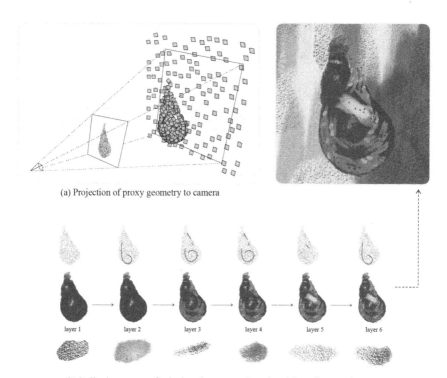

(a) Projection of proxy geometry to camera

(b) Stylization process of point-based geometry: Stepwise adding adjustment layers

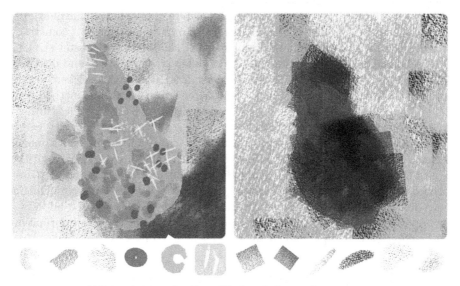

(c) Style variations produced by modifications to adjustment layer parameters

Fig. 9. Dynamic stylization process.

Fig. 10. Different strokes controled by same sketched curves.

5   Conclusions

In this paper, we introduce a system that establishes a non-linear workflow for creating point-based Non-Photorealistic Rendering (NPR) using spatial data. We emphasize the capacity of points, as a form of spatial data, to hold multi-dimensional information, facilitating the creation of diverse rendering styles in point-based geometry. In traditional non-photorealistic rendering of points, where strokes are not generated using parameterized textures, artists often rely on inputting texture images of real strokes. This makes it challenging to effectively create different painting styles for point-based geometry.

Our rendering approach contributes a method to transform the oil pastel painting style into parameterized textured proxy geometry, thus simplifying the creation of oil pastel strokes to a few parameters stored in spatial data. We implement adjustment layers that dynamically modify local proxy geometries, allowing flexible and controllable combinations of different oil pastel strokes. This technique enables the representation of different stroke styles and distributions through a few basic 3D curves. Unlike traditional approaches where artists can only control global styles through parameter adjustments, our method promotes the exploration of a variety of compound oil pastel painting styles through continuous experimentation.

The concept of characterizing variations in stroke appearance with multi-dimensional spatial data can be extended to other rendering styles beyond the oil pastel painting style. By bridging the fundamental observations of point-based data and representation methods with interactivity, we aim to highlight the potential for points to enable diverse expressions. This research addresses the importance of the information attached to points and how this information can be effectively manipulated to enhance visual outcomes.

Acknowledgements. This research is partially supported by National Science and Technology Council, Taiwan, under Grant number NSTC 112-2420-H-A49-002.

References

1. Amenta, N., Kil, Y.J.: Defining point-set surfaces. ACM Trans. Graph. (TOG) **23**(3), 264–270 (2004). https://doi.org/10.1145/1015706.1015713
2. Building 1 / What Homes are Made of: The Architecture of Displacement. https://currentsnewmedia.org/artist/lucija-ivsic/. (Accessed 6 Jul 2023)
3. Chiew, Y. X., Seah, H. S., Montesdeoca, S. E.: Real-Time Art-Directed Charcoal Cyber Arts. In 2018 International Conference on Cyberworlds (CW), pp. 90–95. IEEE (2018, October). https://doi.org/10.1109/CW.2018.00026
4. Cornish, D., Rowan, A., Luebke, D.: View-dependent particles for interactive non-photorealistic rendering. Graph. Interface **1**, 151–158 (2001)
5. Campos, C., Quirós, R., Huerta, J., Chover, M., Lluch, J., Vivó, R.: Non photorealistic rendering of plants and trees. In: International Conference on Augmented, Virtual Environments and Three-Dimensional Imaging, ICAV3D, Greece (June 2001)
6. Colored Pencil Painting NPR Rendering in Houdini Viewport with Photogrammetry Point Cloud. https://www.artstation.com/blogs/hinsunlee/1eXn/colored-pencil-painting-npr-rendering-in-houdini-viewport-with-photogrammetry-point-cloud. (Accessed 6 Jul 2023)
7. Di Blasi, G., Gallo, G.: Artificial mosaics. Vis. Comput. **21**, 373–383 (2005). https://doi.org/10.1007/s00371-005-0292-4
8. Ghost Cell. http://cargocollective.com/antoinedelach/GHOST-CELL. (Acessed 6 Jul 2023)
9. Grabli, S., Turquin, E., Durand, F., Sillion, F.X.: Programmable style for NPR line drawing. In: Rendering Techniques 2004 (Eurographics Symposium on Rendering). ACM Press (2004)
10. Haeberli, P.: Paint by numbers: abstract image representations. In: Proceedings of the 17th annual conference on Computer graphics and interactive techniques, pp. 207–214 (September 1990). https://doi.org/10.1145/97879.97902
11. Hertzmann, A.: Painterly rendering with curved brush strokes of multiple sizes. In: Proceedings of the 25th annual conference on Computer graphics and interactive techniques, pp. 453–460 (July 1998). https://doi.org/10.1145/280814.280951
12. In the Eyes of the Animal. http://intheeyesoftheanimal.com/. (Accessed 6 Jul 2023)
13. Isenberg, T.: Interactive NPAR: what type of tools should we create? In: Proceedings of Expressive, pp. 89–96 (May 2016)
14. Kaplan, M., Gooch, B., Cohen, E.: Interactive artistic rendering. In: Proceedings of the 1st international symposium on Non-photorealistic animation and rendering, pp. 67–74 (June 2000). https://doi.org/10.1145/340916.340925
15. Meier, B.J.: Painterly rendering for animation. In: Proceedings of the 23rd annual conference on Computer graphics and interactive techniques, pp. 477–484 (August 1996). https://doi.org/10.1145/237170.237288
16. Meyer, M.D., Georgel, P., Whitaker, R.T.: Robust particle systems for curvature dependent sampling of implicit surfaces. In: International Conference on Shape Modeling and Applications (SMI 2005), pp. 124–133. IEEE (June 2005). https://doi.org/10.1109/SMI.2005.41
17. Nikander, J., Korhonen, A., Valanto, E., Virrantaus, K.: Visualization of spatial data structures on different levels of abstraction. Electr. Notes Theoret. Comput. Sci. **178**, 89–99 (2007). https://doi.org/10.1016/j.entcs.2007.01.029

18. Philbrick, G., Kaplan, C. S.: Defining hatching in art. In: Proceedings of the 8th ACM/Eurographics Expressive Symposium on Computational Aesthetics and Sketch Based Interfaces and Modeling and Non-Photorealistic Animation and Rendering, pp. 111–121 (May 2019). https://doi.org/10.2312/exp.20191082
19. Praun, E., Hoppe, H., Webb, M., Finkelstein, A.: Real-time hatching. In: Proceedings of the 28th annual conference on Computer graphics and interactive techniques, pp. 581 (August 2001). https://doi.org/10.1145/383259.383328
20. Schmid, J., Senn, M. S., Gross, M., Sumner, R. W.: Overcoat: an implicit canvas for 3d painting. In: ACM SIGGRAPH 2011 papers, pp. 1–10 (2011). https://doi.org/10.1145/2010324.1964923
21. Schütz, M.: Potree: Rendering large point clouds in web browsers. Technische Universität Wien, Wiedeń (2016)
22. Schütz, M., Kerbl, B., Wimmer, M.: Rendering point clouds with compute shaders and vertex order optimization. Comput. Graph. Forum **40**(4), 115–126 (2021). https://doi.org/10.48550/arXiv.2104.07526
23. Schwarz, M., Isenberg, T., Mason, K., Carpendale, S.: Modeling with rendering primitives: an interactive non-photorealistic canvas. In: Proceedings of the 5th international symposium on Non-photorealistic animation and rendering, pp. 15–22 (August 2007). https://doi.org/10.1145/1274871.1274874
24. Semmo, A., Limberger, D., Kyprianidis, J.E., Döllner, J.: Image stylization by interactive oil paint filtering. Comput. Graph. **55**, 157–171 (2016). https://doi.org/10.1016/j.cag.2015.12.001
25. Shaikh, S., Matono, A., Kim, K.S.: A distance-window based real-time processing of spatial data streams. In: 2019 IEEE Fifth International Conference on Multimedia Big Data (BigMM), pp. 133–141. IEEE (September 2019). https://doi.org/10.1109/BigMM.2019.00-33
26. Virtual Depictions : San Francisco. https://refikanadol.com/works/virtual-depictions-san-francisco/. (Accessed 6 Jul 2023)
27. Wagner, R., Wegen, O., Limberger, D., Döllner, J., Trapp, M.: A Non-photorealistic rendering technique for art-directed hatching of 3D point clouds. In: VISIGRAPP (1: GRAPP), pp. 220–227 (2022). https://doi.org/10.5220/0010849500003124
28. Webb, M., Praun, E., Finkelstein, A., Hoppe, H.: Fine tone control in hardware hatching. In: Proceedings of the 2nd International Symposium on Non-Photorealistic Animation and Rendering, pp. 55-ff (June 2002). https://doi.org/10.1145/508530.508540
29. Yan, C.R., Chi, M.T., Lee, T.Y., Lin, W.C.: Stylized rendering using samples of a painted image. IEEE Trans. Visual Comput. Graph. **14**(2), 468–480 (2008). https://doi.org/10.1109/TVCG.2007.70440
30. Yang, C.K., Yang, H.L.: Realization of Seurat's pointillism via non-photorealistic rendering. Vis. Comput. **24**, 303–322 (2008). https://doi.org/10.1007/s00371-007-0183-y
31. Zheng, M., Milliez, A., Gross, M., Sumner, R.W.: Example-based brushes for coherent stylized renderings. In: Proceedings of the Symposium on Non-Photorealistic Animation and Rendering, pp. 1–10 (2017). https://doi.org/10.1145/3092919.3092929

Enhancing Geoscience Communication: Building Virtual Reality Field Trips with the Outcrop Digital Model at Varvito Geological Park of Itu (Brazil, SP)

Douglas B. de Castro[1]([✉]), Jefferson de Lima Picanço[1], Gabriel Santos da Mota[2], and Ítalo Sousa de Sena[3]

[1] Universidade Estadual de Campinas (UNICAMP), Campinas, Brazil
castrodouglasdev@gmail.com
[2] University of São Paulo, São Paulo, Brazil
[3] University College Dublin, Dublin, Ireland

Abstract. This article presents an innovative approach to Geosciences education through the development of a Digital Outcrop Model (DOM) combined with virtual reality. Fieldwork has long been a valuable didactic resource for teaching geological processes and rock outcrop analysis. In this study, a photogrammetric survey was conducted on the rhythmite in the *"Parque Geológico do Varvito,"* and the data was used to create a three-dimensional model. The model was optimized for real-time rendering and adapted to a virtual reality environment. Based on existing literature, a didactic itinerary was designed, allowing users to explore the rhythmite outcrop while receiving educational guidance from teachers on sedimentary formation. The goal is to encourage the use of digital models and gamification techniques to enhance Geosciences teaching, offering an immersive and engaging learning experience for students to understand complex geological processes. The immersiveness of virtual reality aligned with the rhythmite DOM demonstrated to be an engaging solution for Geosciences communication, by supporting virtual fieldworks and accessibility.

Keywords: virtual learning environment · virtual reality in education · photogrammetry · Paraná Basin

1 Introduction

Virtual reality (VR) technology is an advanced human-computer interface that has the capability to simulate realistic virtual environments [1]. It offers users the opportunity to navigate and interact within these environments from various angles and scenarios, providing a highly immersive and interactive experience [1]. VR technology can provide a significant contribution to education, applicable in both remote and in-person teaching settings. The concept of virtual geovisualization holds immense potential across geology,

A. L. Brooks (Ed.): ArtsIT 2023, LNICST 565, pp. 29–45, 2024.
https://doi.org/10.1007/978-3-031-55312-7_3

sparking discussions about its relevance in teaching geological aspects [2, 3]. The field work is crucial to geological knowledge, and the most meaningful approaches are related to the sense of vision [32].

An effective method to enhance didactic aspects within VR is through the implementation of virtual field trips [4]. By combining the visualization of rock outcrops with the immersive experience of virtual reality, students and the general public can engage in investigation and exploration during geological fieldwork. Collecting three-dimensional qualitative information from outcrops becomes essential in various geological segments, proving invaluable for teaching and illustrating geological processes [5].

Through virtual field trips, students can have an experience and understanding of geological phenomena without being physically present at the outcrop locations. This innovative approach enables educators to present complex geological concepts in an interactive and visually captivating manner. By merging technology and Geosciences education, virtual reality not only fosters active learning but also opens doors for more accessible and inclusive learning and communication opportunities.

Based on this context, this work presents a methodological approach for virtual field trips using principles of photogrammetry for the visualization of rock outcrops. The current approach uses images taken by unmanned aerial vehicles (UAV) to create optimized meshes through simplification and retopology. The *Parque Geológico do Varvito*, located in the city of Itu, in São Paulo State, Brazil, was chosen as a use case for the creation of the digital outcrop model (DOM), which would be visited during a virtual field trip.

So that the DOM could be rendered in real-time rendering, two applications were created using C++ language for users to interact with the model: one for Android, designed for VR, and the other for the Windows Desktop version.

Exploratory tests were conducted during the Educational event 'Unicamp Portas Abertas (UPA) 2022,' organized by Universidade Estadual de Campinas, to collect feedback from potential users. UPA is an ideal place for testing this program, as it showcases scientific research developed at the university for elementary and high school students. The tests were carried out in an exploratory, rather than systematic, manner. In future research endeavors, you will have the opportunity to design experiments with a more targeted focus on addressing specific questions, rather than prioritizing application development. The feedback from users was carried out in an exploratory, rather than systematic, manner. For future research, it will be advantageous to conduct experiments with a more precise focus on addressing specific issues, instead of giving priority to application development.

2 Related Work

2.1 Digital Outcrop Model and Virtual Field Work

The outcrops can be scanned and represented by 3D digital models generated from photogrammetric or laser scanning methods [6]. These digitally-generated models are known as Digital Outcrop Models or DOM [6]. The possibility of having the outcrop inside the classroom allows a great optimization of time and money, in addition, it can bring more educational support for vulnerable students. Digital outcrop models can be generated based on photogrammetry surveys, optimized to be visualized in real-time on Android or Windows applications.

In photogrammetric surveys, a significant collection of photographs is employed, which can be acquired using an airborne platform such as a drone or directly with a surface camera (short-range photogrammetry) [7]. Both methods generate a point cloud consisting of 3D spatially positioned points, which must be aligned within the same coordinate system.

Subsequently, the DOM is generated through the 3D interpolation of the point cloud data. This process involves reconstructing the surface and features of the outcrop by connecting the points and creating a continuous representation of the geological features.

By utilizing photogrammetric techniques, the DOM provides a detailed and accurate digital representation of the outcrop, enabling geoscientists and educators to explore and study geological formations in a virtual environment. This approach facilitates the integration of fieldwork data into educational settings, offering an immersive and accessible learning experience for students [4].

The 3D interpolation of the collected points can be generated using a regular network of points or an irregular triangular network (TIN - Triangulated Irregular Network) [8]. A regular lattice of points is a continuously and regularly connected triangular network [8]. Triangular networks are commonly used in mesh generation for representing surfaces, terrains, and geological formations [9]. According to Li [10], the mesh texture can be projected onto the digital outcrop model through five different orthogonal projection mapping methods. Orthogonal projection is the figure formed in a two-dimensional plane, coming from points projected in space [11].

The digital model created by the photogrammetry technique can be inserted into virtual reality software developed inside a graphics engine for games [12]. A graphics engine for games was used as a tool to develop software for interacting with digital outcrops in a much freer and more versatile way compared to web viewers, its documentation was consulted as support during the development.

Virtual reality can be used to simulate different scenarios and also reproduce the physical environment. The generation of an immersive Virtual Field Trip (VFT) can bring a more expository context to the classroom [13], Silva [14] explains that a more expository approach to teaching can enhance the understanding of theoretical content presented in the classroom.

Expository fieldwork has been essential for teaching and forming geological knowledge, and the practical activity is based on the contact with the object of study, the outcrop [15]. Walking through geological time and the formation of an outcrop in VFT can require imagination, the use of an investigative look at details can give important aspects for the interpretation of the outcrop [16]. The VFT can give a simplified path and expand the student's cognitive possibilities using paleoenvironment simulations [17].

2.2 The Didactical Importance of *"Parque Geológico do Varvito"*

The *Parque Geologico do Varvito* (PGV) is a municipal park created in 1995 in the area of a local former quarry. The *Parque Geologico do Varvito* is listed by the State of São Paulo Council for Historical, Archaeological, Artistic and Tourist Patrimony (CONDEPHAAT) as cultural and geological heritage. Since then, the *Parque Geológico do Varvito* has received about 500 thousand visitors, according to the Itu City Hall site [33].

The outcrop situated within the *PGV* displays distinct layers of siltstone and fine sandstone; those layers correspond to the geological dynamics associated with meltwater fluvial systems [21]. The presence of these layers provides valuable insights into the sedimentary processes that occurred in the past, specifically related to the flow of meltwater within fluvial (river-related) systems. The clarity of these layers contributes to a comprehensive understanding of the geological history and environmental conditions in the region.

The term "varvite" refers to a specific type of sedimentary rock formation, while "rhythmite" represents a geological phenomenon characterized by the repeated alternation of distinct layers, often involving sandstone and shale, in the context of varvites and rhythmites [19]. Initially classified as varvite, the rhythmites found in Itu have been redefined as clayey shale rhythmites, these rhythmites consist of alternating layers of clayey shale and finer sandstones to siltstones, often containing fallen clasts [20]. Notably, these layers exhibit a distinct parallel plane lamination pattern, this reclassification reflects a refined understanding of the Itu's region geological characteristics and composition of the rhythmites [20].

Mendes [22] described the seasonality of these rhythmites as an outcome of glacial influence on the sedimentation process. According to this interpretation, during the summer melting period, the resulting water, being denser, would flow into the lake, giving rise to currents carrying turbidity. These turbidity currents would be concentrated near the lake bottom. This seasonal pattern contributes to the distinct layering observed in the rhythmites, as they represent alternating deposits formed during different climatic conditions.

This dynamic transports and deposits coarser fractions to the bottom and subsequently decants finer fractions in winter [23]. In the varvite of Itu, the repetition has pairs of strata (>1 cm) or layers (<1 cm) of light and dark colors, where the light layers correspond to summer and dark layers to winter [24]. The existing layers of sediment on the varvite outcrop are clearly depicted and sufficiently detailed to facilitate an informative explanation regarding sedimentological processes and the formation of rhythmites.

2.3 Education and Expository Didactic Teaching

To create a didactical experience, we used the thoughts of the brazilian education philosopher Paulo Freire, who aimed to offer meaningful training to students [34]. Paulo Freire emphasized the importance of engaging students in a way that resonated with their lived experiences and fostered critical thinking, his approach sought to empower learners by facilitating active participation, dialogue, and reflection, transcending traditional teaching methods [25].

Engaging students with a broad and diverse range of topics, perspectives, and methodologies not only expands their intellectual horizons but also nurtures adaptability and open-mindedness, it prepares them to navigate the complexities of an ever-changing world, fostering a deeper understanding of the interconnectedness of various subjects and realities [26].

Expository classes serve to complement the classroom theory, but effective teacher mediation remains vital to manage the student-knowledge relationship [14]. Gillings [17] suggests that Virtual Reality can be employed to enhance expository classes. Brandão [27] contends that formal education constitutes the phase in which educational knowledge is imparted in alignment with pedagogical principles. This process engenders conducive environments for learning to thrive, resulting in the formulation of methodologies rooted in the student-teacher dynamic.

Non-formal education pursues its objectives through an interactive process where the approach centers around achieving its goals through a dynamic interchange of information and training [28]. Non-formal education often occurs through interpersonal relationships and the exchange of knowledge, leading to a reversal in the traditional flow of information [29].

In summary, formal education takes place within structured academic settings with predefined curricula, while non-formal education revolves around learning during the process of social interaction and engagement [30]. The two educational models are not designed to replace each other but are intended to complement each other in the student's overall learning process [31].

Compiani & Carneiro [18] discuss the application of fieldwork for didactic purposes within geology encompasses a comprehensive classification of field activities based on various parameters. Based on these studies, the geological field trips could be classified as illustrative, inductive, motivational, formative and investigative. The main objectives of the fieldwork activities encompasses subjects as modes of learning, teacher-student relationships, the questioning about the actual scientific models and the logic of the learning process.

3 Methodology

3.1 Virtual Field Trip Creation

This section outlines the primary steps taken to develop and present the immersive virtual field trip at the Parque Geológico do Varvito. The methodological steps can be delineated across three primary stages: first, conducting a survey using photogrammetry techniques; second, processing the DOM and preparing the virtual field trip; and third, conducting tests with participants.

A fieldwork was made on *Parque Geológico do Varvito*, where a UAV captured overlapped images to execute the photogrammetry workflow after. The pictures were processed to create a tridimensional digital mesh, after this the digital model is optimized in order to produce a real-time render. The digital outcrop model of this article was generated using photogrammetry methods and later it was optimized to be rendered in real-time inside an Android and Windows application.

Using the graphical engine Unreal Engine, two softwares were developed using nodes and C++ programming language in order to create a virtual field trip in the *Parque Geologico do Varvito* and a simulated paleoenvironment of a glacial lake environment, one software was coded into a virtual reality using an android system for Oculus Quest 2. The other software developed inside the Unreal Engine was built for Windows Desktop.

A series of steps are outlined to provide further details regarding the methodological advancements of the research:

Fieldwork: A fieldwork expedition was conducted at *Parque Geológico do Varvito*, capturing overlapping images using a drone. These images were then used as input for the photogrammetry workflow.

Photogrammetry Workflow: The captured images were processed using photogrammetry techniques to generate a three-dimensional digital mesh. This digital model represented the geological features of the location.

Optimization: The digital model obtained from photogrammetry was optimized to ensure real-time rendering capability. This optimization enhanced the efficiency and performance of the virtual experience.

Software Development: Two separate software applications were developed using the Unreal Engine. These applications were created using a combination of node-based and C++ programming language. One software focused on creating a virtual field trip experience in *Parque Geológico do Varvito*, while the other simulated a paleoenvironment of a glacial lake.

Platform Compatibility: The software designed for virtual reality was tailored to run on an Android system specifically for the Oculus Quest 2 headset, providing an immersive VR experience. The other software developed within the Unreal Engine was intended for Windows Desktop use.

3.2 Itinerary for Geosciences Teaching

To enhance the educational value of the virtual field trip and promote effective learning in the sedimentology of Varvito, a well-structured itinerary should be developed. It guides users through key geological concepts and processes associated with Varvito formation.

3.3 Exploratory Tests for User Feedback

Conduct user testing sessions with students using the virtual field trip software. After the experience, collect feedback from users regarding the educational value of the simulated fieldwork. Utilize a rating system and a questionnaire to assess the helpfulness of the virtual field trip in enhancing their understanding of Geosciences, particularly sedimentology concepts related to Varvito.

By following these steps, the study successfully achieved its objective of creating an immersive and interactive virtual field trip experience, allowing users to explore the geological aspects of Parque Geológico do Varvito and the simulated paleoenvironment of the glacial lake.

3.4 Fieldwork to Apply the Photogrammetry Technique

The geological park *"Parque Geológico do Varvito"* is located in the city of Itu, São Paulo, Brazil, at latitude coordinates 23° 16′ 4″ S and longitude 47° 19′ 13″ W, approximately 60 km from Campinas.

The data collection was made in the outcrop area of the park to create a detailed digital outcrop model (DOM) of the park's geological formations using the photogrammetry methods. To capture the necessary images, a DJI Mavic Air unmanned aerial vehicle was used, piloted by Dr. Henrique Candido de Oliveira (see Fig. 1). In total, 238 sequential photos were taken, capturing the outcrop from frontal or oblique angles, while an additional 107 photos were captured in a vertical view.

Fig. 1. Doctor Henrique Candido de Oliveira piloting the aerial vehicle in the *Parque Geológico do Varvito.*

Once the image data was collected, it was imported into the Agisoft Metashape Professional software, version 1.8.2. Within the software, the standard photogrammetry processing steps were carried out. The Workstation machine used for this process was equipped with an Intel(R) Core(™) i7–9700 K processor, 80 GB of RAM memory, and an Nvidia GeForce RTX 2060 GPU. These powerful hardware specifications enabled efficient and accurate data processing during the photogrammetry workflow.

The mesh produced by the software contained a substantial number of faces and vertices, specifically 94,430,344 faces and 47,383,505 vertices (see Fig. 2). This level of complexity made the mesh significantly heavy and impractical to perform integrated into real-time rendering software. As a result, it became evident that the model required post-processing work to optimize its structure and reduce its complexity.

Optimizing the model was imperative to enhance its suitability for real-time rendering and other performance-critical applications. The post-processing procedures encompassed a meticulous reduction in the count of faces and vertices, meticulously safeguarding the intrinsic details and overall model accuracy. This strategic optimization endeavor yielded a mesh of diminished weight and enhanced manageability, facilitating seamless integration and swift rendering within real-time software ecosystems.

After applying automated simplification and manual retopology to the model, the newly generated optimized mesh displayed 200,000 faces and 104,817 vertices, representing a remarkable 99.78% reduction in complexity compared to the original mesh.

3.5 Programming the Applications

To build an interactive platform in both virtual reality and Windows desktop versions, the graphics engine used was Unreal Engine 5.0.3 from Epic Games. This powerful engine supported the creation of immersive experiences for both virtual reality and

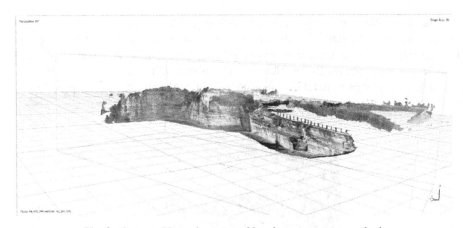

Fig. 2. Outcrop 3D mesh generated by photogrammetry method.

traditional desktop environments. To facilitate the integration with Android devices and specifically target the Oculus Meta Quest 2 hardware, an Android mobile application was made. This Android application would eliminate the need for a computer, once the Oculus Meta Quest 2 already has an internal system to run android applications.

To develop a captivating fieldwork experience through virtual reality, gameplay mechanics for movement and visualization inside the Oculus Quest 2 were made. This was achieved by leveraging the power and versatility of the C++ programming language.

C++ is widely used in game development due to its efficiency and ability to directly access hardware resources, which is crucial for providing a seamless and immersive experience in virtual reality. By utilizing C++, the development could optimize performance and ensure smooth movement and visualization within the virtual environment. They programmed various interactions, such as walking, running, and exploring the fieldwork area, to mimic real-world movements as closely as possible. Additionally, the team implemented visualizations of geological data and features, enabling users to examine and study the park's geological formations and structures in detail.

Through careful programming and leveraging the capabilities of C++, the team created a compelling and immersive fieldwork experience that allowed users to explore the Varvito Geological Park as if they were physically present, enhancing the overall engagement and educational value of the virtual reality application.

For educational purposes, a glacial lake environment was simulated for the users, as shown in Fig. 3. The simulation included several characteristic aspects of a glacial environment, enhancing the realism and educational value of the experience. Notable features added to the paleoenvironment simulation included striated rocks, fallen pebbles, a glacial lake, and glaciers. To create this virtual paleoenvironment, the development team utilized resources from quixel.com/bridge, a web source that provides three-dimensional models specifically designed for real-time rendering in graphics engines. These models were freely available, allowing the team to access a wide range of high-quality assets, thereby streamlining the development process and ensuring a visually stunning and authentic representation of the glacial lake environment.

By incorporating these realistic features and leveraging pre-made 3D models from quixel.com/bridge, the virtual simulation delivered an immersive and informative experience for teaching purposes. Users could explore and interact with the glacial lake environment illustrated on the Fig. 3, gaining insights into the geological processes and unique characteristics associated with glacial regions.

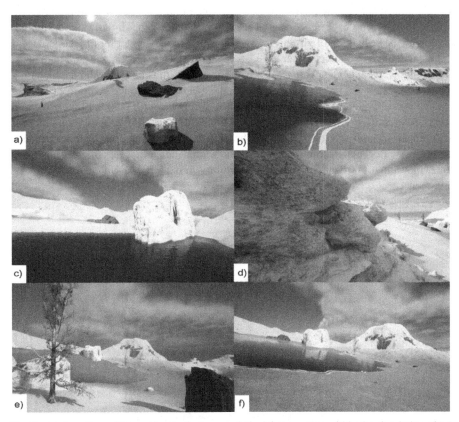

Fig. 3. (a) A section of the meticulously recreated glacial ecosystem within the simulation. (b) A replicated environment segment situated adjacent to the tranquil glacial lake. (c) A representation of a glacial lake adorned with towering glaciers. (d) A rocky landscape simulated with the intricacies of the snow formations. (e) A panoramic vista capturing the expansive surroundings, with the glacial grooves. (f) A simulated portion of the outcrop, recreated to showcase the closest proximity to the eroded region.

3.6   Exploratory Tests for User Feedback

To enhance the educational value of the virtual field trip and promote effective learning in the sedimentology of Varvito, a well-structured itinerary was developed. It aims to guide users through key geological concepts and processes associated with Varvito formation.

A didactic script was carefully developed to test the application's effectiveness in teaching about rhythmites and their formation. The primary goal was to explain the key features of the rhythm in the Itu park. The script was designed to provide users with a comprehensive understanding of the geological formations and processes involved in the geological formation.

After implementing the didactic script into the virtual reality application, a quiz was elaborated for the user. This quiz was designed to understand how immersive and helpful the learning experience was. The incorporation of the digital outcrop model, along with the utilization of virtual reality makes the subject matter more accessible and understandable for students and enthusiasts in the field. User feedback would help identify strengths, weaknesses, and areas for improvement, thus ensuring the application's educational value and promoting a deeper understanding of geological concepts.

The virtual fieldwork project was presented during the 17th edition of the Unicamp Portas Abertas (UPA) program held on 28/08/2022. The UPA program's main goal is to introduce high school students to the public university and assist them in making informed decisions about their future courses for the entrance exams. During UPA 2022, more than 500 visitors had the opportunity to explore the Unicamp Institute of Geosciences (IG) (see Fig. 4).

Fig. 4. User testing the software application during the UPA 2022.

A script was prepared for the visitors to the IG, offering two distinct experiences: one through virtual reality and the other using third-person playable desktop software. Both experiences were centered around the concept of the paleoenvironment and utilized the digital outcrop model. In the virtual reality experience, users could immerse themselves in the outcrop environment of *Parque Geologico do Varvito*. Through the use of VR technology, they were able to interact with the outcrop, examine rhythmite, and look at the fallen pebbles, all of which are characteristic features of the outcrop (see Fig. 5). This context offers an opportunity for a deeper understanding of geological concepts in an engaging and interactive manner.

Fig. 5. (a) Peripheral image of the *PGV* simulation. (b) Outcrop inside the virtual environment, with evident erosion features in the thinnest layers. (c) Enlarged outcrop wall, with darker and lighter layers represented in the texture. (d) Simulated portion of the outcrop where the fallen pebble is located. (e) A broader view of the DOM. (f) Simulated portion of the outcrop, eroded part closest view.

The second experience, in the form of a third-person playable desktop software, provided users with a unique perspective on the paleoenvironment. Players could navigate the environment and experience it from a different viewpoint, interacting with the virtual world and engaging with the rhythm physical samples provided by IG. The primary cause for the development of a desktop iteration was in tandem with the VR counterpart, leveraging its enhanced processing capabilities. This synergy facilitated the dynamic interplay between the map housing the digital outcrop model and the map incorporating paleoenvironmental data.

To assess the effectiveness of the VR experience in teaching sedimentological processes at the varvite outcrop, users participated in a questionnaire. This questionnaire focused on three primary questions, where 0 means a bad experience and 10 corresponds a great learning experience:

1. "From 0 to 10, how was your experience playing the game?" (see Fig. 6).
2. "From 0 to 10, how much did you understand about rhythmite?" (see Fig. 7).
3. "From 0 to 10, was the game important for understanding the local geology?" (see Fig. 8).

We have gathered valuable feedback and suggestions from users, which can be leveraged to enhance the model and its educational concept. These suggestions, alongside positive feedback, included requests for a higher level of immersion within the application and increased rendering details.

4 Results and Discussion

The development and optimization of the DOM, combined with the development of the applications utilizing a graphics engine, has proven to be effective to allow real-time rendering. Based on the rates assigned by users and qualitative feedback collected during the UPA 2022 event, the experience received highly positive responses as a didactic support for Geosciences. The users' feedback indicated that the project successfully enhanced their understanding of Varvito and local geology in an engaging and interactive manner.

A total of 49 randomly selected users tested the virtual field trip, participating during a drop-in event. The questions posed to the users primarily aimed to assess the level of immersion and the quality of their experience in terms of comprehending and interacting with the varvito digital outcrop model, particularly from the perspective of elementary and high school students. In response to the question "From 0 to 10, how was your experience playing the game?", an impressive average score of 9.4 was recorded. This high rating indicates the enthusiasm and positive reception among users when engaging with academic geology content through a technological and gamified approach. The gamification tool had the potential to incorporate elements of interactivity and engagement, aligning with the principles elucidated in Lima's framework for non-formal education [30].

In the question "From 0 to 10, how much did you understand about rhythmite?", the average score was 8.44. While this score slightly dropped compared to the user experience rating, it suggests that users might have faced challenges in fully grasping the content. This feedback is relevant in terms of improving the design of the learning activity. This could be attributed to the event's bustling environment, making it difficult to provide optimal didactic support.

However, in the final question "From 0 to 10, was the game important in understanding the local geology?", the average score increased again to 9.37. This rise in average scores could be attributed to the immersive nature of the software, enhancing users' cognitive ease and investigative engagement. The visual and interactive aspects of the environment seem to have tapped into a potential that not only illustrates theoretical concepts but also sparks users' curiosity and encourages exploration.

The standout aspect, according to user opinions, was the interplay between the outcrop and the paleoenvironment. This aspect of versatility can be correlated with the interactive elements of Virtual Reality, as elucidated by Zheng [1]. This dynamic helped users visualize and comprehend the intricate processes of rhythmite deposition and formation. Qualitative feedback from users also offers valuable suggestions for refining the model and enhancing its educational proposition. Overall, the study underscores the potential of gamification and immersive technology in enriching geology education and piquing students' interest and understanding.

From 0 to 10, how was your experience playing the game?

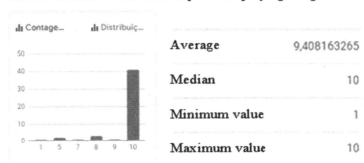

Fig. 6. Graph prepared according to feedback about the game experience from users who tested the application at UPA 2022.

From 0 to 10, was the game important for understanding the local geology?

Fig. 7. Graph prepared according to feedback about the geological understanding from users who tested the application at UPA 2022.

The numerical ratings provided by the users allowed for a quantifiable assessment of their experience, comprehension, and the game's educational impact from the user perspective. The scores from 0 to 10 indicated the level of satisfaction and understanding users felt during the virtual reality and third-person playable desktop software experiences.

From 0 to 10, how much did you understand about the rhythmite ?

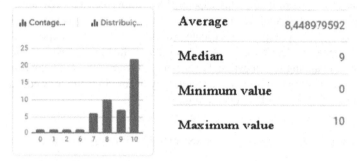

Average	8,448979592
Median	9
Minimum value	0
Maximum value	10

Fig. 8. Graph prepared according to feedback about the understanding of the outcrop from users who tested the application at UPA 2022.

Based on the gathered data, the majority of users awarded high grades, indicating a positive and enjoyable experience while playing the game. The interactive nature of the simulation likely contributed to this positive feedback. Additionally, users reported a significant increase in their understanding of outcrop and local geology, as evidenced by the high scores provided in response to the second question.

Moreover, the users recognized the game's importance in facilitating their understanding of local geology, demonstrating the educational value and effectiveness of the project. Overall, the positive grades and qualitative feedback collected from users at the UPA 2022 event support the initial objective to create a virtual field trip for teaching Geosciences. The interactive and immersive experiences provided by the virtual reality and desktop software left a strong and positive impression on the participants, fostering curiosity and knowledge about the field of Geosciences.

Based on the users opinions, the dynamics between the outcrop and the paleoenvironment emerged as the most significant highlight. This aspect played a crucial role in helping users visualize and comprehend the rhythmite deposition and formation process. Integrating the outcrop data with the virtual paleoenvironment allowed users to bridge the gap between real-world geological formations and their virtual representations. This approach facilitated a deeper understanding of the geological processes involved in rhythmite deposition and how they translate into the formation of the outcrop. This empirical endeavor has paved the way for the establishment of an exceptionally immersive pedagogical setting, aligning with the concepts explored by Klippel [4]. But now specifically tailored to the mechanisms governing the formation of sedimentary rhythmites.

This teaching model, centered around VFT, exhibits significant potential for enhancing both expository and interactive classroom sessions, ultimately leading to improved student engagement and learning outcomes, as posited by Silva [14]. As per Gohn [28], a significant facet of informal teaching lies in its emphasis on socialization and the heightened adaptability of its implementation. Leveraging cost-effective surveys and increased flexibility in its application, the highly motivating VCVi model holds promise for expanding the reach of geology education across diverse segments of society.

The VFT can fulfill several roles with regard to -field trip learning [34]. As far as this experiment is concerned, they are mostly illustrative and motivational, integrating the experience into the motivating field established by Compiani, Mauricio and Carneiro [18]. They are student-centered activities where knowledge comes through interactions with the object of study, aiming to stir up curiosity and induce knowledge. We understand that this is, at the moment, the main function of this type of activity. In the future, with the development of technology, other roles may be added.

The interactive nature of the experience further enriched the learning process. Users could explore the outcrop area and compare it with the virtual representation of the paleoenvironment, thereby gaining insights into the geological features and their spatial relationships. This interactive dynamic provided a more comprehensive perspective, fostering a holistic understanding of the subject matter.

5 Closing Remarks

By following the methodological steps, the study successfully achieved its objective of creating an immersive and interactive virtual field trip experience, allowing users to explore the geological aspects of *Parque Geológico do Varvito* and the simulated paleoenvironment of the glacial lake.

The "Varvito Digital" project at UPA 2022 successfully facilitated an educational and immersive learning experience for visitors. Through innovative technologies and interactive simulations, the project aimed to foster interest in Geosciences and encourage students to explore the field further in their academic pursuits.

The didactical results were highly encouraging, users with no prior knowledge of geology showed increased interest and curiosity after experiencing the simulation. The digital fieldwork cannot replace traditional field trips, but it can effectively supplement and extend the learning experience, allowing students to revisit and interact with outcrops that have been previously visited.

Furthermore, its accessibility and cost-effectiveness make it an excellent tool to reach social strata outside the academic environment. By reducing costs and increasing application flexibility, digital fieldwork has the potential to bring geology education to various layers of society, promoting interest and understanding in Geosciences. In conclusion, the experience proved to be a promising didactic tool for teaching Geosciences, providing an engaging and accessible approach for users to explore and understand geological concepts in the current context. By addressing the observed weaknesses and leveraging teacher feedback, virtual fieldwork can continue to evolve and become an even more effective and valuable resource for Geosciences education. For future research, the focus could shift towards more specific educational and didactic subjects, adopting a more systematic approach.

References

1. Zheng, J.M., Chan, K.W., Gibson, I.: Virtual reality. Potentials. IEEE. **17**, 20−23 (1998). https://doi.org/10.1109/45.666641

2. Queiroz, A.C.M., Nascimento, A.M., Tori, R., da Silva Leme, M.I.: Immersive virtual environments and learning assessments. In: Beck, D., et al. (eds.) iLRN 2019. CCIS, vol. 1044, pp. 172–181. Springer, Cham (2019). https://doi.org/10.1007/978-3-030-23089-0_13

3. Dykes, J., MacEachren, A., Kraak, M.J.: Introduction exploring geovisualization. Exploring Geovisualization **41** (2005). https://doi.org/10.1016/B978-008044531-1/50419-X

4. Klippel, A., et al.: Transforming earth science education through immersive experiences: delivering on a long held promise. J. Educ. Comput. Res. **57**. 073563311985402 (2019). https://doi.org/10.1177/0735633119854025

5. Fantinel, L.M.: O ensino de mapeamento geológico no Centro de Geologia Eschwege, Diamantina – MG: análise de três décadas de práticas de campo (1970–2000). Tese de Doutorado. Universidade Estadual de Campinas. http://repositorio.unicamp.br/handle/REPOSIP/287213 (2005). Acesso 15 Oct 2022

6. Ivan, F.P., Hodgetts, D., Redfern, J.: Integration of digital outcrop models (DOMs) and high-resolution sedimentology - workflow and implications for geological modeling: Oukaimeden sandstone formation, high atlas (Morocco). Petrol. Geosci. Petrol Geosci. **16**, 133–154 (2010). https://doi.org/10.1144/1354-079309-820

7. Nex, F., Rinaudo, F.: LiDAR or photogrammetry? integration is the answer. Italian J. Remote Sens. **43**, 107–121 (2011). https://doi.org/10.5721/ItJRS20114328

8. Jiang, Y., et al.: A comparative experimental study of rill erosion on loess soil and clay loam soil based on a digital close-range photogrammetry technology. Geomorphology **419**,108487 (2022). ISSN 0169–555X

9. Chang, Y.I., Liao, K.Y., Jiang, C.H.: Motion of particle breakthrough curve and permeability reduction in Voronoi and triangular networks. Sep. Purif. Technol. **114**, 38–42 (2014). ISSN 1383–5866, https://doi.org/10.1016/j.seppur.2013.04.006

10. Li, X., Chen, Z., Zhang, L., Jia, D.: Construction and accuracy test of a 3D model of non-metric camera images using agisoft photoscan. Procedia Environ. Sci. **36**, 184–190 (2016). https://doi.org/10.1016/j.proenv.2016.09.031

11. Dabrowski, D., Orponen, T., Villa, M.: Integrability of orthogonal projections, and applications to Furstenberg sets (2021)

12. Le Mouélic, S., et al.: Investigating lunar boulders at the apollo 17 landing site using photogrammetry and virtual reality. Remote Sens. **12**(11), 1900 (2020). https://doi.org/10.3390/rs12111900

13. Fleming, J., Schmidt, N., Cary-Kothera, L.: Visualizando o aumento do nível do mar para examinar o nexo entre mudança climática e segurança socioeconômica. Artigo apresentado no OCEANS 2016 MTS/IEEE Monterey, Monterey, CA, pp. 1–8. Piscataway, NJ. IEEE (2016). https://doi.org/10.1109/OCEANS.2016.77611

14. Silva, C.M.: Educação no ensino superior na contemporaneidade e as metodologias ativas. [Dissertação de Mestrado em Estudos Culturais Contemporâneos]: Universidade FUMEC. Faculdade de ciências humanas, sociais e da saúde. Belo Horizonte, MG, Brasil (2017)

15. Fantinel, L.M.: O ensino de mapeamento geológico no Centro de Geologia Eschwege, Diamantina – MG: análise de três décadas de práticas de campo (1970 - 2000). Tese de Doutorado. Universidade Estadual de Campinas. http://repositoriounicamp.br/handle/REPOSIP/287213 (2005). Acesso 15 Oct 2022

16. Carneiro, C.D.R.: Glaciação antiga no Brasil: parques geológicos do Varvito e da Rocha Moutonnée nos municípios de Itu e Salto, SP. Terrae Didática, [S. l.] (2016)

17. Gillings, M.: Virtual archaeologies and the hyper-real. In: Fisher, P.F., Unwin, D. (eds.) Virtual Reality in Geography, pp. 17–34. Taylor & Francis, London, England (2002)

18. Compiani, M., Carneiro, C.: Os papéis didáticos das excursões geológicas. Enseñanza de las ciencias de la tierra: Revista de la Asociación Española para la Enseñanza de las Ciencias de la Tierra **1**(2), 90–97. 1. 90 (1993)

19. Caetano-Chang, M., Ferreira, S.: Ritmitos de Itu: Petrografia e Considerações Paleodeposicionais. Geociências (São Paulo) **25** (2007)
20. Petri, S., et al.: Grupo Itararé na região de Itu, estado de São Paulo: intensos processos glaciais erosivos e deposicionais. Rev.do Instituto Geológico **40**, 27–48 (2019). https://doi.org/10.33958/revig.v40i3.674
21. Caetano-Chang, M.R., Ferreira, S.M.: Ritmitos de Itu: petrografia e considerações paleodeposicionais. Geociências **25**(3), 345–358 (2006)
22. Mendes, J.C.: Geologia dos arredores de Itu. São Paulo Boletim da Associação Geográfica Brasileira **4**(4), 31–40 (1944)
23. Amaral, S.E.: Nova ocorrência de roche moutonnée em Salto, Estado de São Paulo. São Paulo: Boletim da Sociedade Brasileira de Geologia **14**(1/2), 71–82 (1965)
24. Rocha-Campos, A.C. Varvito de Itú, S.P.: registro clássico da glaciação neopaleozóica. Sitios geológicos e paleontológicos do Brasil. Tradução . Brasília: DNPM (2002)
25. Souza, E., Barbosa, G., Souza, M., Santos, S.: Práticas pedagógicas e educação do campo: paulo freire. Rev. Ibero-Americana de Humanidades, Ciências e Educação. **7**, 1722–1730 (2021). https://doi.org/10.51891/rease.v7i12.3664
26. Almeida, H.M.A.: Didática no ensino superior: práticas e desafios. Rev. Estação Científica. Juiz de Fora, MG, Brasil: n. 14, julho – dezembro (2015)
27. Brandão, C.R.: O que é educação. 19. ed. São Paulo: Brasiliense (1985)
28. Gohn, M.G.: Educação não-formal, participação da sociedade civil e estruturas colegiadas nas escolas.. Rio de Janeiro **14**(50), 27–38, jan./mar (2006)
29. Chassot, A.: Alfabetização Científica: uma possibilidade para a inclusão social. Rev. Brasileira de Educação, 89–100 (2003)
30. Cascais, M., Augusto, F.T.: Educação formal, informal e não formal na educação em ciências. CIÊNCIA EM TELA. **7**, 1–10 (2016)
31. Pereira, W., Silva, J., Deyse, R.: Investigando as relações entre as práticas em espaços de educação não formal e formal. Rev. Cocar. **15**, 1–21 (2021)
32. Frodeman, R.L.: Envisioning the outcrop. J. Geosci. Educ. **44**(4), 417–427 (1996)
33. Itu City Hall Parque Geológico do Varvito. https://itu.sp.gov.br/meio-ambiente/parque-geo logico-do-varvito/ Researched in 09/08/2023
34. Freire, P.: Educação como prática da liberdade. Editora Paz e Terra (2014)

Fostering Collaboration in Science: Designing an Exploratory Time Travel Visualization

Bruno Azevedo$^{(\boxtimes)}$ ⓘ, Francisco Cunha ⓘ, and Pedro Branco ⓘ

Centro ALGORITMI, Universidade do Minho - engageLab, Guimarães, Portugal
d7447@algoritmi.uminho.pt, a84059@alunos.uminho.pt,
pbranco@dsi.uminho.pt

Abstract. Finding important research papers according to a topic in the midst of an exponential growth of scientific publications is a significant challenge for researchers. Digital science libraries interfaces offer inadequate support for effective navigation and exploration, and fail to assist researchers in accurately articulating their queries with their specific interests, by typically offering a massive list of results as visual output. This problem arises not only in terms of interface design, but also in the domains of user experience and information visualization.

Collaboration is a key driver of science, and the collaborative behavior of sharing papers that is based on an individual curation process grounded on researchers' reading experience, can act as powerful social filtering system to find important papers. We present an exploratory visualization structure designed with the aim of mapping and supporting through a temporal perspective a curatorial behavior that already happens in social networks but without a visual communication logic. This article describes the design and implementation process of a temporal visualization structure in D3.js, and which combines a timeline, a node-link diagram, and a force-directed beeswarm algorithm. The findings present preliminary results and set the stage for further investigation into a "time travel" exploratory visualization. The article concludes with a reference to the visualization code, which can be accessed through the provided link.

Keywords: Interface Design · Information Visualization · Timelines

1 Introduction

Finding the most important research papers according to a specific topic in the midst of an exponential growth of scientific publications is a significant challenge, and the relevance of this topic has never been more critical. Evidence of this problematic was recently highlighted by researchers at the height of the Covid-19 pandemic, who had difficulty in tracking, handling and finding important papers [1]. The Association of Scientific, Technical and Medical Publishers (STM) in 2018 reported that three million scientific papers are published annually, meaning that eight thousand, two hundred and nineteen-point eighteen new research papers are published every day [2]. Currently this number must have been exceeded, however STM has not yet released a new update about the number of papers published annually.

© ICST Institute for Computer Sciences, Social Informatics and Telecommunications Engineering 2024
Published by Springer Nature Switzerland AG 2024. All Rights Reserved
A. L. Brooks (Ed.): ArtsIT 2023, LNICST 565, pp. 46–62, 2024.
https://doi.org/10.1007/978-3-031-55312-7_4

Despite the advanced search strategies (e.g., Boolean operators) offered by digital science libraries (e.g., filtering strategies, relevance metrics), the visual output still predominantly comprises an extensive list of results [3]. This can significantly burden cognitive processing, leading to information overload [4, 5], which, in turn, hinders the efficient discovery of results aligned with researchers' specific interests. As a response to the described problematic a recent solution based on social collaborative filtering, visualization and interface design has been proposed by [3, 6, 7]. [3] also points out that this collaborative filtering behavior already happens in social media, however, without a structured communication process. This work aims to make a contribution from a temporal perspective, taking into account the context/scenario described in [3], see Sect. 1.1 for a detailed explanation.

The exploratory visualization framework also must take into account a visual encoding strategy based on marks (e.g., dots, lines) and channels (e.g., size, position, angle, color, shape), but also an interaction strategy (e.g., overview plus details) that makes it possible to present and explore the characteristics of time as a dimension. Therefore, how can interactive and exploratory visualization be used to explore time-oriented data according to the describe context/scenario while considering the structure of time? To address this challenge, we propose Time Travel, an exploratory and interactive visualization structure implemented in D3.js that aims to map and support this collaborative filtering behavior through a temporal perspective. The proposed visualization structure combines a timeline, a node-link diagram, and a force-directed beeswarm algorithm that uses the concepts of forces [8], which is useful for representing distributions of nodes while avoiding overlaps. The visualization design and implementation will be discussed and presented in more detail in Sect. 3. The next section describes the context and scenario for a future usage of the Time Travel technique.

1.1 Future Usage Scenario for the Time Travel

This section provides a brief description of the scenario that shaped the conceptual basis for the Time Travel layout, and it also points out promising benefits in using this structure in a future exploratory interface for science [6].

User Experience (UX) methodologies are fundamental, because designing scenarios demands thoughtful evaluation of context-specific elements, given that the user experience is intrinsically connected with the context (e.g., user's goals, tasks, motivations). A scenario-based design methodology enables the proactive identification of challenges, pinpointing opportunities for enhancement, and the exploration of inventive solutions tailored to users' needs and expectations [9]. The Time Travel scenario is based on the context described in [3] and [7], regarding the collaborative and curatorial behavior that happens in the social dimension of science [10], more specifically in digital social networks. In this particular scenario, it is essential to provide a temporal perspective so that this sharing behavior can be framed on a temporal reference. The aim is to visualize trends, patterns, correlations, bursts, seasonality and set the stage for an analysis focused on the evolution of topics within communities, subareas of knowledge, among other types of analysis [11].

Science progresses continuously over time, which is demonstrated by the rates of variability of the attributes of different science entities [12]. Among other entities, the basic units of science analysis are papers, journals and researchers [12]. The focus of this paper lies on the relationship between these entities, researchers and papers, and the social processes of science, namely the collaborative sharing/filtering of papers [3], where researchers play a key role as curators of important scientific research/results. [3] shows evidences that this behavior also happens on social media (e.g., Twitter and Facebook), namely when the researcher asks for important papers on a specific topic from their peers, and the community is willing to help by responding to the request. These papers (or other documents), are usually stored on researchers' computers and the aim is to use these personal "databases" as a synthesis of important research. This synthesis is the result of a selection process based on the researchers' reading experience. From this point of view, the aim is to develop an interface to support and map a collaborative sharing/filtering behavior based on the curation process described above, however through a temporal perspective, as [3] presents a solution that is based on a relational perspective within a community around a specific knowledge subarea.

Information visualization (InfoVis) design process [13] plays a key role in translating abstract data into a visual language to provide effective cognitive processing [12]. The design of user-friendly interfaces and the improvement of the overall user experience depend fundamentally on user interface design and interaction methodologies. Providing a temporal perspective is essential for understanding the evolution of a subarea, or the evolution of topics. Network visualization aids in recognizing significant connections between the most widely shared pivotal papers and the researchers responsible for disseminating them.

The scenario described above set the stage for the design and implementation of the Time Travel. The next section provides a brief background on the topic of visualizing the temporal architecture of science; Sect. 2.1 presents a brief study and analysis on timelines, and Sect. 2.2 presents the state of the art. The remaining article is divided in the following sections: Sect. 3 outlines the procedure of designing and implementing the Time Travel InfoVis in D3.js, while Sect. 4 introduces preliminary results. The last section presents a brief discussion and future directions. The Observable link to access the dataset, visualization and code can be found at the end of the article.

2 Visualizing Time

Based on the current state of the art, this section presents a brief review of temporal visualizations in the context of science. However, and in order to understand temporal structures, we should first consider the data dimension: time.

Time holds significant value as a data dimension with its own unique attributes. It is found in various domains, including but not limited to medical records, business/finance, biographies, photo collections, history, planning, project management, traffic, mobility, and science, all are defined by temporal information [13, 14]. Time possesses an inherent structure that enhances its complexity, namely a hierarchical arrangement of granularities, from milliseconds to centuries, and it can be aggregated by astronomical time (e.g., hours, years) or cultural time (e.g., semesters). The higher the aggregation the lower the resolution [11]. It also encompasses diverse divisions and relations, one example is the relation between 60 min and one hour [11, 13, 14]. These elements are integrated into design systems (e.g., calendar), and it encompasses natural cycles (e.g., seasonal cycles such as annual temperatures) and recurring patterns such as seasons, as well as social cycles, that are frequently irregular in nature (e.g., school breaks), and trends (e.g., cyclical, seasonal, random) [13, 14]. Therefore, the time structure could be linear, cyclic and hierarchical (e.g., years, months, days).

Temporal data values depend on time and to analyze and visualize temporal data is important to considered a specific visualization framework and time resolution. Temporal analysis aim is to understand the inherent characteristics of events/phenomena mapped trough a sequence of observations, including patterns, trends (e.g., increasing and decreasing tendencies), seasonality, outliers, and bursts of activity [11].

Time-series data consists of a sequence of events and observations and can be categorized as discrete or as continuous [13]. Continuous data is where observations are recorded at regular time intervals, and discrete data is where events occur within milliseconds or extend over extensive periods such as years or centuries [11, 13, 14]. Temporal data can also be categorized as static, involving the examination of historical data, or dynamic, displaying the information flow of data streams like email or news updates [11]. Such example is the work of [15], which combines node link diagrams and circular treemaps to visualize the information flow of sharing behaviors. Another important work is from [16], which presents a summary of a set of important techniques for text data flow, focusing on social networks, more specifically in three areas that are related to the objective of the proposed exploratory visualization structure: events, topics and information dissemination. It also proposes an interactive visualization for depict and analyze anomalous information spreading in social media. It is fundamental to make a reference to the employed technique, specifically the greedy layout algorithm derived from an extended circle packing approach [17]. Other types of analysis to visualize time that should also be considered are: time zones, outliers and time slices [11, 14]. For a more comprehensive analysis, [14] provides an important survey about general time visualizations methods and techniques.

The following section presents the state of the art regarding time visualization in the context of science.

2.1 Temporal Structures in Science: State of the Art

The focus of this analysis is on interactive and exploratory visualizations of science trough a temporal perspective. This section references several important visualizations that make use of temporal structures (e.g., line graphs, stacked graphs, scatter plots, histograms, timelines). [12] provides important milestones in mapping science, from algorithms, visualizations, mapping strategies, tools and books, from 1930 to 2007. This work is used as a starting point for the present analysis. Thus, it is important to highlight the following works: PaperFinder (2004) explores and visualize the history of InfoVis, and it uses the time attribute for the x-axis ordering; The Timeline of 60 years of Anthrax Research Literature (2005) employs three methods to map and visualize a particular topic: a timeline, a technique for clustering papers by time, and the capability to visualize temporal changes within the topic [18]; HistCite visualization of DNA Development (2006), is a tool that automatically generates chronological tables and historiographies from Web of Science searches. The historiography presents documents of a chronological nature and interlinked by their citations [19]; The History Flow (2006)[1] shows the edit history of Wikipedia entries/topic on a timeline [20]; TextArc[2] Visualization of The History of Science (2006) provides a chronological order of the text from the book: *A History of Science* [12]; Science and technology outlook: 2005–2055 (2006) maps uncharted territory of science and technology (S&T) through a timeline from 2005 to 2055 [21]; BiblioViz (2006) is a bibliography visualization over time (Table View) [22]; The 113 Years of Physical Review: Using Flow Maps to Show Temporal and Topical Citation Patters (2007) captures the structure and evolution of physics between 1893 to 2005. The 389,899 documents are arranged in a two-dimensional time-topic reference system. It uses a temporal and topical PACS organization of papers. In the last third of the map, an overlaid layout depicts the citations from every Physical Review paper published in 2005 [23]; Mapping the Universe: Space, Time, and Discovery![3] (2007), time is captured as a time spiral and also maps the sequence of new themes that emerge over time [24]; EdgeMaps (2011), is a tool to visualize influence relations between philosophers using a network graph and a timeline [25]; Citeology (2012) it is an interactive visualization that displays the connections among research publications by analyzing their citations over time [26]; CitNetExplorer (2014), is a tool for analyzing and visualizing citation networks of scientific publications over time. Nodes/publications are displayed on a vertical timeline and colored according to a categorical attribute (e.g., knowledge area) [27];

There are also some interesting web visualizations that should be referenced, although not published in academic papers, such examples are: Science Paths (2015–2016) project,[4] allows the interactive and exploratory visualization of citations of over 10.000 scientists from seven different fields over time by using an histogram [28]; The project Celebrating 150 years of Nature papers[5] (2019) represents an interactive and exploratory reference tree where each ring of the tree represents a cascade of references, and where the papers are arranged in rings by year, grouped and colored by discipline [29]; Searching Covid19[6] (2020) uses the most prominent Covid-19 related search queries starting from January 20th, 2020, and it combines a timeline and a beeswarm [30].

An important work to be referenced, but outside the scope of this review is the work of [31]. The MuzLink uses a linear and chronological layout with three axes of beeswarm connected timelines[7]. However, it does not allow a "time travel" navigation (pan and zoom), because the layout presented is defined by specific time period according to the selected artist. A user study was also conducted to assess the effectiveness of MuzLink with positive feedback regarding the exploration and analysis of musical adaptations and artist relationships.

---

[1] https://whitney.org/exhibitions/idea-line.

[2] http://wbradfordpaley.com/live/#.

[3] http://cluster.cis.drexel.edu/~cchen/projects/sdss/images/.

[4] https://www.youtube.com/watch?v=qlnxM-ld4BU.

[5] https://www.nature.com/immersive/d41586-019-03165-4/reftree-home.html.

[6] https://searchingcovid19.com.

[7] https://muzlink.witify.io/#/artist/1923.

Understanding and interpreting temporal data requires different visual and analytical techniques. In conclusion, this concise overview shows the predominant adoption of the timeline structure and also provides important clues and strategies for the visual coding of temporal data and nodes placement optimization such as the *beewswarm* algorithm.

2.2 Time Lines: Brief Analysis

The focus on this section is to provide a brief analysis and overview regarding linear timeline visualizations, that are designed to present a linear sequence of events [32].

The portrayal of time and time-oriented data through visual representations pre-dates the computer era [14]. Drawing was the primary medium to communicate, and for centuries, timelines conveyed the chronological order of events (e.g., biographies, historical summaries, chronologies) [33–35]. According to [35] timelines are the most common graphical representation of historical time, thus allowing to create sequential and chronological narratives of historical events.

A time-series plot, also called a timeline, chronological or data-distribution plot, plots values over time, revealing the temporal distribution of a data set, such as the first and last time point, any absent values, outliers, values, trends, growths, peak latencies, and decay rates [11, 32, 34, 36]. The spatial layout features temporal intervals and spatial units (e.g., negative space) that indicate "uniform or non-uniform temporal intervals" [35]. A detailed timeline could provide information about the chronological occurrence of events, their duration, and potential overlaps among them. [34].

Succinctly, timelines describe the type, number and order in which events occurred [14], and it can be described as a chronological arrangement of events that provides a historical account [37]. Therefore, a timeline is a visual representation of a series of events in time, namely it represents temporal event sequence data.

[34] provides a design space classification specific to timelines, with the aim of balancing expressiveness and effectiveness. It divides the timeline design space into five categories, namely Linear, Radial, Grid, Spiral, and Arbitrary. It also presents an introduction and analysis of a design space for storytelling with timelines, and identifies three important dimensions for timeline design, namely representation, scale (chronological, relative, logarithmic, sequential, sequential plus intermediate duration), and layout (unified, faceted, segmented, faceted plus segmented). However, these guidelines focus on expressive storytelling and not on exploratory InfoVis. [36] presents a novel study comparing timeline shapes. It assesses how the shape of the timeline influences task performance and whether users have a preference regarding the shape. They found that participants are faster at reading information from linear timelines than from circles and spirals, and that linear shapes are more readable. They also found evidence that linear shapes allow timelines to be read faster than non-linear shapes. It also provides some design recommendations, such as using linear timelines for difficult tasks that requires complicated decision making. This study was conducted with a more limited number of timeline shapes compared to the study conducted by [34].

A more extensive analysis of timeline visualization structures is beyond the scope of this article.

3   Time Travel: Design and Implementation

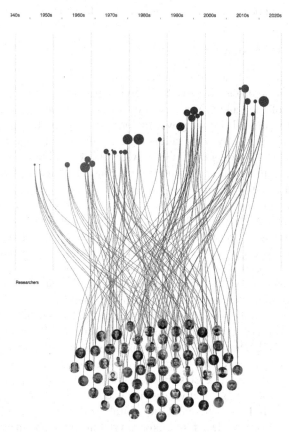

Fig. 1. Time Travel Prototype in D3.js: Clusters on

This section describes the design, implementation and interaction techniques (see Sect. 3.1) of the proposed exploratory Time Travel visualization prototype (see Fig. 1), which will be implemented in a future system/platform according to the scenario described in Sect. 1.1. The Time Travel exploratory visualization is an early-stage prototype and has been implemented in D3.js, an open-source JavaScript library for information visualization and manipulating documents based on data [38]. The visualization conceptualization aimed at supporting a "time travel" navigation was carried out through multiple iterations based on user experience and user interface design methodologies. These iterations have been supported by a focus group: with researchers, visualization, and interaction experts, where no sensitive data was collected. This approach provided guiding advice and evidences for the design of specific tasks, requirements, users profiles and personas, layout design, from early mockups (see Fig. 2) to an early-stage functional protype (see Fig. 1). The aim of Time Travel is to provide researchers with a temporal perspective on the most shared documents in a specific topic and subfield of knowledge (see Sect. 1.1).

The interface layout was divided into three main sections: the bottom section accommodates the researchers; the upper section provides the temporal reference (years and decades); and the middle section displays the shared papers (see Fig. 2). The curved edges/links provide the connections between researchers and papers or between papers and researchers.

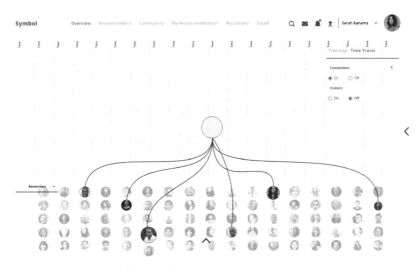

Fig. 2. Time Travel Layout Concept

In the visual encoding strategy, visual marks are used to represent data items, namely circles/nodes and links. Based on a visual abstraction process, the visual mark circle/node is used to represent papers and researchers. Circle size encodes a quantitative variable, namely the number of times a given paper is shared (see dataset), hues and color intensity encode topics, and position the temporal reference. In a future iteration, a weighting factor will be integrated as a quantitative variable, which will act as a constraint between the number of researchers and the number of papers shared. The aim is to avoid/mitigate the visual clutter that can occur as a result of visualizing a high number of links between these two entities. However, a discussion of the weighting factor is beyond the scope of this paper.

A dataset in JSON was fabricated based on the described scenario, where we can find three different entities, namely a standard user profile (personas design), a set of shared papers (scenario) and the links between researchers and papers (network). The entity name of researchers is `person`, the name of papers is `object` and the links between the entities is `links`. A set of attributes defines the three items (ordinal and quantitative variables). For more details, see the dummy dataset in the Observable page. The link can be found at the end of this article.

A beeswarm visualization structure works similarly to a histogram, except it allows individual data points to be displayed. When compared to histograms, beeswarm plots offer several advantages, namely to interpret dense regions by highlighting individual data points within a distribution instead of binning them [39]. Additionally, beeswarms are effective at visualizing data gaps and irregularities. The data is plotted along a singular axis, with points vertically displaced while maintaining their horizontal positions [39]. Each data point is represented by circles, and the data points are placed along the axis based on their corresponding values. Beeswarm charts present various benefits, including clarity by allowing clear observation of individual data points and their distribution, thus simplifying the identification of patterns and outliers [39]. They can also be used for categorical and quantitative data, which makes them versatile for various types of data analysis. For extensive datasets, employing additional techniques, such as grouping or filters, becomes essential as a mitigation strategy for dense layouts [39]. In contrast to simple scatter plots that illustrate the correlation between two numerical variables, beeswarms depict the distribution of an individual numeric measure across either a singular category or multiple categories [39]. Another feature, is to visualize different scales (linear and logarithmic), transitions between scales, and hue or size can add additional dimensions [39].

In the Time Travel layout, the points are nodes of a network, and the relative position of each circle/node is based on D3.js forces[8]. The position of the papers and researchers are defined by the implementation of two force-directed beeswarm algorithms[9], which employs a simulation of physical forces acting on nodes to achieve the intended arrangement without overlapping [39, 40]. The collision force `collide` uses nodes as circles with a given radius, and avoids nodes from overlapping. The `tick` function runs n iterations of a force simulation, and on each iteration of the simulation the "tick" function is executed. This function associates the nodes array with circle elements and then adjusts their positions according. The `forceX` and `forceY` induce the movement of elements towards designated position(s). The used forces are based on the following code:

```
//SimulationForce
var simulation = d3.forceSimulation(data.object)
.force('x', d3.forceX((d) => xScale(parseTime(d.date))))
.force('y', d3.forceY((d) => y(d.domain)))
.force('collide', d3.forceCollide(d => d.radius))
.on("tick", tick);
console.log(data)
var simulation1 = d3.forceSimulation(data.person)
.force('forceX', d3.forceX(width / 2).strength(3))
.force('forceY', d3.forceY(height2 + 250).strength(10))
.force('collide', d3.forceCollide(30).strength(1))
.on('tick', ticked)
//ForceSimulation Ticks
function tick() {
svg.selectAll('#circle1')
.attr('class', 'dataPoint')
.data(data.object)
.attr('cx', d => d.x)
```

---

[8] https://d3js.org/d3-force.
[9] https://observablehq.com/@harrystevens/force-directed-beeswarm.

```
.attr('cy', d => d.y);
}
function ticked() {
svg.selectAll('#circle2')
.data(data.person)
.attr('cx', d => d.x)
.attr('cy', d => d.y);
}
```

The data-driven concept implies that the dataset should have a specific structure [20], namely a relational dataset structure (network) in order to compute and draw the links between the papers and the researchers. As previously mentioned, the dataset network structure encompasses attributes. These attributes make it possible to establish links between persons and objects. The links have a target attribute and a source attribute, where the target corresponds to the id of specific objects and the source corresponds to the id of specific persons. In this way we have a pair of person and object (bi-partite network). Based on the ids of the object and person nodes, the x/y position of the node in the layout is determined, which is then used to create a vertical link (path) by using d3.linkVertical().

Color was used to simplify category identification, and is based on Lch/HCL color space interpolation that is both intuitive and perceptually uniform [49]. The transformation from RGB to Lch/HCL interpolation space, was performed in the Leonardo color webtool[10], and the color scheme was generated in Coolors[11] webtool. A further discussion about the color topic is beyond the scope of the current article.

3.1 Interaction

The process of data interaction, exploration and visual analysis involves the active participation of the user. The aim is that the interaction techniques implemented allow users to effectively analyze and explore the data.

A linear arrangement is characterized by the progression of time moving in a straight line from the past to the future. However, the linear arrangement implemented starts on the most recently shared papers. The main objective is to align with the reasoning inherent in a systematic literature review procedure, in which a time period of five years from the present year is delimited. Thus, the position of the nodes is related to the year of publication of the papers, and the hue represents a specific topic taking into account a knowledge subarea. The timeline interactivity (zoom and pan) is based on code already implemented[12], and the aim is to navigate through the shared papers over a time reference. The panning function enables navigation both backward and forward along the timeline, while zooming provides the capability to focus on specific dates, such as moving between decades and years (see Fig. 3).

The interaction buttons allow users to select a set of functions, namely the grouping function, where nodes are arranged vertically according to specific topics and are

---

[10] https://leonardocolor.io/#.

[11] https://coolors.co.

[12] https://observablehq.com/@notanaccent/timeline.

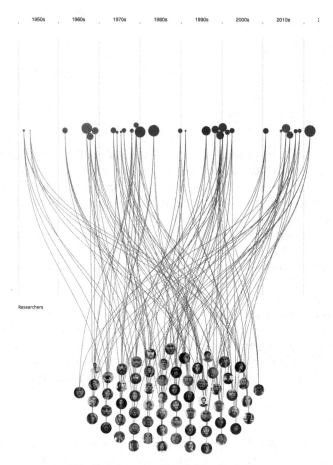

Fig. 3. Interaction Detail: Distribution on

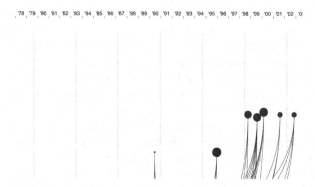

Fig. 4. Zoom and Pan Navigation

distributed over a time reference (see Fig. 4). The distribution button allows users to visualize the horizontal distribution of papers over a time reference (years) (see Fig. 5).

The connection button allows users to enable or disable links between researchers and papers (see Fig. 4). The color gradient aims to convey visual information about the researcher's knowledge domain and the paper's topic.

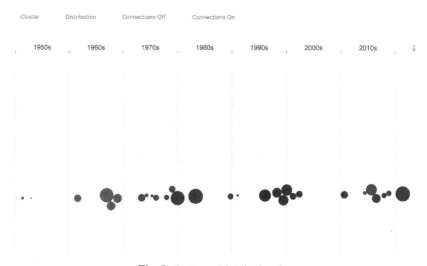

Fig. 5. Buttons: Distribution On

A mouseover event triggers a tooltip/infotip to display detailed information/ metadata related to papers (e.g., Date, Title, DOI among other metadata), and to the researchers (see Fig. 6 and Fig. 7). Another visual feature is the tooltips opacity that is used to prevent users from losing sight of the context (see Fig. 6 and Fig. 7). Opacity with a value of 0.85 was used to make the tooltip translucent.

Another interaction feature is the possibility to individually select researchers and therefore check details/metadata related to their profile (see Fig. 6). Another important technique is to visually " deactivate " the context to reduce visual clutter. This technique makes it possible to focus on a node and its respective links without letting users lose sight of the context, and thereby deliver a seamless and foreseeable user experience during transitions in the interface context. Opacity with a value of 0.85 was used, see Fig. 6.

It is important to note that the buttons (connections on, connections off, distribution and clusters) are not the final design, a new design will be proposed in a future iteration (Fig. 5).

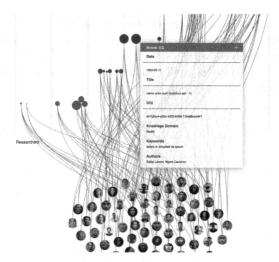

Fig. 6. Papers Metadata Tooltip

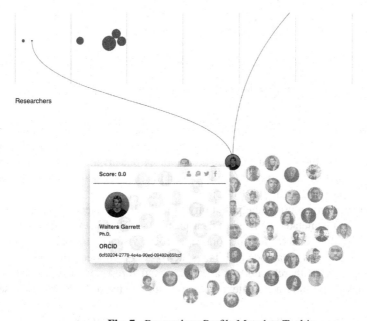

Fig. 7. Researchers Profile Metadata Tooltip

4 Preliminary Results

The main goal of this paper was to describe the designing and implementation of an interactive Time Travel visualization structure in D3.js, given a specific scenario. This section introduces initial findings, covering aspects such as user interface design, interaction techniques, algorithms and implementation strategies.

Section two provided important guidelines and points to some techniques and approaches for visualizing the temporal structure of science, namely the use of linear structures and some important algorithms, such as the beeswarm algorithm. In addition, it was concluded that timelines are a predominant structure used in almost all the projects previous presented. Based on the referenced works, linear representation proves to be the most suitable visual approach to assist the chronological placement of items in a sequential order.

The Time Travel layout strategy was devised through the blending of a timeline, a node-link diagram, and a force-directed beeswarm algorithm. (see Sect. 3). The connections among the nodes involved the transformation of the distribution points into a network of nodes (see dataset on the provided link) and also the work of [7]. The aim was to establish links between two different entities, namely researchers and papers. In order to avoid overlap between nodes, a force-directed beeswarm was used. [41] points out a drawback in this approach, which is its tendency to represent slightly incorrect values in the placement of nodes at the x position. However, this topic needs further exploration.

Succinctly, Time Travel can interactively and aesthetically accommodate a social cooperation behavior based on a curation process that is grounded in the researchers' reading experience through a temporal perspective. Interaction techniques allow users to interact and navigate, but also mitigate visual clutter. Examples of this are the use of buttons to activate or deactivate links, the use of zoom in and zoom out functionalities, as well as panning for linear navigation through sequences of events. While we have preliminary results, the aim is to reveal intriguing patterns of social sharing in a future iteration /increment of the propose visualization structure.

5 Discussion and Future Work

This paper presents a solution to the problem of science information overload. It proposes a temporal visualization that supports and maps a social filtering behavior that already happens on social platforms, but without a visual language and communication logic. Time Travel is an interactive and exploratory visualization that aims to support a collaborative and curatorial process based on the reading experience of researchers. The proposed visualization layout incorporates a set of visualization structures/algorithms, and the aim is to provide an effective visual interface to time exploration/navigation according to a specific scenario. Time Travel remains in its initial prototype stage, and some issues need to be addressed in forthcoming iterations. An existing issue relates to the interaction of zoom and pan and the automatic updating of node placement. Whenever a user engages in zooming or panning, the positions of nodes must be readjusted. This update can be carried out using either the buttons (distribution and cluster buttons) or by refreshing the page. It is through these buttons that the node positions and the aforementioned forces are recalculated. The second issue involves the hue gradient employed within the lines, which is reversed in certain links (See Fig. 3). The purpose of using a gradient between hues is to establish a perceptual connection between the researcher's sub-area of interest and the paper's topic. This issue will be addressed in a future iteration.

There is also room for further research on the proposed visualization, such as the integration of the edge bundling technique in order to reduce visual clutter when a larger number of links are presented. And, it is also important to conduct a user study/assessment based on performance and satisfaction metrics to understand how researchers interact and navigate. The study will be conducted with researchers playing 3 types of roles: beginner level (e.g., PhD students), intermediate level (e.g., postdoctoral researchers) and expert (e.g., established researchers and professors), in a total of 15 testers and according to a set of task scenarios, using the Think Aloud protocol.

As future work, it is intended to design and implement a time visualization perspective from an individual point of view, that is, from the selection of a specific researcher. The objective is to visualize a temporal perspective based on the papers shared by a selected researcher. A cyclic arrangement can be used to carry out other types of analysis, such as visualize time granularities allowing users to discover trends, or low frequency variations in the data, or bursts and other temporal dynamics of papers sharing/information spreading.

The presented work represents a proof of principle and the current results are still at a preliminary stage, so more research is needed on the main points highlighted. The visualization and code can be found in the following link: https://observablehq.com/d/03f2b04695e9947a.

Acknowledgements. This work has been supported by FCT – Fundação para a Ciência e a Tecnologia within the R&D Units Project Scope: UIDB/00319/2020.

References

1. Brainard, J.: Scientists are drowning in COVID-19 papers. Can new tools keep them afloat? Science (1979) (2020). https://doi.org/10.1126/science.abc7839
2. Johnson, R., Watkinson, A., Mabe, M.: The STM Report: An overview of scientific and scholarly journal publishing. https://www.stm-assoc.org/2018_10_04_STM_Report_2018.pdf
3. Azevedo, B., Branco, P., Cunha, F.: Design for science: proposing an interactive circular 2-level algorithm. In: Brooks, A.L. (ed.) In Creating Digitally: Shifting Boundaries: Arts and Technologies - Contemporary Applications and Concepts, pp. 445–472 (2023)
4. Khazaei, T., Hoeber, O.: Supporting academic search tasks through citation visualization and exploration. Int. J. Digit. Libr. **18**, 59–72 (2017). https://doi.org/10.1007/s00799-016-0170-x
5. Wurman, R.S.: Information Anxiety 2. QUE (2001)
6. Azevedo, B., Baptista, A.A., Oliveira e Sá, J., Branco, P., Tortosa, R.: Interfaces for science: conceptualizing an interactive graphical interface. In: Brooks, A.L., Brooks, E., Sylla, C. (eds.) ArtsIT/DLI -2018. LNICSSITE, vol. 265, pp. 17–27. Springer, Cham (2019). https://doi.org/10.1007/978-3-030-06134-0_3
7. Azevedo, B., Cunha, F., Branco, P.: Designing an interactive 2-level circular algorithm to visualize and support collaboration in science. In: Brooks, A.L. (ed.) ArtsIT 2022. LNICSSITE, vol. 479, pp. 576–590. Springer, Cham (2023). https://doi.org/10.1007/978-3-031-28993-4_41
8. Cheong, S.-H., Si, Y.-W.: Force-directed algorithms for schematic drawings and placement: a survey. Inf. Vis. **19**, 65–91 (2020). https://doi.org/10.1177/1473871618821740

9. Courage, C., Baxter, K.: Understanding Your Users: A Practical Guide to User Requirements Methods, Tools, and Techniques. Morgan Kaufmann (2015)
10. Leydesdorff, L.: The Challenge of Scientometrics: The Development, Measurement, and Self-Organization of Scientific Communications. Universal Publishers / uPUBLISH.com, USA (2001)
11. Börner, K.: Atlas of Knowledge Anyone Can Map. The MIT Press (2015)
12. Börner, K.: Atlas of Science: Visualizing what We Know. MIT Press (2010)
13. Börner, K., Polley, D.: Visual Insights: A Practical Guide to Making Sense of Data. MIT Press (2014)
14. Aigner, W., Miksch, S., Schumann, H., Tominski, C.: Visualization of Time-Oriented Data. Springer, London, London (2011)
15. Viégas, F., Wattenberg, M.: Google+ ripples: a native visualization of information flow. In: Proceedings of the 22nd International Conference on World Wide Web, pp. 1389–1398. Republic and Canton of Geneva, Switzerland (2013)
16. Zhao, J., Cao, N., Wen, Z., Song, Y., Lin, Y.-R., Collins, C.: #FluxFlow: visual analysis of anomalous information spreading on social media. IEEE Trans. Vis. Comput. Graph. **20**, 1773–1782 (2014). https://doi.org/10.1109/TVCG.2014.2346922
17. Wang, W., Wang, H., Dai, G., Wang, H.: Visualization of large hierarchical data by circle packing. In: Proceedings of the SIGCHI Conference on Human Factors in Computing Systems, pp. 517–520. ACM, New York, NY, USA (2006)
18. Morris, S.A., Boyack, K.W.: Visualizing 60 years of anthrax research. In: Proceedings of the 10th International Conference of the International Society for Scientometrics and Informetrics (2006)
19. Garfield, E.W., Paris, S.G.: Stock: HistCiteTM: A Software Tool for Informetric Analysis of Citation Linkage. http://garfield.library.upenn.edu/papers/histcite2006.pdf
20. Viégas, F.B., Wattenberg, M., Dave, K.: Studying cooperation and conflict between authors with *history flow* visualizations. In: Proceedings of the SIGCHI Conference on Human Factors in Computing Systems, pp. 575–582. ACM, New York, USA (2004)
21. Pang, A.S.-K., Pescovitz, D.: Science and technology outlook: 2005–2055. https://legacy.iftf.org/our-work/people-technology/technology-horizons/science-technology-outlook-2005-2055/
22. Shen, Z., Ogawa, M., Teoh Tee, S., Ma, K.-L.: BiblioViz: A System for Visualizing Bibliography Information. (2006)
23. Herr, B.W., Duhon, R.J., Börner, K., Hardy, E.F., Penumarthy, S.: 113 Years of physical review: using flow maps to show temporal and topical citation patterns. In: Proceedings of the International Conference on Information Visualisation, pp. 421–426 (2008). https://doi.org/10.1109/IV.2008.97
24. Chen, C., Vogeley, M.S., Gott III, R.J., Juric, M., Kershner, L.: Coordinated Visualization and Analysis of Sky Survey Data and Astronomical Literature. http://cluster.cis.drexel.edu/~cchen/projects/sdss/images/
25. Dörk, M., Carpendale, S., Williamson, C.: EdgeMaps: visualizing explicit and implicit relations. Presented at the January 23 (2011)
26. Matejka, J., Tovi Grossman, G.F.: Citeology: visualizing paper genealogy. In: CHI EA '12 CHI '12 Extended Abstracts on Human Factors in Computing Systems, pp. 181–190. , Austin, Texas (2012)
27. van Eck, N.J., Waltman, L.: CitNetExplorer: a new software tool for analyzing and visualizing citation networks. J. Informetr. **8**, 802–823 (2014). https://doi.org/10.1016/j.joi.2014.07.006
28. Albrecht, K.: Science Paths. http://sciencepaths.kimalbrecht.com
29. Grishchenko, A., Martino, M., Gates, A., Ke, Q., Varol, O., Barabási, A.-L.: The project Celebrating 150 years of Nature papers. https://www.nature.com/immersive/d41586-019-03165-4/index.html

30. Schema Design: Searching Covid19. https://searchingcovid19.com
31. Lévesque, F., Hurtut, T.: MuzLink: Connected beeswarm timelines for visual analysis of musical adaptations and artist relationships. Inf. Vis. **20**, 170–191 (2021). https://doi.org/10. 1177/14738716211033246
32. Aigner, W., Bertone, A., Miksch, S., Tominski, C., Schumann, H.: Towards a conceptual framework for visual analytics of time and time-oriented data. In: 2007 Winter Simulation Conference, pp. 721–729. IEEE (2007)
33. Rosenberg, D., Grafton, A.: Cartographies of Time: A History of the Timeline. Princeton Architectural Press; Illustrated edition (2012)
34. Brehmer, M., Lee, B., Bach, B., Riche, N.H., Munzner, T.: Timelines revisited: a design space and considerations for expressive storytelling. IEEE Trans. Vis. Comput. Graph. **23**, 2151–2164 (2017). https://doi.org/10.1109/TVCG.2016.2614803
35. Meirelles, I.: Design for Information: An Introduction to the Histories, Theories, and Best Practices Behind Effective Information Visualizations. Rockport Publishers (2013)
36. Di Bartolomeo, S., et al.: Evaluating the effect of timeline shape on visualization task performance. In: Proceedings of the 2020 CHI Conference on Human Factors in Computing Systems, pp. 1–12. ACM, New York, NY, USA (2020)
37. Yan, R., Wan, X., Otterbacher, J., Kong, L., Li, X., Zhang, Y.: Evolutionary timeline summarization. In: Proceedings of the 34th International ACM SIGIR Conference on Research and Development in Information Retrieval, pp. 745–754. ACM, New York, USA (2011)
38. Bostock, M., Ogievetsky, V., Heer, J.: D^3 data-driven documents. IEEE Trans. Vis. Comput. Graph. **17**, 2301–2309 (2011). https://doi.org/10.1109/TVCG.2011.185
39. Kirk, A.: Data Visualisation: A Handbook for Data Driven Design. AGE Publications Ltd (2019)
40. Heinz, M.: BeeSwarm Plot. https://martinheinz.dev/blog/27
41. Trimble, J.: Accurate-Beeswarm plot. https://github.com/jtrim-ons/accurate-beeswarm-plot

Enhancing Scientific Communication Through Information Visualization: A Proposal for a Multimodal Platform

Mariana Pereira[1]([✉]) ⓘ, Bruno Azevedo[2] ⓘ, and Sílvia Araújo[1] ⓘ

[1] Universidade do Minho, Braga, Portugal
pg43613@uminho.pt, saraujo@elach.uminho.pt
[2] Centro ALGORITMI, Universidade do Minho - engageLab, Guimarães, Portugal
brunomiguelam@gmail.com

Abstract. This article delves into the realm of science communication and data visualization, presenting a platform designed to enhance the dissemination of scientific knowledge. Rooted in the context of the DIAL4U project, the study investigates the creation of online resources to bridge the gap between scientific experts and diverse audiences. The integration of multimodal data and advanced visualization techniques underscores the platform's point in fostering engaging interactions with the public. This endeavor facilitates the creation of resources that cater to varied scientific communication goals, ranging from sharing recent findings to influencing public opinions. Through science communication and visualization, this article contributes to the evolving landscape of effective scientific discourse and knowledge dissemination.

Keywords: Science Communication · Data Visualization · Multimodal Content · Digital Platforms · Academic Literacy

1 Introduction

This research focuses on the development of a scientific communication tool to address and facilitate the sharing of scientific information with different audiences. Effective communication of scientific knowledge to wider audiences remains a key challenge in our evolving digital information landscape. In which new formats and actors arise in the area, in part due to social media innovation leading to a diversification in the field. As members of the scientific community at the forefront of knowledge generation, it is our responsibility to engage in knowledge transfer, develop creative and practical approaches to education, and engage on the creation of resources to understand and improve education. This research investigates the creation of online resources as a potential solution to this challenge, offering a clear, interactive, and highly engaging medium for the public based on an information visualization user interface.

A. L. Brooks (Ed.): ArtsIT 2023, LNICST 565, pp. 63–71, 2024.
https://doi.org/10.1007/978-3-031-55312-7_5

1.1 Science Communication

On this section we'll analyze the multifaceted panorama of science communication. Effective scientific communication is a complex process that requires a differentiated understanding of the various target audiences [1]. Whether it's the general public, specific interest groups, the media, political decision-makers, or individuals with no scientific background, starting a dialog about science requires direct interaction between scientists and a heterogeneous public [2]. To achieve this, leveraging resources such as analogies and visual aids, while respecting the audience's existing knowledge, becomes pivotal for sustaining engagement and enhancing comprehension [3].

To effectively communicate science, it's crucial to prioritize the objectives of sharing recent findings and generating excitement for the field. It's also essential to promote public appreciation for science, deepen understanding of scientific concepts and viewpoints, and incorporate diverse perspectives when tackling complex societal issues [4]. The goal is harnessing a range of strategies and resources, to develop cutting-edge and dynamic approaches to science communication, and foster engagement and bridge gaps [5].

It is essential to highlight that the landscape of science communication is multifaceted, encompassing various models and approaches [5]. While some scholars advocate for a unidirectional flow of scientific information from experts to the public as an effective strategy, alternative models prioritize the inclusion of dialogue and deliberation among the public, experts, and decision-makers [1]. The progression of these models traces back to the Deficit Model in the 1960s, transitioning to the Public Understanding of Science (PUS) paradigm in the 1980s and 1990s [6]. Both models inherently acknowledge the necessity to address gaps in public knowledge. However, despite the waning popularity of the deficit model and its empirical limitations, a deficit-oriented approach continues to persist in science communication [1].

The enduring influence of deficit-focused approaches can be attributed to factors such as the presumption among academics that audiences are eager to learn [1]. Moreover, the simplicity of attributing gaps in understanding to the public rather than the science itself has contributed to the persistence of this approach. Notably, the PUS paradigm posits that enhancing the general public's scientific knowledge will correlate with increased support for science [6]. Nevertheless, this paradigm's longevity underscores the necessity for a more comprehensive and nuanced approach to science communication. Consequently, the emergence of models like Public Engagement with Science (PES) signals a shift toward bidirectional discourse between science and the public, fostering active participation and collaborative interactions [7]. This transformative shift aims to empower the public and promote equality in scientific discussions and decision-making [8].

2 Information Visualization

The fundamental role of Information Visualization (InfoVis) in the effective transmission of scientific knowledge as a communication tool with great potential in today's information-oriented society, since it can present information in an engaging, interactive, and accessible way [9, 17]. Information visualization, a fusion between science, art, and design, amplifies the perception of visual patterns, trends, and anomalies, redefining how reality is apprehended [10].

Serving as a complementary tool to scientific communication, InfoVis assists in elucidating textual information through graphical representations. By effectively conveying complex data in a visual format, it ensures objective communication across diverse audiences and promotes the dissemination of scientific knowledge [10]. Visualizations are asserted to be the most effective means for constructing and conveying evidence-based public policy recommendations on climate change, vaccines, and policing [9].

The foundational principles of InfoVis draw from various intellectual knowledge domains, including, Design, Art, Cartography, Human Computer Interaction, among other knowledge areas [10]. This combination maximizes the generation of reliable, replicable, and representative results. By analyzing the historical, conventional, and practical facets within observational, descriptive, hermeneutic, normative, and critical approaches, visualization optimizes its potential to convey intricate scientific concepts effectively [10].

Within both academic and civil society spheres, InfoVis serves as a conduit for communication. It underscores three vital aspects of scientific communication: the scientific knowledge dissemination, the scientific experience dissemination, and the scientific community dissemination [11]. Notable examples include the University of Oxford's "COVID-19 Pandemic: Visualizing the Global Impact"[1] which disseminates scientific knowledge, and "Mapping Antibiotic Resistance"[2] which employs interactive maps to depict antibiotic resistance levels worldwide. Similarly, Stanford University's Palladio[3] project (2021) illustrates the integration of graphical interfaces with humanistic methodologies, enabling users to create historical visualizations pertinent to their research.

This paper introduces an exploratory InfoVis approach within the DIAL4U project, focusing on seamlessly integrating interactive layouts with knowledge discovery to enhance the scientific communication process. Our contributions aim to bridge the gap between scientific knowledge and the general public, making usage of InfoVis a valuable resource to engage the public. The impact of this research on the public's understanding will be quantified through upcoming user tests, allowing us to measure the effectiveness of these approaches.

2.1 Hierarchical Structures Visualization

Within the realm of information visualization, an array of techniques exists for depicting hierarchical structures. These techniques include nesting, stacking, indentation, and node-link diagrams. Among the various tree visualization approaches, such as treemaps, layered icicles, indented plots, and hierarchical node-link diagrams [18] (the latter of which will be employed in this work), each possesses distinct advantages and disadvantages tailored to specific dimensions of data exploration and other aspects [12].

---

[1] https://ourworldindata.org/coronavirus.

[2] https://resistancebank.org/.

[3] http://hdlab.stanford.edu/palladio/.

Node-link diagrams, treemaps, and sunburst layouts are all layout fitable to hierarchical data [12, 14]. Node-link diagrams perform well to show the overall structure and parent-child relationships in a hierarchy. Making it easy to see how nodes are connected with edges, which is good for complex and interconnected hierarchies. However, they can overlap and take up a lot of space as the hierarchy gets bigger [13]. Treemaps and sunburst layouts are more compact and space efficient [18]. They use nested shapes or circular segments to represent nodes. This makes it easier to summarize and compare hierarchical data. Sunburst layouts display efficiently the hierarchy in-depth and the area of each node on the representation [14]. However, they may not be as good at showing complex parent-child relationships and interconnectedness as node-link diagrams.

For instance, treemaps efficiently use available display space through a space-filling approach, allowing for effective data representation [12]. However, they fail in conveying intricate hierarchical structures, especially when dealing with sizable datasets characterized by complex branching and depths [18]. In contrast, node-link diagrams adeptly communicate hierarchical relationships, yet may exhibit inefficiencies in utilizing display space as hierarchies grow deeper, particularly in situations with exponential growth [12]. The aim of these visualization techniques is to offer users an intuitive and interactive means to investigate, comprehend, and interact with the content effectively.

In the context of graph theory, a tree is defined by nodes and links. Crucially, trees feature hierarchical relationships among nodes, categorizing them into distinct roles: "above" or "parent", "below" or "child", and "at the same level" or "sibling" [12]. Within this framework, a tree represents the collective assembly of nodes and links within a hierarchical dataset. Tasks centered around the complete tree visualization constitute tree-level tasks, focusing on the overarching presentation of the entire tree.

3   Methodology

3.1   The DIAL4U Project

In response to the evolving landscape of education, particularly language instruction, the DIAL4U[4] project emerges as an initiative co-funded by the European Union's Erasmus + program, targeting the advancement of digital language teaching, and learning methodologies to cater to dynamic educational demands.

This multifaceted endeavor aspires to enhance language education through a comprehensive approach. By equipping language educators with knowledge, crafting tools for language acquisition, disseminating learning across diverse contexts, and promoting the use of open educational resources, the project envisions a holistic access to language learning resources. These efforts collectively work towards enhancing the effectiveness and accessibility of language learning experiences while empowering learners to take an active role in shaping their educational journeys.

One of the facets of this project centers on the creation of an interactive guide for educators. This guide is designed to facilitate the development of language skills within virtual and multimodal contexts. Recognizing the vital role of mediation interaction in

---

[4] https://www.dial4u-uni.eu/.

foreign language acquisition, this guide aims to empower educators in access language instruction materials.

The content is organized in a logical sequence, starting with teaching modalities and progressing through methodologies, strategies, learning paths, and concluding with various resources, including a collaboratively created glossary. This structured approach allows for user-friendly navigation and enables individuals to explore the interface at their own pace, fostering an accommodating learning environment.

Aligned with the overarching goals of the DIAL4U project, this research introduces a model for science communication through an interactive exploratory visualization. The objective is to empower users with academic expertise to proficiently communicate research content and structural elements. By fostering effective communication through visual representation, this project not only complements the DIAL4U initiative but also contributes to the broader landscape of science communication.

3.2 Data Collection

In crafting a visualization tool for language educators, the project harnesses the capabilities of multimodal data. The construction of data structures is a collaborative effort involving DIAL4U partners, particularly language teachers who curate pertinent information and hierarchical arrangements reflective of the language landscape. This data, accessible online via a designated link, is organized in a JSON file format – ideal for accommodating large datasets with hierarchical structures. The JSON file encapsulates both the hierarchical order of data and the corresponding content for each data point within the multimodal data, including text, video, podcast, or images.

3.3 Interface Development

In this project, we propose a visualization and interface to communicate and share knowledge regarding specialized languages utilizing a node-link diagram. We also choose to use the radial node-link diagram as a visualization method. This method is versatile and helps users to easily explore and clearly represent large hierarchical structures, making it easier to compare different elements within the tree. This visualization technique is especially effective when used with the zoom and collapse functionalities, as they allow more targeted exploration of specific parts of the hierarchy [14]. This type of approach is also used in both 2D and 3D formats and can even incorporate hyperbolic surfaces [12] for enhanced complexity [16].

The used radial tree layout provides a clear and concise overview of the project components and their interdependencies in an effort to facilitate user interaction with InfoVis. The design of the DIAL4U webpage[5] involved a meticulous process aimed at seamlessly integrating advanced visualization techniques with user-friendly content dissemination. The webpage design was designed to accommodate the multifaceted needs of language educators and learners alike. Drawing on web development technologies such as HTML5, CSS, and JS, and the Echarts.js library further enriched the interface's

---

[5] https://www3.elach.uminho.pt/dial4u/index.html or
https://dial4u.000webhostapp.com/home.html.

dynamic capabilities, enabling the creation of an interactive visualization that has the potential to engage users and facilitate comprehension. Notably, the page layout adopts a dual-column structure, where on the left, a radial tree visualization offers an intuitive representation of hierarchical relationships. Adjacent to this visualization, the right-hand column shows content intrinsically linked to each node in the tree, triggered by clicking on the node in the visualization. This arrangement leverages the synergy between data visualization and content dissemination, embodying the project's commitment to effective science communication (Fig. 1).

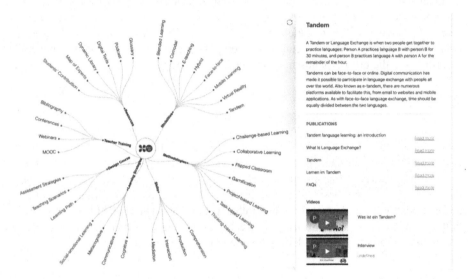

Fig. 1. Web page interface presenting a radial tree visualization and content related to nodes. Developed within the framework the DIAL4U project.

4   Amplification and Future Directions

This project evolved to a new perspective [15], the solution presented in this paper, in the context of the DIAL4U project, not only demonstrates its potential for wider replication, but is also promising for a broad impact. Taking into consideration the DIAL4U context, a dynamic platform has been conceptualized in order to empowering users to create hierarchical visualizations linked with multimodal content created and curated by the users. This innovative platform will serve as a catalyst for communication, facilitating the articulation of research structures and results with greater public involvement.

The transition between DIAL4U to Lang2Science [15], were catalyzed by the synergy between the goals and objectives of both projects. In the evolution of the DIAL4U into Lang2Science [15], a significant stride is taken toward a vibrant community of practice dedicated to scientific communication. Fostering collaboration, shared learning, and the collective enrichment of knowledge dissemination practices. This step signifies not only a cultural shift in the sense of a stronger and more interconnected scientific community but also a technical advancement. Due to the expansion of capabilities and functionalities offered by the platform. This transition implies the development of a more sophisticated and robust interface, data management systems, and interactive visualization tools.

The Lang2Science [15] platform draws inspiration from the design principles of the DIAL4U project, aiming to empower individuals to generate their own interactive resources for knowledge dissemination. By leveraging the foundation established by DIAL4U, Lang2Science seeks to amplify resource offerings, including features like users profiles associations and links connected to the project's main page. Since the content associated with each project is of substantial value, users can readily access and exploit this collective repository of research materials. The new project uses the same DIAL4U node-link diagram layout, a radial tree, to compose the interface design. The choice was based on positive feedback from users. Users can exercise control over the visualization by toggling labels on or off for each node, thereby enhancing the user experience and facilitating navigation within the visualization structure.

Aiming to empower users to create their own interactive resources and facilitating the sharing of knowledge. The project Lang2Science, a funded initiative, endeavors to establish a repository of resources that collectively support scientific researchers and the community, fostering citizen science, and open science. As part of this global effort, a platform is being actively developed to provide interactive communication resources specifically designed for the effective dissemination of scientific knowledge.

5 Conclusion

Our research has highlighted the potential of online and interactive resources as a valuable tool to bridge the gap and enhance science communication. The replicability potential demonstrated by the DIAL4U project indicates that similar initiatives can take advantage of InfoVis alongside multimodal content to improve user engagement. Given the abundance of high-quality online resources available, these can be seamlessly linked to the platform, further facilitating communication and knowledge dissemination. Increasing communication effectiveness.

As future work, the Lang2Science platform will offer a repository of resources that meet the needs of scientific, academic, and technical literacy. With the integration of specialized language resources, Lang2Science has the potential to make an effective contribution to the way students and educators communicate science. The next crucial step involves subjecting the DIAL4U to obtain a broader perspective on the user experience of the DIAL4U project website requires user testing based on satisfaction and performance metrics. This iterative process will ensure that our design aligns perfectly with the user's expectations in order to improve their experience. By grounding our efforts in user feedback, we aspire to refine and amplify the impact of our platform on the landscape of science communication and education.

Furthermore, the project's future perspectives include the use of advanced AI, powered by a Language Large Model (LLM), to automate the creation of JSON structures for specific knowledge domains. This automation will facilitate the easy generation of visualizations, benefiting academic users who can take advantage of this technology to create educational resources more efficiently and effectively.

References

1. Kessler, S.H., Schäfer, M.S., Johann, D., Rauhut, H.: Mapping mental models of science communication: How academics in Germany, Austria and Switzerland understand and practice science communication. Public Underst. Sci. **31**(6), 711–731 (2022). https://doi.org/10.1177/096366252110657

2. Schäfer, M.S., Kessler, S.H., Fähnrich, B., Leßmöllmann, A., Dascal, M., Gloning, T.: Analyzing science communication through the lens of communication science: reviewing the empirical evidence. Handbooks Commun. Sci. **17**, 77–104 (2019). https://doi.org/10.1515/9783110255522-004

3. van Geenen, D., Wieringa, M.A., Engebretsen, M., Kennedy, H.: Approaching data visualizations as interfaces: An empirical demonstration of how data are imag (in) ed. (2020). https://doi.org/10.1515/9789048543137-013

4. Kappel, K., Holmen, S.J.: Why science communication, and does it work? a taxonomy of science communication aims and a survey of the empirical evidence. Front. Commun. **4** (2019). https://doi.org/10.3389/fcomm.2019.00055

5. Volk, S.C., Vogler, D., Fürst, S., Schäfer, M.S., Sörensen, I.: Role conceptions of university communicators: a segmentation analysis of communication practitioners in higher education institutions. Public Relat. Rev. **49**(4), 102339 (2023). https://doi.org/10.1016/j.pubrev.2023.102339

6. Entradas, M.: Public communication at research universities: Moving towards (de) centralised communication of science? Public Underst. Sci. **31**(5), 634–647 (2022). https://doi.org/10.1177/0963662521105830

7. Bucchi, M., Trench, B.: Science Communication Research. Routledge Handbook of Public Communication of Science and Technology, pp. 1–14 (2014). https://doi.org/10.4324/9781003039242

8. Haklay, M.M., Dörler, D., Heigl, F., Manzoni, M., Hecker, S., Vohland, K.: What is citizen science? the challenges of definition. Sci. Citizen Sci. **13** (2021)

9. Franconeri, S.L., Padilla, L.M., Shah, P., Zacks, J.M., Hullman, J.: The science of visual data communication: what works. Psychol. Sci. Pub. Interest **22**(3), 110–161 (2021). https://doi.org/10.1177/15291006211051956

10. Engebretsen, M., Kennedy, H.: Data Visualization in Society. Amsterdam University Press (2020).https://doi.org/10.5117/9789463722902

11. Yansong, C., Zhigang, W., Hantian, X., Xiaoxi, L., Luhan, W., Fei, G.: Exhibition design of the thematic science popularization space based on scientific visualization. In: 2020 International Conference on Culture-oriented Science & Technology (ICCST), pp. 269–273. IEEE (2020). https://doi.org/10.1109/ICCST50977.2020.00058

12. Pandey, A., Syeda, U.H., Shah, C., Guerra-Gomez, J.A., Borkin, M.A.: A state-of-the-art survey of tasks for tree design and evaluation with a curated task dataset. IEEE Trans. Visual Comput. Graphics **28**(10), 3563–3584 (2021). https://doi.org/10.1109/TVCG.2021.3064037

13. Burch, M., van de Wetering, H., Klaassen, N.: The power of interactively linked hierarchy visualizations. J. Vis. **25**, 857–873 (2022). https://doi.org/10.1007/s12650-021-00818-3

14. O'Handley, B., Wu, Y., Duan, H., Wang, C.: TreeVisual: Design and evaluation of a web-based visualization tool for teaching and learning tree visualization. In: Proceedings of American Society for Engineering Education Annual Conference (2022)

15. Azevedo, B., Pereira, M., Araújo, S.: Designing a multilingual, multimodal and collaborative platform of resources for higher education. In: Brooks, A.L. (eds) ArtsIT, Interactivity and Game Creation. ArtsIT 2022. Lecture Notes of the Institute for Computer Sciences, Social Informatics and Telecommunications Engineering, vol. 479. Springer, Cham. (2023). https://doi.org/10.1007/978-3-031-28993-4_27

16. Munzner, T.: Visualization Analysis and Design. CRC Press (2014). https://doi.org/10.1201/b17511

17. Hasegawa, Y., Sekimoto, Y., Seto, T., Fukushima, Y., Maeda, M.: My city forecast: urban planning communication tool for citizen with national open data. Comput. Environ. Urban Syst. **77** (2019). https://doi.org/10.1016/j.compenvurbsys.2018.06.001

18. Burch, M., van de Wetering, H., Klaassen, N.: Multiple linked perspectives on hierarchical data. In: Proceedings of the 13th International Symposium on Visual Information Communication and Interaction, pp. 1–8 (2020). https://doi.org/10.1145/3430036.3430037

Fragments of Fungi: Eliciting Dialogue Through a Virtual Experience

Astrid Pedersen$^{(\boxtimes)}$ ⓘ, Clara Mc Nair ⓘ, Morten Jørgensen ⓘ,
Sebastian Evangelista ⓘ, Olga Timcenko ⓘ, and Rolf Nordahl ⓘ

Aalborg University, Copenhagen, Denmark
{aekp19,cnair21,mojarg19,sevang22}@student.aau.dk,
{ot,rn}@create.aau.dk
https://melcph.create.aau.dk/

Abstract. Both the arts, natural phenomena and technologies hold possibilities for profound emotional experiences. This interdisciplinary project explores the world of mycelium networks through an artistic virtual reality (VR) experience, as a medium to communicate the awe-inspiring natural phenomenon. The VR application was conceptualised and designed through various creative activities and participatory design workshops with external participants. The final VR experience *Fragments of Fungi* was brought into a physical forest setting in the north of Copenhagen, Denmark, and evaluated by four participants engaging in a two-hour collaborative workshop, including individual exercises and collective discussions. The findings suggests the potential of virtual reality as an awe-eliciting medium, and highlights the benefits of collective spaces for reflection and dialogue when presenting such self-contained and emotional experience inherent to virtual reality. Finally, the paper encourages VR artists to consider the holistic experience, rather than focusing solely on the VR aspect.

Keywords: Collective Experiences · Awe · Virtual Reality

1 Introduction

The act of *experiencing* is a fundamental part of living, however the notion of having *an experience* is something else. An experience is something we recall, something that had a profound impact on us, and we retain them with a certain character [10]. Such an experience may arise from the arts, which is known for lending itself to emotional responses, sparking conversations, and captivating people's attention [17]. Other experiences arise from nature, which can leave us in awe, by connecting us with something much larger than one can comprehend, both physically and mentally [24].

While searching for such natural experiences in the preliminary stages of this project, we came across mycelium networks, a phenomenon which left us

A. L. Brooks (Ed.): ArtsIT 2023, LNICST 565, pp. 72–90, 2024.
https://doi.org/10.1007/978-3-031-55312-7_6

struck by its inherent intricacy and complexity; leaving a feeling of being awe-struck. This emotional response can envelope individuals when they encounter something vast, captivating, or beyond their usual experiences. It is a feeling of reverential wonder, a sense of being simultaneously humbled and inspired by the grandeur or beauty of the world around us [9]. Recognising the innate capacity of artful experiences to evoke similar affective states, inspiration was found in combining these elements - an artistic representation of the awe-inspiring mycelium phenomenon. By harnessing the potential of art to ignite conversations, spaces can be created where individuals can connect and collectively reflect on their experience. When people experience art together and engage in dialogue about their interpretation and emotions, it can foster a sense of connection and shared experience, allowing for the exchange of ideas, challenging assumptions, and gaining new perspectives.

Though both nature and the arts can undoubtedly give rise to profound experiences, novel technologies, such as virtual reality (VR), allow for entirely new experiences to be had; we can travel to places otherwise impossible, experience life from the perspective of others, and immerse ourselves in fictional worlds. Such opportunities give rise to new areas of study, but is still in its adolescent state in the realms of academia in regard to what emotions and conversations might arise from such experiences. Literature reviews in the field of VR often focus on its applications in diverse domains such as gaming, education, and social learning. These reviews also investigate the effects of VR on improving task performance and usability. [2, 4]. Exploring VR as a medium for artistic expression and prioritising the emotions of the audience opens up new frontiers of creativity.

As we dwell on the opportunities arising from the arts, nature, and technology, we present this interdisciplinary study, which introduces the development of an artistic VR experience that unfolds the natural world of mycelium. The development is based on artistic practises as means to create an emotional experience that would encourage dialogue and reflection. The experience acts as the foundation for an experimental workshop that took place within a forest north of Copenhagen, Denmark. The project aims to unfold the impact that the unification of art, nature, and technology has on fostering impactful experiences and conversations. As such, the means of this project are as follows; *facilitating a conversation about nature through a virtual experience.*

2 Background

When entering the human experience and the fluctuating emotions that come with it, it is beneficial to present the academic work that has directed this project. Therefore, this section will investigate the realms of the arts, awe, and digital media in the following.

2.1   Academia and Art

The arts have captivated the human species throughout times; it compels our attention through its mesmerising and mysterious nature, which remains ambiguous and up for interpretation. It does not hold one truth, but rather stimulates multiple meanings. Many scholars have dwelled on the application of art in academia and highlighted its importance in different regards. In *The Participatory Museum*, Simon stresses the importance of participation in the arts and focusses on the conversations that might arise from engaging in artistic experiences and how this is often neglected in museum contexts. She calls attention to people who wish to participate, rather than passively consume, and demonstrates how cultural institutions can actively cultivate a shared space to engage in dialogues between visitors about the material exhibited [20]. Leavy has also highlighted the conversation-provoking quality of the arts, speaking about the emotional connections formed through art and how they may spark conversations and community among people. She highlights how such conversations are crucial in facilitating and reconfiguring our understanding of others [17].

Other academics have reinforced the representational value of artistic performances and the emotional and bodily engagement that may arise: *By expanding the possibilities of representation- using theatre, art, or multimedia, for example- scientists invite a more fully embodied response from their audiences* [11]. However, the arts not only offer an alternate form of mediation, but can also act as the subject of enquiry throughout the course of a research project [17]. This has been emphasised within the area of artistic research, which merges artistic practises with research and theory. Such research approach can allow for a deeper understanding of one's artistic work, as well as a more unified relationship with one's research. [13]. Although the area of artistic research remains rather ambiguous in its form, it develops an experimental and playful attitude toward research, which allows alternatives and speculations to thrive [3]. In Halberg's reflections on her own artistic research practise, she draws focus to *experiential reflections* that arise from her own experiences and the experiences of others. These are documentations of aesthetic experiences, which reflect both an empirical- and an artistic value, arising from the interplay of reflection and experience [12].

2.2   Awe

Within the realms of profound experiences lies awe. In the paper titled *Approaching Awe: A Moral, Spiritual, and Aesthetic Emotion*, the authors propose a conceptual framework to understanding the complex emotion of awe. Drawing from a literature review of works across various disciplines, including religion, sociology, philosophy, and psychology, the authors identify two predominant themes that are crucial for experiencing awe: vastness and accommodation. Vastness refers to encountering entities, phenomena, or experiences that surpass an individual's sense of self. While accommodation denotes the cognitive processes of adjusting mental structures that cannot assimilate a new experience, an appraisal to awe becoming necessary when faced with encounters that exceed one's prior knowledge, requiring an adaptation of their cognitive framework [16].

Applying the framework of awe to the domain of art, the authors argue that vastness can be represented in physical forms, such as large-scale objects, or in conceptual forms, such as depictions of god-like figures or capturing exceptional moments resulting from powerful forces. Additionally, artworks that elicit awe are not easily comprehended, but rather possess an element of obscurity and challenge the viewer's interpretation, creating a need for cognitive accommodation to fully experience the sense of awe [16].

The experience of awe has been found to evoke a shift in one's self-perception, wherein attention is directed towards entities or phenomena that surpass one's own magnitude, thereby diminishing personal concerns and objectives. Research indicates its potential to help people perceive themselves as integral components of a larger social framework, fostering an augmented sense of empathy and consideration toward others [18]. With these transformative potentials of awe, researchers have investigated the potential of synthetically eliciting the emotion by presenting participants with immersive VR environments, depicting nature scenarios. Findings indicated that the researchers achieved significantly higher reports of awe and presence under their experimental conditions [8].

In the study by Quesnel et al., the fundamental concept of awe was the foundation of a VR application specifically developed to improve mental well-being [19]. This application drew inspiration from the overview effect, a cognitive phenomenon encountered by astronauts when observing the Earth from outer space. The experiment was physically located in a mediation lounge, while the VR experience consisted of serene forest landscapes and underwater scenarios, ultimately transitioning into a view of the Earth from outer space. Using physiological measures, the designers observed that approximately half of the participants experienced goosebumps, with greater reports of awe. In addition, a significant majority of participants reported a positive sense of connection with nature after engaging with the application [19]. Therefore, inducing awe is not an unexplored realm in the field of VR, as previous research has demonstrated its feasibility. Furthermore, the work by Quesnel et al. suggests considering the surrounding physical environment before the VR experience initiates.

2.3 The Virtual Experience

Human escapism is a deeply rooted inclination. The earliest concept of VR can be dated back to 1935, in the novel Pygmalion's Spectacles by the science fiction writer Stanley G. Weinbaum [23]. The novel stands as a testament to our inherent desire to transcend the confines of reality and immerse ourselves in captivating alternate worlds. Almost ninety years later, the virtual world imagined in the book has gradually become a reality. In the virtual world, the body's presence is felt but not seen. Yet, there remains a sense of agency and the ability to perceive independently of the outside world. Thus, one can briefly escape the real world, as if they were fully engaged in the illusion before them [14].

This phenomenon above written has been the subject of classification by Slater as *immersion* and the *illusion of presence* [22]. A classification in which the level of immersion might correspond to different levels of the illusion of being in the virtual world. The illusion of presence presents the essence of an interesting paradox: the feeling of 'being there', despite one's conscious knowledge of its unreality. It is a perceptual illusion, but not a cognitive illusion, where the perceptual system instinctively reacts to stimuli. Although the cognitive system may later comprehend the illusory nature of the experience, the initial reactions have already transpired. Therefore, the powerful capacity of virtual reality lies in its ability to evoke genuine perception and elicit authentic responses, unfettered by conscious acknowledgment of its illusory nature [22]. This quality makes VR a powerful medium for evoking authentic and impactful emotional responses, providing a valuable avenue for creating art. Starting in 2017, La Biennale di Venezia launched the first competition for work in VR. However, despite the artistic merit of VR becoming clearer, it is still in its adolescent phase [25].

Exploring the concept of embodied cognition, Hsin-Chien's VR installation *Samsara*, allows the audience to experience the six realms of existence in Buddhism, by reincarnating the audience into the bodies of different persons and creatures [7]. He claims that *it is when we feel this world in different bodies that we may truly appreciate thoughts of others, empathize with them, and comprehend our existence in full.* Similarly, virtual worlds grant us the ability to inhabit impossible spaces. *To the Moon*, by the same artist, allows the individual to explore the surface of a new moon using imagery and tropes from Greek mythology, literature, and science; commissioned by the Louisiana Museum in Denmark [6]. This duality of embodied cognition and transport to impossible spaces exemplifies the captivating nature of VR as a medium. *In the Eyes of the Animal*, an immersive installation created by Marshmallow Laser Feast, explores the realm where virtual and real-world experiences converge. The experience presents an artistic interpretation of the sensory perspective of three species of animals, blending virtual and physical environments within a real-world forest setting. By anchoring the virtual experience in parallel with the real world, the installation provides a unique opportunity to investigate the interplay between human perception, the constructed virtual environment (VE), and the authenticity of the natural world [1].

Although VR has allowed a variety of artworks that cater to the sense of wonder and curiosity, there are works such as Taryn Simon's 'An Occupation of Loss' - a provocative VR piece exploring the anatomy of grief [21]. The artistic value of this piece is not determined by its experiential pleasantness, and the artistic aspiration transcends mere aesthetics, seeking to evoke emotions and provoke contemplation, surpassing immediate comfort.

3   Design Development Process

It was clear already from the initial stages of conceptualisation, that this project would take a creative approach with participatory elements. The creative design process in art is complex and multifaceted and can be approached in different ways depending on the individual style of the designer and the stage of the design process [5,15]. The activities undertaken throughout the project were not viewed as a linear succession of stages, but rather as an iterative process.

The beginning phases were shaped by independent research and artistic expression exploring the phenomenon of mycelium networks. Throughout the remainder of the process, collective discussion and diverse forms of creative expression remained essential. This was supported by dynamic activities such as body- and brainstorming sessions on the whiteboard, excursions to museums and the forest, as well as engaging with tactile mediums like clay and drawings, as shown in Fig. 1. This collaborative and multidimensional approach aligned with our commitment to artistic, participatory, and exploratory practises, paving the way for a rich exchange of ideas, insights, and emotions.

Fig. 1. Creative work with tangible mediums

3.1 Initial Participatory Workshop

As part of the early development stages, a group of four participants were invited to partake in a two-hour design workshop, as an instantiation of the participatory design methodology. Since the goal of the project was to provide a meaningful experience, the importance of focusing on the users was essential; hence, it would not seem rational to enter a design process that neglected the individual and their experience. As we wanted a high level of user participation, the participatory design approach was deemed highly applicable for creating the framework of the cooperative workshop. Furthermore, since the overall theme of the workshop deals with abstract topics of underground mycelium networks, it was important to facilitate a common space wherein participants felt comfortable to coopera- tively thinking outside the box, and challenge their own preconceptions.

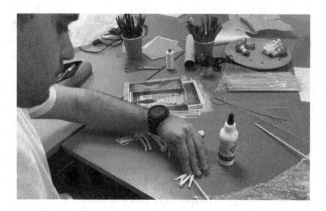

Fig. 2. Participant working with the mushrooms in the creative warm up exercise

The workshop was held in three parts; an initial introduction, a creative warm-up exercise as seen in Fig. 2, and finally, the evaluation of three different noninteractable VEs following a *think aloud* protocol. The workshop provided the groundwork for our participatory understanding, new design requirements related to both interaction and VE design, and validated pre-established conceptions related to the emotional arc of the experience, across the three acts. The framework for this initial workshop would shape the methods for the final evaluation; which itself would resemble a collaborative workshop. This will be further described in Sect. 5

4   Fragments of Fungi

The target audience for *Fragments of Fungi* includes young people with an interest in cultural and artistic experiences. The viewer embarks on a journey, traversing the hidden realms of the subterranean across three acts, as depicted in Fig. 3. Each act was designed and developed with the requirements acquired from the aforementioned participatory design workshop, along with internal discussions and design activities. Throughout the whole experience, a narrator accompanies the viewer, directing their interactions while also supplying narrative context and information.

PROLOGUE
The stage is set with an introductory speak about a network hidden in plain sight and introduces the viewer to the journey that lays before them; a journey into the enchanting world of mycelium.

ANTICIPATION
The viewer is presented with a forest at dusk and is asked to take a deep breath and settle into the surrounding environment. They are then prompted to pick up the glowing mushroom within reach, which shortly after disintegrates into particles in their hand, before fading out to the next act.

Fig. 3. From top: prologue and initial scene, middle: anticipation and mycelium scene, bottom: anti-space and outro scene

ANTI-SPACE

The viewer is situated underneath the forest floor, looking up at the roots of the trees they were previously surrounded by. As the narrator speaks about the phenomenon of mycelium networks, mycelium threads gradually begin to form in the distance, enveloping the visitor. Branches of mycelium then slowly grow towards the viewer, and as they touch the branches, the network further entangles them and light up - visualising the complexity of the network. As the viewer moves their hands through the mycelium network, the interconnected branches illuminate sequentially, extending into the distance. Simultaneously, reactive auditory feedback propagates from these branches.

Two distinct branches of mycelium emerge, and the viewer is prompted to be the bridge that unites them. As the branches are connected, a symphony of light and sound unfolds within the network and the trees above, which extends into the dark vastness. Fruiting fungal bodies appear besides the viewer, and as these begin to grow, the viewer grows with them, moving them up towards the surface, and the next act is initiated.

RELIEF

Surrounding the viewer are high-reaching mushrooms, standing tall as guardians of the delicate balance of the forest. The narrator speaks of viewing the world with refreshed eyes, attuned to the subtle miracles that exist beneath the surface.

The objectives of this project encompass not only the immersive VR experience, but also the emotional outcome that emerges from it. In addition to taking participants on a journey through the virtual realm of mycelium, the experience was extended to the physical world. The VR experience was to be held in a physical forest setting, followed by a collective discussion that would spark reflections and foster meaningful conversations. The combination of virtual exploration and real-world discussions would allow for a deeper connection with the subject matter and a shared exploration of its significance, enriching the experiences beyond the virtual realm. The group discussion was recorded and later transcribed for further analysis. The qualitative data from interviews and observations were analyzed thematically, identifying recurring themes, emotional responses, and patterns of understanding.

4.1   Technical Implementation

The implementation of the virtual experience was carried out using Unity version 2021.3.11f1, a powerful game engine that facilitated the integration of interactive elements and the creation of intricate VEs. The 3D objects were modeled and optimised in Blender version 3.4.1 and exported into Unity using the .fbx format. The application was targeted a Meta Quest 2, and the Oculus Integration package for Unity was used as the main interaction framework. As tracking outdoors was needed, the use of hand tracking was omitted, as this feature was more susceptible to overexposure in the sunlight compared to tracking the Touch controllers. The interaction techniques was simple; using just the trigger button to select and manipulate the virtual objects, along with collider-based physics to detect collision between controller and interactive objects.

The experimental feature of Application SpaceWarp by Oculus was used through a custom render pipeline (URP) providing increased computational overhead for standalone Oculus Quest development. In Unity, optimisation strategies as Level Of Detail on terrain objects and offline light rendering with lightmap textures were utilised. For visual feedback when touching the mycelium, the implementation of shaders was done in Unity ShaderGraph 12.1.6, while various particle effects were done in Unity VFX graph 12.1.6.

The audio-engine of choice was FMODUnity, using the Oculus Spatializer plugin for spatialisation and reverberation. The audio SDK provides spatialisation of monophonic sounds using generalised head-related transfer functions (HRTFs). It was important to include spatialised audio in order to guide the user's attention to the main action. During the process, it was of interest to compose our own soundscape for both the ambience and the auditory feedback triggered when touching the virtual mycelium. The sound design took inspiration from samples of fungi sonification, using authentic biodata. The soundscape was designed and synthesised using an Arturia MiniBrute 2 combined with Ableton

Live. To increase the expressivity of the auditory feedback, the hand/controller tracking velocity was captured and used as threshold values, enabling different audio samples with various tempo and pitch; high velocity would increase tempo and pitch. This reactive feedback would propagate from the location of the specifically touched mycelium branch, facilitated by the Oculus Spatializer plugin.

5   Framing a Collective Experience

With the experience being at centre for the entirety of this project, this was to also be reflected in the final evaluation; thus blurring the boundaries of the central artistic experience and the evaluating parts. This resulted in an almost performance-like excursion to a forest north of Copenhagen, Denmark, which involved four individuals over the course of two hours. The participants had varying backgrounds; an actor, a graduate in Diversity and Change Management, a graduate in Medialogy, and a neuroscientist.

The workshop was situated in a reclusive area of the forest amidst towering trees, following the structure illustrated in Fig. 4. Gathered in a circle on the forest floor, participants were given a general introduction to the project and what was to unfold during the workshop; additionally, all facilitators and participants took a moment to present themselves and complete a formal consent form. Finally, a brief introduction to VR interaction was demonstrated on the Touch controller. Afterward, as a means to attune the participants' senses to the surrounding forest, they were asked to spend a few minutes taking a picture of something that caught their attention. These pictures were saved and reintroduced later during the discussion part of the workshop. After completing the picture task, each participant was accompanied to a designated spot in the forest, scattered around the central space, where they would experience *Fungi fragments*. Facilitators helped initiate the experience and equip the participants with the head-mounted display, headphones, and controllers. Upon completing the experience, the participants would hand over the devices and find a pillow with a clipboard next to their spot. On the clipboard, a piece of heavy art paper, a pencil, and a brief instruction were attached. The instruction prompted the following: *Document how you are feeling in this moment through writing or drawing. Return to the group when you are finished.* The participants were left alone for this task and would return to the group when they wished. The three activities - photography, VR experience and documenting - are depicted chronologically in Fig. 5.

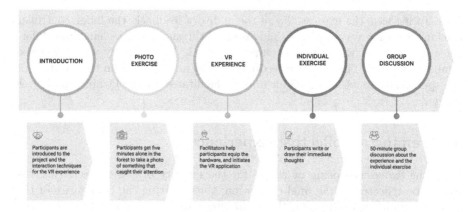

Fig. 4. Evaluation workshop structure

Fig. 5. From left: photo-exercise, middle: VR experience, right: individual writing exercise

Once all participants had returned to the group, a 50-minute group discussion was held, predominantly focused on their emotions during the experience. The discussion was recorded and later transcribed using an intelligent verbatim approach. To initiate the discussion, participants were asked to present their immediate experience as they had documented on paper. The continued discussion grew out of these initial manifestations, and subtle prompts from facilitators continuously guided the conversation without too much interference. In addition, printed pictures of the three VEs were laid out for the participants to have more concrete points of reference during the discussion. Toward the end of the discussion, the participants were asked to revisit the pictures, which they had taken in the beginning of the workshop, and asked what made them take the pictures in the first place and if anything had changed now looking back at it. As a finalisation of the group discussion and workshop as a whole, participants were asked if there were parts of the experience that they would have changed and their experience of the overall format of the workshop.

Throughout the entire workshop, it was important for us/the facilitators to provide a safe poetic space, in which the participants would feel comfortable to share their emotions. This was reflected in both the workshop activities but also in the physical setup, which catered to the simple details and having the group discussion unfold in a circular position on the forest floor submerged in the greenery of the forest and the soundscape of the birds. The setting is captured in Fig. 6.

Fig. 6. Picture of the group post-experience discussion setting

6   Findings

This section is structured around two main themes of investigation, wherein the results will be presented and discussed. Subsequently, these findings will be challenged in light of the inherent bias within this project. For readability purposes, throughout the remainder of this paper, the four participants will be referred to by an alias. All participants, except Bastian, had little to no prior experience with VR. Firstly, relevant quotes related to *the illusion of presence* will be presented, as this is an important element to generate genuine emotional responses in a VE. The feeling of "being there" in the virtual environment was explicitly conveyed by Carl when asked about the transition from VR back to the real world:

> I felt a little sad because I was leaving, like waking up from a nap, but also in a nap in which you had a really good dream. And you wake up and you're like, if I could just fall asleep right now, just again, I could maybe continue it for a little bit longer.

When asked about presence within the VE, Luna expressed this as: *(. . .) like being in two places at once; the real world, and what felt very like the real world; the VR (...) I guess I was quite present during the seven minutes.* All participants agreed to this sense of presence throughout the whole experience. The physical setting of the forest further added to this, as Mark noticed: *you have the leaves underneath feet. And the wind,* followed by Bastian: *and also when you walk around in the experience, (...) the leaves are kind of crunching, that just adds to the whole experience.* It can therefore be suggested that all participants had indeed experienced a high sense of presence. The presence was increased by the use of headphones, providing immersive audio, and the sensation of tactile stimuli from the surrounding forest environment, which would serve as a foundation for evoking intense emotions such as awe.

6.1 An Awe-Inspiring Experience

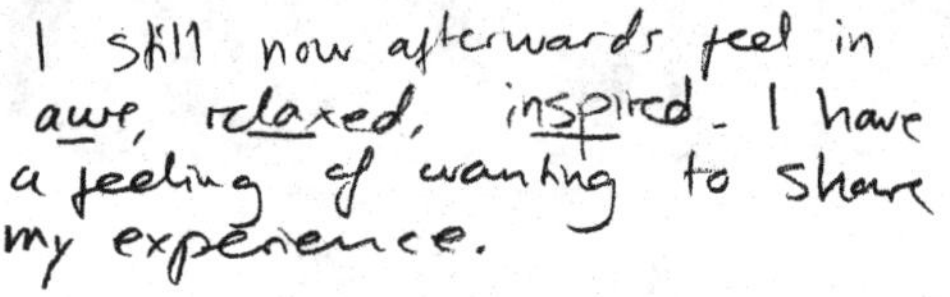

> I still now afterwards feel in awe, relaxed, inspired. I have a feeling of wanting to share my experience.

Fig. 7. Luna's individual exercise

Though being a nebulous and transcendent concept, the participants did describe their experience as awe-inducing. This observation was prominent without directly imposing the word on the participants and was initially broadened in the discussion by Luna, who highlighted the term in her individual reflective documentation, as depicted in Fig. 7. Upon further elaboration, Luna explained:

> I often get it [the feeling of awe] when I see nature; like when you come to the Grand Canyon. In the same sense that it was beautiful, I felt small (..) I feel like the awe-feeling is when you realise you are small (..) and one day you will be gone; but the world is still here

The remaining participants consented to the same feeling and highlighted its presence towards the second and last scenes of the experience. Prominent in the further discussion of awe were elements of feeling small or being in the presence of something much larger than one self, as initially described by Luna. Bastian directly linked the scale of the mushrooms to the feeling of awe and described the last scene as follows: *This was also a very explicit representation of this feeling [awe].* Within the same topic, Bastian spoke to the complex nature of awe as the feeling of being scared; *it is a bit of an ambiguous feeling when you are in awe, and I think being scared is part of that.*

In the talks of a larger entity, associations to god-like creatures and tribal encounters came to mind. Mark elaborated on this association: *I felt that the scale, with small birds and huge mushrooms, really made it feel like they were gods (. . .) like big Greek statues.* In continuation Carl adds: *(. . .) the God aspect of it is its complexity.* Both notions speak to the scale and complexity as elicitors of the religious, and both connect to the idea of vastness related to awe.

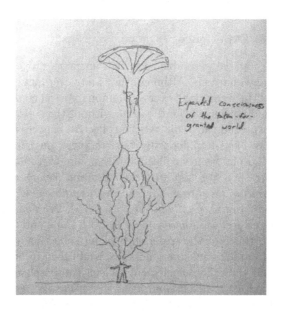

Fig. 8. Carl's individual reflection exercise

Drawing upon the conceptual framework outlined in Sect. 2.2, the emotion encompasses two primary themes: vastness and accommodation. The concept of vastness was prominent in the discussion, but understanding the cognitive process of accommodation of participants presents a greater challenge. However, Carl expressed a transformation in how he understood the surrounding forest after the experience, whilst pointing to the drawing on this individual reflection exercise depicted in Fig. 8:

> Coming to a new understanding of a world that was taken for granted. (. . .)
> I forget to see how complex they really are. (. . .) With this experience it
> becomes easier to understand how complex it actually is.

This newfound understanding was expressed in a similar way by Bastian, as he stated *I felt it was a very meditative journey that helped me realise the subjective scale of human consciousness and I was introduced to a new state of interconnectedness.* Luna describes another shift happening when she entered the second scene: *that is when my brain stopped thinking about if it was artificial or not, because it was so new and different.* In the abstract mycelium world she felt more perceptive, whereas the initial forest scene had her being analytical. This shift conveys the captivating power of the arts - whereas the forest is to be experienced in the physical world, the imaginative power of art and VR allowed Luna to enter realms otherwise unseeable. Carl talks about the same experience and describes the shift as: *shifting from rational to perhaps emotional or interpretive.* These views resemble the idea of accommodation, in the sense of acquiring a new cognitive framework due to a new experience.

6.2 A Collective Experience

Through the workshop, we were able to bring four people together in conversation about nature and emotions. Although they had never met each other before, participants shared their emotions over a common experience. It was noted that the experience would not have been the same without the collective elements, as the participants expressed they enjoyed the collective aspects of the experience and would not have liked it differently. Carl reflected on what it would have been like if there had not been a collective discussion: *it would have been super lonely (...) and then you are with all these ideas and feelings in your head and you cannot communicate them to anyone or anyone who would understand.* This speaks to a phrase by Luna: *can't help but notice (...), we come out from a short experience, and we all feel like talking about it.* Both Luna and Carl's statements underline the importance of the experience being collective and the collective potentials arising from artistic experiences, as established in Sect. 2.1.

Luna enjoyed all the different elements which composed the entire experience: *having this [the group discussion] as a part of the experience, to reflect by yourself, doing everything in parallel, I think it wouldn't have been the same if it had just been me, out here by myself.* This aligns with the initial goals set for the workshop and supports the deployment of more participatory evaluations. It was clear that all participants felt all elements as parts of the experience: *us connecting afterwards really adds to the experience* (Mark). The resulting product of this project was not just about going through the virtual environment, but all other aspects played a significant part in the experience as a whole; from being scattered around the forest and dwelling on ones own reflections to gathering on the forest floor and sharing thoughts and perspectives. Mark specifically recalled the act of physically bringing your own documented thoughts into the collective space and how such small details enriched the flow of the experience.

In further reflection, the participants spoke to a calmness and a meditative state rising from the experience: *it was a kind of meditation experience*(Luna), *As you're coming out of that experience, also really calmed me down and made me just really relaxed and grounded* (Mark). These feelings were related to both the virtual experience and the physical setting in the forest, as Luna put it: *it was a good choice to be in a real forest. If it was on the main plaza in Copenhagen having people talk and bicycle, it would have been kind of weird.* These notions once again emphasise the importance of pertaining to the whole experience and how the forest brings a different layer to the discussion: *Looking around at the forest, I feel like I wish I could see the level of detail that I was able to see in the scenes.*

6.3 Final Discussion

It's important to highlight that the above findings stem from a thoughtfully orchestrated combination of factors: the forest backdrop, the evocative tone set by facilitators, and group conversations on the forest floor. The workshop's overall atmosphere likely primed participants to delve into these abstract themes. Moreover, the very notion of engaging with mushrooms could align with spiritual and abstract concepts for some. Looking at the findings related to awe, all participants explicitly stated this emotion. In relation to the conceptual framework introduced by Keltner and Haidt, the most apparent element was the perceived encounter with something bigger than themselves, whether it be in size or complexity [16]. As for accommodation, it can be suggested that the transformation in how the participants viewed the forest after the VR experience could imply the occurrence of the cognitive process, though this cannot be validated. However, given the highly subjective nature of emotions, the most reliable indications of awe lie within the first-hand accounts provided by the participants. It is relevant to acknowledge that awe is made up of more elements than accommodation and vastness, such as sense of time, self-loss, and connectedness, although the above findings speak to multiples of these variables. A summary of the qualitative results can be seen on Fig. 9.

It is important to recognise that a central element for this project was the experience of being located in the real world forest. In addition to providing an important point of reference for discussion, the surroundings offered multisensory stimuli furthering presence within the VE. To ascertain the contributions of the forest setting to the overall experience, conducting a comparative study in a more controlled environment, like a laboratory, could provide valuable insights. Same applies for the impact of post VR experience group discussion atmosphere, to shed light on its contribution to the overall experience and emotional outcome.

It was prominent that all participants enjoyed the collaborative aspects of the workshop, which made an otherwise individual experience, that is, VR, collective. As affirmed by [20], humans are eager to participate, and this is truly noticeable within the collective workshop. These collaborative efforts were not only a reflection of the final workshop, but are also a reflection of the entire process of the project. We have continuously valued new thoughts and ideas, and as such these are reflected in the final outcomes. *Fragments of Fungi* allowed the participants to enter the awe-inspiring world of mycelium networks, and afterwards share this with others. As participants accentuated the need for more experiences that provide space for reflection and collective dialogues, this underscores the potential of more holistic VR experiences, and suggests that designers should be mindful of both before and after the VR experience has occurred.

The reliance on voluntary participation of participants in the study introduces bias, as people who were mentally prepared to disclose and express themselves have been more likely to participate. Similarly, had the participants been more introverted and less eager to do so, the same result might not have been achievable. The framework of the workshop was highly dependent on the collaboration of the participants, which could have led to less successful outcomes.

Findings	Description	Relevant Quotes
Awe-inspiring experience	Participants reported feeling awe during the VR experience, especially in the second and last scenes	"I feel like the awe-feeling is when you realize you are small (..) and one day you will be gone; but the world is still here." "The God aspect of it is its complexity." "It is a bit of an ambiguous feeling when you are in awe, and I think being scared is part of that."
Collective experience	Participants valued the collective aspects of the workshop, especially the group discussion after the VR experience	"Having this [the group discussion] as a part of the experience, to reflect by yourself, doing everything in parallel, I think it wouldn't have been the same if it had just been me" "Us connecting afterwards really adds to the experience."
Transformation in understanding	Participants' perception of the forest environment transformed after the VR experience, leading to a new understanding of its complexity	"Coming to a new understanding of a world that was taken for granted. (…) I forget to see how complex they really are." "Helped me realise the subjective scale of human consciousness and I was introduced to a new state of interconnectedness" "Shifting from rational to perhaps emotional or interpretive"

Fig. 9. Summary of Qualitative Findings

Furthermore, although the participants had different backgrounds, they all had their appreciation of art and culture in common. And though these qualities align with our target audience, it introduces the risk that attaining the same findings would likely be compromised with participants exhibiting different degrees of pre-existing art appreciation. Given that the workshop was not conducted under controlled conditions but was taken out of a laboratory setting, this strongly advocates reproducing the setup with a different set of participants to ascertain the reliability of the results.

7   Conclusion

This project explores the interplay of the arts, nature, and technology, and how the union of these can allow new means of artistic expression that facilitates emotions and dialogue. From this position we present an artistic virtual reality experience that uncovers the hidden world of mycelium networks, an experience titled *Fragments of Fungi*. As part of an evaluative workshop, four participants were brought to a physical forest setting, explored the virtual world, and engaged in a collective conversation. The findings suggest that the participants indeed experienced awe and related emotions, such as the feeling of a smaller self and a higher connectedness.

It is believed that these results would not have surfaced without the collective parts of the workshop. We have continuously, throughout this study, emphasised the importance of enveloping the virtual experience in a prolonged holistic experience, that caters to further reflection and dialogue among individuals. This paper should not be regarded as a definitive guide for developing awe-inspiring VR experiences or a framework for eliciting conversations. Instead, it serves as an illustration of the advantages and the desire for such reflective spaces, and peoples' desire to engage collectively and share their otherwise lonesome VR experience. Furthermore, it emphasizes how facilitating this sense of connection afterward can potentially amplify emotional experiences within the virtual environment. In light of the findings, we propose a greater emphasis on harnessing the potential of deploying a comprehensive artistic experience induced through virtual experiences.

References

1. In The Eyes of the Animal kernel description. http://intheeyesoftheanimal.com/behind-the-scenes Accessed 22 May 2023
2. Ankomah, P., Vangorp, P.: Virtual reality: a literature review and metrics based classification. Comput. Graph. Visual Comput. (CGVC) **2018**, 173–181 (2018)
3. Arlander, A.: Artistic research does 1 - Investigação em Arte e/como Interdisciplinaridade — Artistic research and interdisciplinarity (01 2016). https://doi.org/10.24840/978-989-98745-8-9
4. Asad, M.M., Naz, A., Churi, P., Tahanzadeh, M.M.: Virtual reality as pedagogical tool to enhance experiential learning: a systematic literature review. Educ. Res. Int. **2021**, 1–17 (2021)
5. Bonnardel, N., Wojtczuk, A., Gilles, P.Y., Mazon, S.: The creative process in design. The creative process: Perspectives from multiple domains, pp. 229–254 (2018)
6. Chien, H.: Chien's new media art projectto the moon vr (Sep 2018). https://hsinchienhuang.com/pix/_3artworks/i_ToTheMoonVR/p0.php?lang=en
7. Chien, H.: Chien's new media art project samsara (Sep 2021). https://hsinchienhuang.com/pix/_3artworks/i_samsara/p0.php?lang=en
8. Chirico, A., Ferrise, F., Cordella, L., Gaggioli, A.: Designing awe in virtual reality: an experimental study. Front. Psychol. **8**, 2351 (2018)
9. Chirico, A., Yaden, D.B.: Awe: a self-transcendent and sometimes transformative emotion. The function of emotions: When and why emotions help us, pp. 221–233 (2018)
10. Dewey, J.: Art as experience. In: The richness of art education, pp. 33–48. Brill (2008)
11. Gergen, M.M., Gergen, K.J.: Performative social science and psychology. Historical social research/Historische sozialforschung, pp. 291–299 (2011)
12. Hallberg, G.W.: Artistic research in sisters academy-sisters academy in artistic research. Peripeti **18**(34), 44–63
13. Hallberg, G.W.: Sisters academy as a space and time for experimentation and healing. In: Periskop-Forum for kunsthistorisk debat, pp. 136–155. No. 24

14. Huang, M.H., Tsau, S.Y.: A flow experience analysis on the virtual reality artwork: la camera insabbiata. In: Proceedings of the International Conference on Machine Vision and Applications, pp. 51–55 (2018)
15. Huber, A., Leigh, K., Tremblay, K.: Creativity processes of students in the design studio. Coll. Stud. J. **46**(4), 903–913 (2012)
16. Keltner, D., Haidt, J.: Approaching awe, a moral, spiritual, and aesthetic emotion. Cogn. Emot. **17**(2), 297–314 (2003)
17. Leavy, P.: Social research and the creative arts: An introduction. Method meets art: Arts-based research practice, pp. 1–24 (2009)
18. Piff, P.K., Dietze, P., Feinberg, M., Stancato, D.M., Keltner, D.: Awe, the small self, and prosocial behavior. J. Pers. Soc. Psychol. **108**(6), 883 (2015)
19. Quesnel, D., Stepanova, E.R., Aguilar, I.A., Pennefather, P., Riecke, B.E.: Creating awe: artistic and scientific practices in research-based design for exploring a profound immersive installation. In: 2018 IEEE Games, Entertainment, Media Conference (GEM), pp. 1–207. IEEE (2018)
20. Simon, N.: The participatory museum. Museum 2.0 (2010)
21. Simon, T.: Taryn Simon – tarynsimon.com
22. Slater, M.: Immersion and the illusion of presence in virtual reality. Br. J. Psychol. **109**(3), 431–433 (2018)
23. Weinbaum, S.G.: Pygmalion's spectacles. Simon and Schuster (2016)
24. Yang, Y., Hu, J., Jing, F., Nguyen, B.: From awe to ecological behavior: the mediating role of connectedness to nature. Sustainability **10**(7), 2477 (2018)
25. Yoo, K., Gold, N.: Emotion evoking art exhibition in vr. In: Proceedings of the 25th ACM Symposium on Virtual Reality Software and Technology, pp. 1–2 (2019)

Creating a Methodology to Elaborate High-Resolution Digital Outcrop for Virtual Reality Models with Hyperspectral and LIDAR Data

Douglas B. de Castro$^{(\boxtimes)}$ and Diego Fernando Ducart

Universidade Estadual de Campinas (UNICAMP), Campinas, São Paulo, Brazil
castrodouglasdev@gmail.com

Abstract. Outcrops are essential for geological interpretation on the surface, with the advancement of technology, outcrops can also be studied in the office through three-dimensional outcrop models. Photogrammetry and laser scanning techniques allow the creation of a three-dimensional digital outcrop model (DOM) for out-of-field analysis. Virtual reality technology is a human-computer interface that allows direct user interaction with a virtual environment. In geosciences, virtual reality (VR) is currently widely used to overcome difficulties encountered in the field, the quality of the analysis increases by allowing the collection of a greater number of qualitative and quantitative data in the office. The combination of DOM and RV is used in geosciences to make data extraction more accurate and intuitive for the geologist, but photogrammetry and LIDAR data can reach millions of points and present an excessive density for processing when collected on large scales by airborne platforms. Although some areas of science such as engineering have techniques for optimizing 3D models, geosciences do not yet have an optimization protocol that allows integration between DOM and VR without loss of relevant resolution aspects for geological interpretation. Given the present scenario, the proposal of this project is to develop a valid and replicable method for creating and optimizing a high-resolution three-dimensional outcrop model, using LIDAR and hyperspectral readings of the mining front of the Au Córrego do Sítio deposit as a database.

Keywords: geotechnology · geoprocessing · scan data · geo interpretation · digital interpretation

1 Introduction

Outcrops are bodies of rock that are exposed on the surface by natural events on the planet or anthropological events. Outcrops in general are a way of visualizing and interpreting surface and subsurface geology [1]. Collecting quantitative information in three dimensions (3D) in the field is a fundamental requirement in many geological, structural, stratigraphic, sedimentological, geomorphological, and geotechnical studies [2].

© ICST Institute for Computer Sciences, Social Informatics and Telecommunications Engineering 2024
Published by Springer Nature Switzerland AG 2024. All Rights Reserved
A. L. Brooks (Ed.): ArtsIT 2023, LNICST 565, pp. 91–93, 2024.
https://doi.org/10.1007/978-3-031-55312-7_7

Rock exposures can be represented by 3D digital models generated from photogrammetric or laser scanning methods [3]. These models are known as Digital Outcropping Models or DOM [4]. The spatial analysis process can be carried out from measurements in the same field or from DOM in an office [5].

DOM has become much more accessible and used in modern geosciences [6]. Although it does not replace geological field activities, DOM is a powerful additional tool for geosciences. The possibility of "bringing" the outcrop to the laboratory allows a great optimization of time and money, in addition to allowing more complex analyses to be carried out with a greater level of detail [7].

Many areas of geosciences have adopted the use of the digital outcrop model [5]. In the hydrocarbon industry, it is used to provide a qualitative check that validates reservoir models [8]. The field of use of Digital Models has also expanded to geological mapping, with the aim of defining patterns of lithological, mineralogical and chemical alteration characteristics [2]. Collecting data from three-dimensional models allows measurements to be carried out remotely in the office.

DOMs can be generated from photogrammetric or laser scanning (LIDAR - Light Detection and Ranging) methods [4]. When DOM is fused with hyperspectral images collected from the same outcrop, the result can help with lithological and mineral mapping. Hartzell [9] combined hyperspectral reading data with models generated via LIDAR, thus being able to identify different lithologies after orthogonally projecting hyperspectral images onto the digital outcrop model.

DOMs can be analyzed and interpreted with the support of Virtual Reality (VR) tools. Digital files generated by these methods may have sizes that are unfeasible for processing in virtual reality, overloading computers when placed on viewing platforms for data collection [6]. However, in the construction area, some mesh optimization techniques are used to reduce the need for processing the 3D model [10].

According to Chen [10], 3D models generated from point clouds can be optimized through different reconstruction techniques. Although there are appropriate proposals for digital model optimization processes, none have yet been applied and validated for digital outcrop models. This project aims to test and validate different methods of creating an optimized DOM for geological studies through virtual reality, preserving precision for its application within geosciences. For this, hyperspectral and LIDAR data already collected on a mining front of the Au Córrego do Sítio deposit will be used. The Córrego do Sítio Gold Deposit is situated in the eastern region of the Brumal district, within the municipality of Santa Bárbara, Minas Gerais, Brazil.

2   General Results

2.1   Main Results

This project's general objective is to create and test a valid and replicable method for creating and optimizing digital outcrop models. The method must allow the extraction of qualitative and quantitative geological information with precision and high resolution from virtual reality.

2.2 Specific Results

Between the specific objectives, we can find the creation of hyperspectral and LIDAR data processing with interactive protocols between hyperspectral data. The test of different mesh optimization methods on the DOM, so that the model becomes lighter and still has a level of detail and accuracy sufficient for data collection through virtual reality. The data must be validated in optimized models within virtual reality with field data. Using C++ mesh programming we expect to train a GPU code so that the optimization process can be done in an automated manner.

3 Methods

The methods used for data processing are parallel processing techniques such as MPI, CUDA, and OpenMP. The model used will be the SIMT (Single Instruction Multiple Threads) type, where SM (Stream Multiprocessors) is used to share memory. In the memory sharing model envisaged, the CPU and RAM store variables and code. RAM memory is slow compared to the CPU and creates the need for cache memory.

References

1. Trinks, I., et al.: Mapping and analysing virtual outcrops. Int. Rev. Econ. (2005)
2. Bistacchi, A., Massironi, M., Viseur, S.: 3D digital geological models: from terrestrial outcrops to planetary surfaces (2022)
3. Assi, A.: Light detection and ranging (2016)
4. Fabuel-Perez, I., Hodgetts, D., Redfern, J.: Integration of digital outcrop models (DOMs) and high resolution sedimentology - workflow and implications for geological modelling: Oukaimeden Sandstone Formation, High Atlas (2010)
5. Keogh, K., et al.: Data capture for multiscale modelling of the Lourinha Formation, Lusitanian Basin. An outcrop analogue for the Statfjord Group, Norwegian North Sea, Portugal (2014)
6. Nesbit, P., Boulding, A., Hugenholtz, C., Durkin, P., Hubbard, S.: Visualization and sharing of 3D digital outcrop models to promote open science. GSA Today **30**(6), 4–10 (2020)
7. McCaffrey, K.J., et al.: Unlocking the spatial dimension: Digital technologies and the future of geoscience fieldwork. J. Geol. Soc. (2005)
8. Jones, R., et al.: Extending digital outcrop geology into the subsurface (2011)
9. Hartzell, P., Glennie, C., Biber, K., Khan, S.: Application of multispectral LiDAR to automated virtual outcrop geology. ISPRS J. Photogrammetry Remote Sens. (2014)
10. Chen, Z., Ledoux, H., Khademi, S., Nan, L.: Reconstructing compact building models from point clouds using deep implicit fields. ISPRS J. Photogrammetry Remote Sens. (2022)

Games

Hearing Sounds Through Different Ears:
A Video Game Case Study

Gabriel Dargains Gonzaga[(⊠)]

Federal University of the State of Rio de Janeiro, Rio de Janeiro, RJ 22290-255, Brazil
`Gabriel.dargains@edu.unirio.br`

Abstract. This article seeks to demonstrate that different listening vantage points and types of listeners are established in open world video games to create meaning and immersion. Using *Dragon Age: Inquisition* (2014) as a case study, I analyze its sound image through Smalley's concept of space-form (2007) to find these points. Even though all sound is spatialized in the sound-image, not all sounds are located in the 3D navigation space. Positioned and non-positioned sounds (in reference to the player character) lead us to understand there's a combination of both peripatetic and fixed listeners during the gameplay. In the case study, by soundwalking in one of the game's zones we conclude there is an overlap of listening vantage points that do not refer to the same space. This helps the players have a holistic view of the soundscape, making it more favorable to express meaning through the ambience's sound design.

Keywords: ludomusicology · ambient sounds · space-form

1 Introduction

Through this article I'll present different listening perspectives in open world video games, using *Dragon Age: Inquisition* as an example. While there are articles about soundscape and soundwalk in video game context[1], I couldn't find literature addressing listening perspectives that we can find while playing a navigable game. I hope to contribute with the argument that these perspectives mean something and that meaning can be manipulated when composing video game ambience. For the identification of these perspectives, that from now I shall call vantage points[2], I will analyze the space-form in the game's sound image and find that some of the sounds are rendered in relation to the character's location and orientation, while others are ambiguous to both.

---

[1] Hambleton's take on Westerkamp's Soundwalk opens the interpretation to soundwalk in virtual spaces. She uses soundwalk as a method to map out the soundscape [9]. For other articles using soundscape as a method in analyzing video game sounds, see Galloway [8] and O'Hara [13].

[2] This is the term used by Smalley in [15], to delineate a point from where the listener will turn his attention to the sounds in space. The reason to not choosing a term like "point of listening", analogous to 'point of view', can be a way to reinforce that the vantage point is also about perceiving or observing space and not simply hearing.

© ICST Institute for Computer Sciences, Social Informatics and Telecommunications Engineering 2024
Published by Springer Nature Switzerland AG 2024. All Rights Reserved
A. L. Brooks (Ed.): ArtsIT 2023, LNICST 565, pp. 97–110, 2024.
https://doi.org/10.1007/978-3-031-55312-7_8

The first chapter will introduce open world video games, vantage points and the concept of sound image. In the second chapter, we will discuss space representation in object-based audio and Stockburger's categorization of sound objects [18] applied to *Dragon Age: Inquisition* (that I'll refer to as DAI for brevity). Then in the third chapter after a brief explanation of Smalley's space-form, I'll present an analysis using some of the concepts to identify the different vantage points in a zone within the game.

1.1 Open World and Vantage Points

Open world video games have players navigating 3D maps through avatars [10]. The player controls a character in that world, through a first person or third person view - in the first person the world is shown as if through the eyes of the character; in third person the player can see the controlled character as an observer, external to its body. Both first person and third person imply a point of view, one where the camera orientation is attached to the character's and another where the camera orientation is independent of the character's orientation. In this video game genre, the receiver of sounds is attached to the camera, so if the camera moves, panoramic visual and sound space also does. In both cases, the camera functions as eyes and also ears into the game world.

Since we have a point of view, we will have a listening vantage point too. Listening vantage points, as described by Smalley, can be fixed, variable or peripatetic in location and orientation [15]. The ones that interest us for this matter are the fixed and peripatetic, fixed being the position (location and orientation [11]) of the listener. Peripatetic meaning the listener has a moving position. The camera-sound receiver attached to the controlled character moves his point of view in the navigable space, so it can be considered as a peripatetic listener. We can't forget the player sitting in front of the screen is playing from a fixed vantage point. Those are the first immediate vantage points in the game experience, they both coexist as long as some sounds are directed to the character and some sounds are directed to the player.

1.2 Space in Sound

Movement and space are characteristics we can find regardless of the listening vantage point. Every sound carries some degree of spatial and causal qualities and information, we naturally imagine or 'feel' these qualities as they appear. While some qualities can be understood through spectromorphological [16] analysis, others are quite difficult to establish in an objective way. The feeling and imagination of the qualities do not belong to the auditory domain, they occur through transmodal perception of the sound[3] - when we listen to a sound, we are also virtually feeling it through other senses and that happens with every sensory experience.

Our transmodal perception of the sounds might hint on vision and touch as we often describe sounds relative to its size or texture - where I come from it's not uncommon to

---

[3] 'Transmodal' is not the only term possible, but it is the one Smalley uses. Lars Elleström [7] will refer to it by a sensorial multimodality in the first chapter in the book Media Borders, Multimodality and Intermediality. Chion will call it transsensorial [5]. I also recommend reading Rodolfo Caesar's writings on sound's transsensoriality [2] and [3].

hear people calling higher pitch sounds 'fine' or 'tiny' sounds, but we see transmodal links happening even with commonly accepted terms like grain, wet, dry, etc.

With audio recording (or through memory) we can attempt to inspect the image that forms in our imagination as the transmodal links occur. This image is referred to as sound-image or *i-son* as coined by François Bayle[4]. When discussing space-form, Smalley will break down this sound-image into spatial forms to be able to discuss space within acousmatic music. In video games, as with every audiovisual media, sound and image is produced separately, so all the spatial qualities in sound comes from sound recording, editing and rendering.

2 Representation of Space in Games and Music

Sound in both video games and recorded music will often be heard in similar conditions - except when in concert or performance situations - with audio output devices such as headphones, loudspeakers or television. The sound image in both cases have the same 'depth' and dimensions, even with different mixing methods – game engines that deal with 3D space have multichannel audio, what makes the directionality of sound more refined but the frame is still the same, a pair of headphones or loudspeakers. Although the listening conditions are similar, we find that there is a striking difference on how we perceive space through hearing. Spatialization[5] is quite a complex topic to discuss in detail here, but there is some basic distinction we have to make before proceeding.

Nicolas Tsingos [14] can help us start the discussion by bringing up Roberta Klatzky's [11] distinctions on spatial representation. He mentions that in order to represent space, we first need a frame of reference and explains them briefly:

> [...]An egocentric frame of reference encodes object location relative to the position (location and orientation) of the observer or "self." An allocentric frame of reference encodes object location using reference locations and directions relative to other objects in the environment. [...]

> For interactive rendering, video game applications programming interfaces (APIs) generally express the positions of audio objects as allocentric world-space Cartesian coordinates. The coordinates of the objects may be converted to a listener/egocentric frame of reference at rendering time depending on the player's position, in particular if a single perspective has to be rendered (e.g., on headphones). (pp.244–245)

By 'player's position' he means the player-character (PC, the character controlled by the player). Since the player themselves doesn't move much during gameplay, they'll

---

[4] For a more original reference, see [6] chapter 9, footnote 14 at page 259. Chion has a transcription of the definition of the term sound-image.

[5] *Immersive Sound: The Art and Science of Binaural and Multi-Channel Audio* [14] has various articles on spatialization and you can find a more in-depth discussion there. The point is that spatialization creates a more refined space by using virtual outputs in various directions, but the output devices used are mostly the same – the same 'frame' or boundary is used.

look at their screens and stay still for the most part. Sounds help create the sense of causal relation between sound object and its visual object, but since that causality is artificial, the proximity of the sounds may be used to simulate distances and virtual locations. For example, an audio object can be near the PC-listener in cartesian coordinates but have qualities of a sound being heard far away.

It's important that he mentioned perspective, because orientation is mostly what tells players where objects are in the surrounding space. In first person view this is literal, but in third person even while the camera moves with the character, both might have independent orientations. In *Dragon Age: Inquisition* the camera's field of view is approximately 100° and it is independent of the character's orientation. The character's egocentric reference is important because the camera can't stray too far away from it, but the camera's reference are the eyes and ears that allows players to experience the game world.

Being able to face the character and the camera in opposite ways helps navigation and a better assessment of the surrounding space, but most of the time the player will want to see where the character is heading. The merging of character and camera's orientation reinforces roleplay, the player recognizes the intimacy between the character and themselves more readily[6].

2.1 Positioned and Non-positioned Sounds

Ideally, in most cases the listener and player's position will be between both audio outputs and facing a direction perpendicular to the axis left and right in the stereo image. This way both music and video game listener's 'vantage points' makes the sound image an allocentric representation. To simplify the matter, I'll be referring to the audio objects positioned (located and oriented) in world-space's cartesian coordinates and those that are not positioned, as positioned and non-positioned sounds. Positioned sounds are the ones in the world-space that are rendered in reference to the player-character or camera's position (egocentric to the PC-Camera) and non-positioned sounds are the ones that have no relation to the player-character nor camera location and orientation in the game world-space (Allocentric to the player). In DAI's case, the reference to positioning is always the camera.

Positioned sounds can still be acousmatic - sounds whose sources are not seen - if their sources are not in the camera's field of view. They expand the proximity by using the PC's circumspace[7] and hinting at objects directionality and distance. It becomes easy to spot an enemy if their sound effects are present even if the player isn't directly looking at them. Positioned sounds are the indication of a peripatetic listener and non-positioned indicates a fixed listener. Non-positioned sounds share the same orientation reference as the player, since they are facing the screen and not navigating themselves. They coexist and serve different purposes.

---

[6] This intimacy can be thought of as immersion or as Modena and Parisi argues, a *process of becoming*. It is relevant to understand that camera and character's orientation can create or tear down the distance between player and avatar, but the distance is always present causing friction. Perhaps this is a good point to further the topic.

[7] "Space around the listener" p.48 – Smalley [15]

The main purpose of the positioned sounds is creating depth and directionality, making navigation possible (directionality includes height too). The main purpose of non-positioned sounds is presenting information directly to the player instead of making it go through the camera reference[8]. But as we'll see, non-positioned sounds often create ambiguities, not making it clear as to which listener the sound is being addressed to. When there is visual navigation but some sounds do not accompany the movement, we can't say that the camera and the player hold the only vantage points.

Camera and character position means nothing if the game world has no objects to be perceived by them. Tsingos refers to audio objects, but in video games they are often attached to entities in the game world, be it a creature, a non-player character (NPCs), representation of inanimate objects or even structures and zones. As he mentions, these objects are generally located in the game world-space and are rendered in relation to an ego (which in this case is the avatar or camera, not the player). We'll see now what kinds of sounds there are in the game.

2.2 Sound Categorization

Stockburger [18] has recognized five types of sound objects in video games according to their function: speech sound objects, effect sound objects, zone sound objects, score sound objects and interface sound objects. I shall explain the categories as we see examples of how they are used in Dragon Age: Inquisition.

Speech Sound Objects
The speech sound objects can be found either located as NPC's greetings and short phrases as the character walks by or in cinematic cutscenes and dialogues. The use of the speech objects in cutscenes are not located in the map's coordinates, the voice is centralized in the sound image for the sake of clarity just as in cinema [5]. In greetings, short phrases and dialogues that are not cinematic, they are subject to the PC-camera's position, so they will have more volume as the characters approach each other and will change directionality in the sound image as the player character turns in different directions. During dialogues, player's frequently have to choose how to reply, creating the feeling of choice and participation in the narrative. The answers are set and there might be additional options depending on how you created or progressed the character. Being able to perceive directionality in dialogues makes the players feel as if they are part of the conversation circle, but having to choose how to reply brings them back to the roleplay, as if they were between observing and controlling the PC.

Effect Sound Objects
Effect sound objects refer to the sounds made by objects and characters in the game world-space. They might be sound effects from actions like attacking, jumping, footsteps in different terrain, blacksmiths striking the anvil, a creature's growl and objects like

---

[8] In DAI, the camera is the egocentric reference for the sound receiver, but in some other games this might not be the case. *Assassin's Creed Syndicate (2015)*, for example, utilizes the PC orientation as reference until the camera crosses a certain angle, then the camera serves as reference.

campfires, torches, forges – these are mostly positioned. In this game, during combat and roaming, you control a team of four members and even though there is a protagonist, the main PC's actions are heard from the camera's vantage point[9]. This reinforces the strategist playstyle, letting the players distance themselves and looking at each member from outside.

Object's sound sources are often used to give perspective while navigating. In most positioned sounds you can hear its approach and recession through volume. If the player only hears (non-positioned) sounds in a distance they never seem to reach, the world-space feels empty or inanimate quite fast. DAI uses mostly speech, sources of fire, non-hostile creatures and enemies to fill those gaps - the PC won't likely be walking for more than a few seconds without listening to positioned sounds.

Zone Sound Objects

Zone sound objects are often the ones that have more non-positioned sounds because they are also mostly ignored since the players have a lot of tasks and goals ahead of them. They are sounds from the environment, which is more ambiguous than it seems - a bird singing is part of the environment but if its audio object is attached to a source that is a NPC, it would make it an effect sound object. In this category we put the sounds of wind, rivers, rain and other ambient features. Some acousmatic sounds are never revealed and have ambiguous position, as I'll demonstrate during the analysis. The player hears some of them at roughly the same volume and fixed directionality, making it seem as if they are always at the same distance relative to the PC.

It happens with crowd chatter in a zone called Haven, for example, the sounds come from fixed directions in the sound image and do not adapt according to the PC's position (and this is why I haven't put this among the other speech sound objects). Wind is an exception too, because there is a non-positioned layer that keeps playing, but in open field and elevated areas you notice stronger winds shifting sides as you look around[10]. Of course, the player will hardly notice all of this unless they are focused on listening. Nonetheless, the zone sound objects are also the ones that evidentiate the different listening vantage points in a peculiar way since they are the only diegetic sounds that are not positioned (besides cutscene speech).

Score and Interface Sound Objects

Score and Interface sound objects are both directed to the player as they are non-positioned and not diegetic. The former sets the mood, creating narrative accompaniment and the latter consists of notifications, warnings, cues for action and indication of an action being executed. They reinforce the player's fixed vantage point but are not as interesting for the discussion since they do not create any tension over different types of listeners.

---

[9] You can hear the PC's positioned bow attacks coming from the left side of the image in Loopy Longplays' video at 14:02–14:10, for example. < https://youtu.be/hvH1Hy7HEwg?t=842 >

[10] See the How Big is the Map? – Walk across DAI video at 1:38:38–1:38:45 < https://youtu.be/3nqyCpkJV18?t=5918 >

Sound Objects Positioning

While looking at the categorization, it didn't seem fitting to group sound objects by their cause when zone and effect sound objects can easily become one another depending on their position. The player's navigation interacts with the sound image and that hints on the sound objects' positioning.

For example, critter sounds could either be environmental or tied to NPCs that are roaming around in the world-space. This has an implication on which portion of the ambient is interactive and which portion is not, rather than having sounds coming only from non-acousmatic sources. NPC sounds will change with listener's or their own positioning. Non-positioned sounds can have randomized spatialization but won't react directly to player's navigation, making the sound out of their reach and therefore indicating interactive boundaries of the world-space. We could incorporate the division between positioned and non-positioned sounds among Stockburger's categories of sound objects to differentiate the modes of spatial representation (Fig. 1).

	Positioned	Non-Positioned
Speech Objects	Short NPC lines	Cutscene/some dialogues
Effect Objects	Actions, attacks, sfx	Internal Sounds
Zone Objects	Birds, spatialized wind, water	Crowd chatter, non spatialized wind
Score Objects	Diegetic music, if any	Soundtrack
Interface Objects	Diegetic interface sounds	Regular interface sounds

Fig. 1. Stockburger's sound objects categorization divided by positioning with a few examples.

After going through this brief list, I realized Zone sound objects are the most variable since they're easily taken for Effect objects if they're positioned (or even speech in the case of a crowd chatter). Because of that, they're also more likely to convey different vantage points in ways that are not explicit – as it happens with NPC lines and cutscene speech, they can be positioned or not but their representation is clear for the player-listener. For the sake of the investigation, we will address these Zone sound objects more closely in our analysis to point out vantage points created by their positioning.

3 Game Sound Analysis

Elizabeth Hambleton's soundscape analysis [9] of the navigable narrative *Leaving Lyndow* (2017) does quite a good job in mapping out sound sources and it should add to what I'm proposing here, since this kind of game deals mostly with Zone sounds. It is not fit for discussing listening vantage points or orientation but some kind of mapping is important because we will always be looking out for any information on distance, depth and direction. She used schaferian soundscape analysis, which is concerned more with soundmarks (place signatures) and tonal profile. Along with Porteous and Mastin's style of mapping of SPL (sound pressure level) that is more accurate in determining distances.

What I did to investigate the positioning of DAI's sounds was a space-form analysis. By approaching the sound image, it is obvious that it only makes sense if we actively use our other available senses - we can only perceive a positioned sound from the peripatetic listener, if we stand still in both game and life, all of the vantage points are fixed. I started the research watching videos of gameplay walking across maps and free roaming before I decided to choose this game. The videos were good for listening to sonic positioning from different zones and looking out for differences in ambient sound design, especially at the parts where they move the camera around the character to show the surroundings. For confirmation of ambiguous sounds, I had to play the game myself and try out soundwalking and static listening. I played until the PC gets access to a camp called Haven, exploring the camp and the space outside of it was the focus of this analysis.

Even though I mentioned space-form back in the introduction, I couldn't actually bring it up until now. Smalley describes it as "An approach to musical form, and its analysis, which privileges space as the primary articulator[...]" He makes a quite thorough conceptualization of space in the sound image, but it is all based on the fixed listener. Some of his categories will merge, transform and behave differently when we're talking about the game[11], but it is still useful since most of the times we can only access its sound image and not the project from the developers.

3.1 Perspectival Space

There's a glossary of terms at the end of Smalley's article and we won't use all of it. The terms best used were the types of space that deals with direction and distance. Regarding direction we have the ones that constitute the perspectival space: prospective, panoramic and lateral. Prospective space is formed in the frontal image, the central axis. Panoramic is the extension of the prospective space to the sides. Lateral is "the extension of panoramic space towards the rear of the listener."

Then we have the proximate and the distal space, which is a way of grouping sounds by distance from the vantage point. The distance is often understood by volume change but mainly the transformation that occur to sounds far away - difference in frequency, intelligibility of the sound, echo, response time, etc. The differences that constitute the distance in the perspective space creates a sense of scale in the sound image [15]:

> With perspectival space there is a kind of contract between me, the viewer, and what is viewed. That my perceptions are rooted in me as a physical and spatial being means that all I perceive is 'sized up', located, and put into perspective in relation to the human scale – in relation to egocentric space. (p. 48)

Perspectival space concerns an egocentric reference position, so there will be the PC-camera and the player perspectives. As I mentioned before, an object's position does

---

[11] To mention the ones left out: There is no use for performed space and its components, nor the spaces that are related to cause because all of the ambience's sounds have implied causes. The movements of approach and recession is given by the moving of the PC in relation to source or vice versa and were used to understand volume and positioning but they are obvious. Ouverture and enclosure can be noticed by entering and leaving buildings or when hearing loud thunder noises but besides that they were not very important for the analysis either, the PC's navigation will naturally create approach, recession, ouverture and enclosure in the sound image.

not necessarily imply on its spatial qualities. The spatial forms and position are separate things, but they can be used together to compose the sound image in a creative way. There is no actual distance besides the one between audio output and the player's ears, every distance is simulated in the sound image, so there can be sound objects near the camera-listener that sounds like events occurring in the distal space or the other way round.

The scale in the perspectival space is managed through those simulations and differences between positioned and spatial form's distance. It will combine the positioning and the perspectival space in the sound image to achieve the scale intended for the ambient. Thus, making the positioned distance an egocentric reference to the camera-PC and the perspectival space egocentric to the player.

Vectorial Space
There is also the vectorial space, that is "the space traversed by the trajectory of a sound, whether beyond or around the listener, or crossing through egocentric space" [15] p. 56. The main tool to differentiate positioned from non-positioned sounds was moving the camera around paying attention to the vectorial space of each sound. Trying to shift a positioned sound from panoramic to prospective space and then trying to hear them crossing over to the other side of the headphones was the way to confirm that. Non-positioned sounds do not form a vectorial space that follows the camera's panoramic space, so one can hear which sounds are vectorized and which aren't. Sometimes non-positioned sounds can feel as if they are positioned when another sound with shared frequencies has a vectorial space that crosses its fixed space.

3.2   Game Context

Dragon Age: Inquisition is a CRPG (combat role playing game), which is a genre of RPG that favors strategic real time combat. One of the reasons I chose the game is that because of this strategy interest, the player controls a party of four characters, navigating and issuing commands one at a time. The genre usually features a pause function and strategic view, so the player can step back and decide how to proceed whatever time needed. For DAI, this means that the eyes and ears of the player become the camera and not the PC, even though it can't stray away too far from the party members. The player can choose, however, to play as one of the characters throughout the combat (as the difficulty scales higher, players are encouraged to micro manage each character simultaneously with the strategic view). This is in itself a clear indication of the different vantage points, the camera attached to each of the four members and the one attached to the strategic view - all of them peripatetic.

Dragon Age franchise has a long history through video games, books, tabletop rpgs, comics and a movie. It depicts a high fantasy world setting called Thedas that portrays adventures, political conflicts, romance and religious themes. Not only humans live in Thedas, there are dwarves, elves, qunari and many fictional creatures. *Dragon Age: Inquisition* starts with a nightmarish cutscene showing the character you create swarmed by monsters while a luminous entity calls for you. The PC wakes up imprisoned with a mark on his/her left hand being interrogated by the authorities as the prime suspect to a catastrophic event, the opening of a portal called The Breach. It spawns demons and hellish hostile creatures that threaten life in Thedas. PC is then led by Cassandra Pentaghast, with the intent to use the mark to close the portal. The beginning of the gameplay consists of a tutorial area, where the player learns how to control the character, attack, loot items, and interact with objects. It ends when the PC manages to stabilize the Breach after slaying a huge monster spawned by it. The Chantry - a powerful religious organization - seeks to execute the PC for the suspicion of being responsible for these portals. Cassandra and other military characters break allegiance to the Chantry in order to promote an Inquisition against these demons and portals, using the PC's mark to restore order to Thedas.

After these events, the game's logo appears and the player is free to roam in Haven, the camp where the Inquisition was set. It is a snowy region, surrounded by mountains in the distance, it has roads leading to both east and west. There's a river that comes from the east, under a bridge and ends in a lake in the middle of the map, both river and lake are frozen. There is a big hall in a stone construction, there are wooden cottages that serve as lodging, infirmary, there's the apothecary, the forge with the blacksmith anvils, the soldier's tents where they also spar. There are some trees in the northeastern part of the map.

3.3   Field Notes

My analysis began soundwalking through and around the camp to listen to whatever sounds I could find. After having heard the sounds that are part of the zone, I had a good holistic view of the soundscape and started taking notes to organize them between positioned and non-positioned sounds. I thought there were four non-positioned sounds: birds, wind, dogs and crowd chatter. Then I chose a spot to try to pick them apart, away from positioned speech since they capture our attention and space in the sound image.

Fig. 2. In front of the apothecary's lodge, facing north.

The first spot was in front of the apothecary's lodge, right between two other wooden houses and facing the wooden palisade (Fig. 2). In the proximate space I could hear crackling fire of the torches in front of the houses, birds (lateral space while facing north) and wind (as a permanent texture, centralized in prospective space). In the distal panoramic space, there was crowd chatter[12] and dogs barking. When turning the camera around to force vectorial spaces, I noticed the birds were actually positioned sounds, probably attached to the middle or top part of the trees. They are misleading because they can be heard often at the same volume, making it seem as if they were really close in the proximate space. Here the birds and the dogs are acousmatic (at least I couldn't find any). The chatter is indeed non-positioned, during the search for vectorial traces of single vocal sounds I noticed they were fixed in the sound image, in the prospective but mostly on the edges of panoramic space. Wind is always centralized in the image, but depending on where the PC is, you can hear another layer of positioned gales. This happens usually out in the open and at elevated places like top of hills and mountains, as I found out walking outside of the camp site.

The dogs were left for last because they were the most elusive. I went on another walk to try to pinpoint anything I could find regarding their direction and got to a small pier north of the frozen lake (Fig. 3). To the south you could see the camp gates, where the dogs should be coming from. And to the north, behind my PC was a few trees that had birds chirping. From time to time there was a sound of something big made of wood creaking. Not many sounds compared to the first spot, so I could test out the position of the dogs in a more precise way. I stood by the end of the pier facing a direction until I heard the dogs bark, after I exhausted the four directions, I noticed their direction was fixed. I could hear dogs barking from somewhere to the south and somewhere from the east. This sound resists positioning to some degree as it is always in the distal space, never in the proximity. If you follow the trail leading to the direction from where you heard the bark, you end up hearing the same dogs barking from other directions.

I deduced the dogs' barking sounds are part of the zone's texture, always playing within the map but with fixed directions when played. It is understood through abstraction that there are dogs somewhere, but they are never found. They are always in the distal space, following the PC around the zone at roughly the same perceived volume level. The opposite happens with the birds, we often hear them as if more proximate than they really are.

If the dogs are part of the zone the PC is walking through, positioned but always at an unreachable relative distance, it addresses a vantage point that is not the PC's - maybe a point fixed in space or at a different scale. If the chatter is non-positioned, the addressed vantage point is not clear either. It is fixed as from a listener that doesn't move along the navigation of the PC - could it be an imagination of what would be heard in the area? Even though those sounds are clearly diegetic, non-positioning could refer to the presence of a listener that is neither the PC-camera nor the player, that has a fixed listening vantage point to be able to hear them.

---

[12] Like the wind, can also be understood as a permanent texture, but the chatter is confined to the camp's limits while the wind's texture can be heard throughout the whole map zone.

Fig. 3. Map distance between the apothecary's lodge and the spot in the frozen lake. (Straight line between the two circles).

We can also see the distance of these specific vantage points as a distal interpolation that doesn't rely on closure of the proximate space[13]. Some vantage points could be representing listeners that are listening from elsewhere, like a stethoscope [17], while others are positioned in a clearly represented space. This external listener adds up to the other vantage points, creating a virtual space that convey different distances being heard.

On a different key, there is a mood set by the crowd chatter in Haven, indicating that even though in a cold place with people recovering from wounds, they are engaging in social activity, the place is filled with movement. For the context of the start of the game, Haven is a boiling pot for social and political activities. The sonic atmosphere sets the feeling of having a lot of work ahead, while the PC travels around the continents in political missions. Its purpose is figurative and does not need to abide to causal relations the same way as other positioned sounds, making use of an external listener to deliver the mood to the player efficiently.

4   Conclusion

The tension between space in world-space positioning and space in the sound image create multiple vantage points and types of listeners. These vantage points are not always coherent in terms of where the sounds are being heard from, but they have an effect on how the players understand the zones. Even though most players won't hear in a detailed manner, the details are meaningful and can be used to help compose ambient sounds that describes the zone creatively.

---

[13] Smalley defines it as "a temporary rupture in ongoing proximate space thereby permitting access to a distal view".

The combination of different listeners with their own vantage points creates a virtual result of the listening experience, assembling an entity with many ears, listening to sounds from different locations at the same time. There is the peripatetic camera, the fixed player, the crowd listener, the non-oriented wind listener, the dog's listener – even though they all express different vantage points, they are seen as one since the sound image mediates them all.[14]

Stockburger's categorization could be expanded upon by including positioned and non-positioned sounds as a way to separate sounds that are represented in the world-space and those that aren't. That could help sound designers think about sound and spatial representation within their projects, contributing to the structure design and helping shape the sound identity of the game.

Also, Smalley's space-form analysis has proved to be a useful method for investigating ambient sound design processes in video games and could be further developed by condensing the core concepts and techniques in a way that's easier to apply without in-depth understanding of terms that might not be relevant when talking about a video game's sound image. Being able to draw information from sound image might assist case studies since most game projects are not shared publicly (what makes audio implementation studies somehow obscure and reliant on hearing and personal knowledge on the audio engines used in the process).

Besides that, sound takes an important role in the way we understand our surroundings, both real and virtual space [16], so investigating spatial representation and cognition is important in order to create environments that can express feelings and narrative through either matching or expressively rejecting expectations of causality. The crowd sound loop and the barking dogs are nowhere to be seen and still they are largely responsible for how the mentioned zone feels active and lively for the player. If these sounds were positioned, the player would certainly lose the effect of the lingering texture throughout the zone – making it more realistic, but also imparting the absence of activity at more spots around the map. On one hand, the world-space feels empty if the player only hears non-positioned sounds since there isn't perspective being formed in the sound image. On the other hand, it feels equally empty if the player only hears positioned sounds and there is not enough activity to fill the perceived space.

This discussion is relevant to any kind of open world game and it would be interesting to elaborate and apply it on navigable narratives (also called 'walking simulators'[15]) as one of the facets of environmental storytelling [4]. Ambience can create or reinforce meanings and gameplay features through causal and figurative uses of sound. The mood set in Haven is one of the many possible examples and it's only because we have different vantage points at the same time. There is a lot of room for combining different vantage

---

[14] It's possible to add and remove listening vantage points while dealing with the non-positioned and acousmatic ones, but the overlapping of virtual non-positioned listeners makes them sound as if they are one since they belong to the same sound image and share the same reference. There is also the possibility of adding vantage points through using extra audio output devices, offering distinguishable fixed listeners (as it occurs with the inbuilt speaker of PS5's controller).

[15] There is a debate on the pejorative use of the term to refer to this genre. Elizabeth Hambleton [6] proposes 'navigable narratives' as an alternative and after reading other sources on the subject, this is the one I preferred.

points and listeners in order to create meaning in the holistic view of the sounds. The virtual multi-eared player gets to manage all of the vantage points, making the auditory experience of the game unique among other media.

References

1. Bioware.: Dragon Age: Inquisition. EA, [Windows/PlayStation 3/ PlayStation 4/Xbox 360/Xbox One/Xbox Cloud Gaming] (2014)
2. Caesar, R.: The bandwidth of soundness: between sound and image. XXIII Congresso da Associação Nacional de Pesquisa e Pós-Graduação em Música. Natal (2013)
3. Caesar, R.: Sound isn't a thing in itself, but a carrier for leaking things. Dossiê Sonologia 297–308 (2020)
4. Chang, A.: Games as environmental texts. Qui Parle **19**(2), 56–84 (2011)
5. Chion, M.: Audio-Vision: Sound on Screen. Columbia University Press, New York Chichester, West Sussex (1994)
6. Chion, M.: Sound: an Acoulogical Treatise. Duke University Press, Durham (2016)
7. Elleström, L.: Media Borders. Palgrave Macmillan, Multimodality and Intermediality (2010)
8. Galloway, K., edit. Gibbons, W., Reale, S.: Music in the Role-Playing Game : Heroes & Harmonies (chapter 10) Soundwalking and the Aurality of Stardew Valley: An Ethnography of Listening to and Interacting with Environmental Game Audio. Routledge, New York (2020)
9. Hambleton, E.: Gray areas analyzing navigable narratives in the not-so-uncanny valley between soundwalks, video games, and literary computer games. J. Sound Music Games **1**(1), 20–43 (2020)
10. Jørgensen, F., edit. Svensson, D., Saltzman, K., Sörlin, S.: Pathways: Exploring the Routes of a Movement Heritage (chapter 9) Walking and Worlding: Trails as Storylines in Video Games. White Horse Press (2022)
11. Klatzky, R.: Allocentric and Egocentric Spatial Representations: Definitions, Distinctions, and Interconnections. Springer, Spatial Cognition An Interdisciplinary Approach to Representing and Processing Spatial Knowledge (1998)
12. Modena, E., Parisi, F.: Exploring stories, reading environments: flow, immersion, and presence as processes of becoming Elisabetta modena. Cinergie – Il cinema e le altre arti. **19** (2021)
13. O'Hara, W.: Mapping sound play, performance, and analysis in proteus. J. Sound Music Games **1**(3), 35–66 (2020)
14. Roginska, A., Geluso, P.: Immersive Sound: The Art and Science of Binaural and Multi-Channel Audio. Routledge, New York (2018)
15. Smalley, D.: Space-form and the acousmatic image. Organised Sound **12**(1), 35–58 (2007)
16. Smalley, D.: Spectromorphology: explaining sound-shapes. Organised Sound **2**(2), 107–126 (1997)
17. Stankievech, C.: From stethoscopes to headphones: an acoustic spatialization of subjectivity. Leonardo Music J. **17**, 55–59 (2007)
18. Stockburger, A.: The game environment from an auditive perspective. In: Proceedings of the Level Up, Digital Games Research Conference. NL (2003). http://www.stockburger.at/files/2010/04/gameinvironment_stockburger1.pdf

A Review of Game Design Techniques for Evoking and Managing Curiosity

Ying Zhu[(✉)]

Georgia State University, Atlanta, USA
`yzhu@gsu.edu`

Abstract. Curiosity is essential for learning and discovery and is a key motivational factor for playing games. As a complex psychological phenomenon, curiosity has been studied extensively in psychology, education, and other fields, and many theories have been proposed. In the field of game design, many techniques have been developed to evoke players' curiosity. However, there is a gap between theoretical insights and practical application. This paper aims to bridge that divide by providing a comprehensive review of game design techniques for evoking and managing curiosity. We attempt to connect practical game design techniques with curiosity theories to help people understand the psychological mechanism of curiosity in the context of games. Our first contribution is to develop a theoretical framework for systematically categorizing game design techniques for evoking and managing curiosity. Our second contribution is to use this framework to classify and analyze many game design techniques. Our work not only provides a deeper understanding of how curiosity can be managed in games but also provides a framework for game designers to develop new techniques.

Keywords: Curiosity · Game Design · Game Mechanics · Artifacts

1 Introduction

Curiosity, the strong desire to learn about something or seek new experiences, is an integral part of human cognition and essential for learning and discovery. Curiosity is important for game design because it motivates players to engage with the game, leading to a higher level of enjoyment [16]. Researchers in psychology, education, communication, and behavioral science have proposed various theories on the causes and natures of curiosity [1, 4, 10–12, 14, 15, 17, 19, 20, 24], and some of the theories have been adapted to game design [3, 6–8, 18, 26]. Despite the advances in theoretical studies, there still lacks a comprehensive review of game design techniques that affect players' curiosity. From a game designer's perspective, the key question is, "How to evoke players' curiosity in games?" While theories may offer high-level concepts like "create information gaps" or "create novelty," the existing literature falls short of providing specific, actionable guidance needed for practical implementation. For this reason, a comprehensive

© ICST Institute for Computer Sciences, Social Informatics and Telecommunications Engineering 2024
Published by Springer Nature Switzerland AG 2024. All Rights Reserved
A. L. Brooks (Ed.): ArtsIT 2023, LNICST 565, pp. 111–126, 2024.
https://doi.org/10.1007/978-3-031-55312-7_9

examination of game design techniques can bridge the gap between theoretical concepts and practical applications.

In this paper, we provide a comprehensive review and categorization of game design techniques for evoking and managing curiosity. For this review, we have developed a theoretical framework by integrating and expanding the existing theories on curiosity in games. Based on this framework, we have identified and analyzed a large number of game design techniques that can evoke players' curiosity.

Our work makes two novel contributions. First, we have developed a new framework for categorizing game techniques for curiosity. Specifically, we divide game design tasks into three layers: storytelling, game mechanics, and artifacts. Within each layer, we first categorize game design techniques based on the two primary curiosity motives: deprivation and discovery. We further classify the techniques based on the types of uncertainties or curiosities they evoke. This theoretical framework helps analyze existing game design techniques and can be used as a brainstorming tool for game designers to develop new ideas. This framework may also help people design serious games for learning or for studying human cognition.

Second, this is the most comprehensive review of game design techniques for curiosity compared to previous works. Although many techniques in our study have been well-known in the game design field, they have rarely been analyzed through the lens of curiosity theories. Therefore, we bring a new analytical perspective on these techniques for game designers.

The rest of the paper is organized as follows. In Sect. 2, we briefly review the theories of curiosity and previous works on curiosity in games. In Sect. 3, we present our theoretical framework for classifying curiosity-driven game design techniques. In Sect. 4, we discuss storytelling devices for evoking curiosity. In Sect. 5, we discuss the various game mechanics for evoking and managing curiosity. In Sect. 6, we discuss using game artifacts to evoke curiosity. In Sect. 7, we discuss secondary game design techniques that may change the intensity of curiosity. The last section is the conclusion and future work.

2 Background and Related Work

2.1 Theories of Curiosity

Curiosity is a psychological state characterized by a strong desire to explore, seek information, and learn about the unknown. Many theories about curiosity have been proposed and debated [4,14,15]. These theories provide a foundation for understanding how different game design techniques evoke and manage curiosity.

Loewenstein, et al. [14,15] provide comprehensive reviews of the research on curiosity, focusing on its psychological underpinnings, measurement, dimensionality, and situational determinants. Dubey and Griffiths [4] pointed out that previous theories about curiosity generally fall under two different theoretical approaches: novelty-based theories and complexity theories. The novelty-based theories hypothesize that gaining information to satisfy curiosity brings reward and pleasure [4,5,12–14,20]. This means that curiosity may also lead many people to seek out irrelevant or negative information [19,20]. These theories suggest that curiosity is a key motivational factor that drives people to keep playing games, despite repeated failures in gameplay. The complexity theories posit that curiosity is caused by an intermediate level of complexity [4,14]. This group of theories is consistent with one of the main game design principles, that a game should be neither too simple nor too complex.

In addition, Dubey and Griffiths [4] and Wojtowicz and Loewenstein [28] have attempted to explain curiosity using economic concepts.

Some researchers have proposed different classifications of curiosity, such as perceptual curiosity, epistemic curiosity, specific curiosity, diversive curiosity, etc. [14]. Some of these classifications have been adapted to game design [26], as discussed in the next subsection. Noordewier and van Dijk [20] argued that how it feels to be curious depends on whether people have a deprivation or discovery motive. We have adopted these two motives as the basis for our classification of curiosity in games.

A relevant question is whether the intensity of curiosity is affected by external stimuli, and several previous works have addressed this question. There is evidence that the intensity of curiosity may be affected by the size of the information gap, the time to close the information gap, and the specificity of the missing information [14,20]. In other words, people tend to experience a higher level of curiosity when they feel close to finding the missing information, when the resolution appears imminent, or when the nature of the missing information is less specific. Markey and Loewenstein [15] observed that, in an educational setting, curiosity increases in supportive environments, when questions are answered effectively, on topics of importance, when information gaps are made salient, and when students are surprised.

Finally, it's important to understand that curiosity is dependent on personality [14,20]. Different people have different levels of tolerance for uncertainty and confusion. People with higher openness and a lower need for structure appreciate novel external stimuli more than people who prefer structure and clarity. As a result, the same game design technique may evoke different levels of curiosity for different players.

2.2   Previous Work on Curiosity in Game Design

To, et al. [26] analyzed the relationship between curiosity and uncertainty [2] in games and proposed a theoretical framework for studying curiosity in game design. At the center of this framework are five types of curiosity: perceptual curiosity, manipulatory curiosity, curiosity about complex or ambiguous, conceptual curiosity, and adjustive reactive curiosity. They point out that while they provide proof of concept, there is a need for systematic work that can reveal patterns and gaps in game design techniques. Specifically, they "believe it is important to collect and catalog existing techniques for curiosity management in games, even if those techniques are not intended by the designers to address curiosity as such." Our work directly addresses this issue.

Costikyan [2] identified nine types of uncertainty that help enhance the enjoyment and engagement of players in games, such as luck-based uncertainty, decision uncertainty, tactical uncertainty, etc. Given the intrinsic link between curiosity and uncertainty, Costikyan's framework significantly influenced the research conducted by To et al. [26] and our current work.

Gomez-Maureira and Kniestedt [7,8] conducted a survey to analyze which game genres and titles invoke curiosity in players. Their results showed that exploration and social simulation games ranked high in triggering curiosity. They identified curiosity-related level design patterns within these genres, such as reaching extreme points, encountering out-of-place elements, and understanding spatial connections. Later, Gomez-Maureira, et al. [9] developed a 3D open-world game to test how these level design patterns affected players' curiosity-driven exploration.

In other related works, Muscat and Duckworth [18] described the design and evaluation of a first-person exploration game and presented six design strategies for creating ambiguous exploration environments that evoke curiosity. Dahabiyeh, et al. [3] showed that curiosity was a key motivator for playing online games despite cybersecurity risks. Power, et al. [23] described a method to measure player uncertainty in games, which is closely related to curiosity.

2.3   Curiosity, Interest, and Suspense

Curiosity, interest, and suspense are often used interchangeably in causal language. Therefore, it is helpful to clarify the differences between them.

According to Hidi and Renninger [10], curiosity and interest share some common characteristics, psychological state, and physiological responses but are not the same. Curiosity is usually short-lived. Once the information gap is closed, the curiosity is over. On the other hand, interest may last a long time. Although curiosity and interest may be triggered by uncertainty, complexity, and novelty, interest may be triggered by factors not associated with curiosity. For example, a person's interest in aviation may be triggered by the heroic story of a pilot. A player's interest in gaming may be triggered by a sense of mastery and elevated social status within the gamer community.

Curiosity and suspense [21,27] are closely associated with uncertainty, but some key differences exist. Suspense is the feeling of anxiety or excitement over what will happen next. Although curiosity is an integral part of suspense, suspense is not necessarily part of curiosity. For example, a player may explore certain parts of the game world for the sake of exploration. This behavior is driven by curiosity but not suspense because there is no anxiety. While suspense is more emotional, curiosity is more intellectual.

3   Theoretical Framework

We have developed a framework to classify game design techniques for evoking and managing curiosity. The framework serves four purposes. First, it helps game designers look for techniques that evoke and manage curiosity. Second, it helps game designers understand the psychological mechanism behind each technique for evoking or managing curiosity. Third, game designers can use this framework to brainstorm and develop new techniques for evoking and managing curiosity. Fourth, game designers can use this framework to analyze a game through the lens of curiosity, identify the types of curiosity triggered by the game, and decide whether to add more triggers for different types of curiosity.

This framework consists of the following components.

- **Layers.** We group game design techniques by these three layers: **story, game mechanics**, and **artifacts** [22]. Such division helps game designers focus on a specific aspect of game design at a time. In addition, stories, mechanics, and artifacts are often handled by different teams. Grouping game design techniques by layers can help different teams focus on their tasks.
- **Motives.** For each layer, we further classify the game design techniques by two curiosity motives: **(information) deprivation** and **discovery** [20]. Each game design technique is aimed at creating a particular curiosity motive.
- **Curiosity triggers.** Each game design technique is connected with one or more curiosity triggers. For the techniques in the deprivation motive group, each trigger is a type of uncertainty. For the techniques in the discovery motive group, each trigger is a type of novelty. We use Costikyan's classification of uncertainty [2] as the basis for our framework. These uncertainties include the following.
  - Luck-based uncertainty
  - Hidden information uncertainty
  - Decision uncertainty
  - Skill uncertainty
  - Strategy uncertainty
  - Environmental uncertainty
  - Metagame uncertainty
  - Tactical uncertainty
  - Social uncertainty
- **Curiosity types.** Each curiosity trigger is then connected with a curiosity type. The connection between a game design technique, the trigger, and the curiosity type reveals how the game design technique triggers curiosity in players. In this framework, we adopt the curiosity types proposed by To, et al. [26].
  - Perceptual curiosity
  - Manipulatory curiosity
  - Curiosity about complex or ambiguous
  - Conceptual curiosity
  - Adjustive-Reactive curiosity

– **Secondary game design techniques.** Secondary game design techniques may not directly evoke curiosity, but they can modulate the intensity of curiosity [14,15,20]. Examples include revealing player progress, strategically dispensing information to adjust informational gaps, or adjusting the complexity of the tasks.

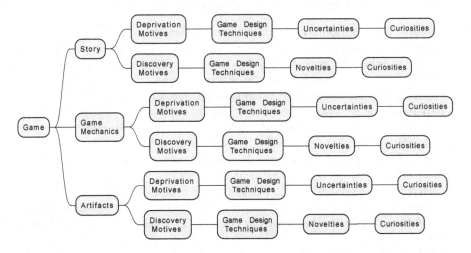

Fig. 1. This figure shows our framework for classifying the game design techniques for evoking curiosity. The secondary design techniques are not included here to keep the picture simple.

Figure 1 shows our framework for classifying the game design techniques for evoking curiosity. This framework allows us to systematically classify various game design techniques and connect them with different curiosity types based on some of the theories discussed earlier. With this framework and the classified techniques, a game designer can ask various questions, such as the following.

– What techniques can I use to evoke perceptual curiosity?
– What kind of curiosity is evoked if I allow users to randomly pick up and play with many game objects?
– How to use the camera to evoke curiosity?
– How to use game mechanics to evoke conceptual curiosity?
– What specific types of curiosity are evoked within this level? How can a broader range of curiosity triggers be incorporated for a more diverse experience?
– Can I develop a novel game design technique based on the mechanism of curiosity?

The discussion below is organized as follows. First, we divide our discussion into three layers: story, game mechanics, and artifact. Within each layer, we further divide the discussion based on the two curiosity motives: deprivation and discovery. Within each motive group, we list and discuss the game design techniques for evoking curiosity and how each technique triggers a specific type of curiosity. Finally, we discuss how secondary techniques can be used to modulate the intensity of curiosity.

4   The Story Layer

The story layer of a game is similar to the traditional storytelling medium such as film, TV, and literature. Therefore, many conventional storytelling devices can be applied to game storytelling. In this layer, information can be passed to players via cutscenes, dialogues with Non-Player Characters (NPCs), voiceover, and artifacts such as documents, videos, text messages, voice messages, signs, etc.

4.1   Deprivation Motive

Information deprivation [20] means that a player realizes certain information is missing, which is called an information gap [14]. The desire to close the information gap motivates the player to seek the missing information. To create a deprivation motive, game designers must first provide players with partial information, making them aware of the existence of knowledge they currently don't possess. The key questions are how much information should be disclosed to the players and how to present that information. We have identified the following groups of design techniques for evoking curiosity.

- **Forward or backward referencing.** Forward referencing, also called foreshadowing, informs players of a future event, often a significant one, such as a big showdown, a big decision, a big award, etc. Certain NPCs may give players hints about a future or past event. For example, the game may begin with an unresolved conflict between the player and an enemy, with the enemy promising to return for revenge. Or let the player know that a certain NPC possesses some key information.
  Or the designer may adopt a reversal narrative structure [12], where an event is presented at the beginning, and then the story goes back and starts from an earlier time. Other common techniques include flashforwards, flashbacks, an oracle-like figure making predictions, etc.
  Forward or backward referencing creates hidden information uncertainty [2]. The desire to know the details or outcome of future or past events triggers conceptual curiosity [26].
- **Mystery.** Mystery also creates hidden information uncertainty, which triggers conceptual curiosity. Here are some common design techniques.
  - Create a mystery character where the character's identity, history, whereabouts, capability, intention, motivations, or any combination of them is unknown. For example, Andrew Ryan is a mysterious character in BioShock.

- Create a mystery object whose characteristics, history, whereabouts, or usage are unknown. For example, Majora's Mask is a mysterious object in The Legend of Zelda: Majora's Mask.
- Create a mystery event where some details are unknown.
- Create cryptic messages. For example, the game Fez is filled with mysterious symbols and encrypted languages.
- Create a locked room mystery. This technique is often used in crime stories where a crime is carried out in a way that seems impossible to explain. We use the term "locked room mystery" to describe any mystery that defies the usual logical explanations.
- **Multiple options.** Providing players with multiple options can create decision uncertainty [2] that triggers curiosity about complex or ambiguous [26]. Here are some common design techniques.
  - Create a branching story and let the player choose which branch to follow. Players will be curious to know what each ending is like if the game has multiple endings, such as in BioShock.
  - Create parallel stories and switch between them. This technique is often used to create cliffhangers that make readers want to know what happens next in the other stories.
  - Create moral dilemmas. For example, in The Witcher 3: Wild Hunt, players need to make decisions often involve moral dilemmas.
  - Use ambiguous language that can be interpreted in different ways. For example, the game Braid often presents texts that are intentionally vague and open to multiple interpretations.
  - Provide contradictory information. For example, in the game Soma, the main character Simon Jarrett is often given contradictory information.

4.2 Discovery Motive

Discovery motives are related to people's desire to attain new knowledge [20]. Based on the curiosity classification by To, et al. [26], this type of conceptual curiosity is triggered by novelty, not by uncertainty. There are many techniques to add novelties to a story, and here are some examples.

- **Create unusual situations and unusual events.** For example, in Japan World Cup, characters ride on pandas, fake horses, and other weird things
- **Create an unusual environment.** For example, BioShock is set in an underwater city (Fig. 2), a very unusual environment.
- **Create unusual characters.** In the game Dropsy, the main character is a clown exploring a strange world with unusual characters.
- **Create surprises.** For example, BioShock Infinite contains a major plot twist.
- **Tell the story from an unusual perspective.** For example, the game Spec Ops: The Line is told from unusual perspectives, as the main character's mental health gradually deteriorates.
- **Use novel narrative structures.** For example, the game Her Story is based on an unusual narrative structure based on found footage.

- **Use non-linear storytelling.** For example, BioShock contains multiple endings.
- **Subvert common tropes.** For example, the game Undertale subverted the common RPG tropes by allowing players to befriend monsters instead of fighting them.

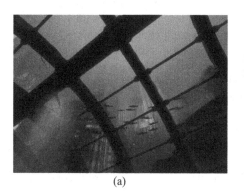 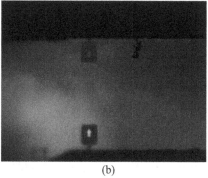

(a)   (b)

Fig. 2. a. The underwater city of Rapture in BioShock; b. The reverse-gravity mechanics in Limbo

5   The Game Mechanics Layer

The terms "game mechanics" and "gameplay" are often interchangeable. But we differentiate the two terms as follows. Gameplay is about looking at a game from the player's perspective. Game mechanics is about looking at a game from the game designers' and developers' perspectives. Game mechanics include gameplay and everything a player cannot see or control, including goals, rules, and behavior of game objects. Some games, such as The Last of Us, have substantial story and game mechanics layers. But some games, such as Super Mario Bros and Pacman, focus primarily on the game mechanics layer, with little to no story. Even without a story, there are still many ways to evoke curiosity through game mechanics.

5.1   Deprivation Motive

In the game mechanics layer, deprivation-based curiosities are triggered by the uncertainty and complexity of game mechanics. Here are the common design techniques for evoking curiosity.

- **Information hiding.** Naturally, most games only provide players with partial information about the opponents. If a player does not know when or how the opponent will attack, this creates hidden information uncertainty [2] that triggers conceptual curiosity [26].

- **Creating moderately complex gameplay.** Complex gameplay can be created by providing players with different weapons or tools, weapon upgrades, or customization. In chess, each player can use 16 pieces to generate enormous gameplay. Game designers can also introduce rules to create complex gameplay. For example, in Shogi (Japanese chess), players can incorporate captured pieces into their own army, resulting in unusually complex gameplay. Giving players many options can also create complex gameplay. The board game Go might look like a simple game with only a single type of piece. But its 19×19 board (and 361 initial options) enables extremely complex gameplay.

  Complex gameplays create decision uncertainty, skill uncertainty, tactical uncertainty, and strategic uncertainty [2] that trigger curiosity about complex and ambiguous [26]. However, it should be noted that curiosity theories suggest that the relationship between curiosity and complexity resembles a reversed U-shaped curve. Very low or very high complexity will lead to diminished curiosity. Therefore, it's important for game designers to tune the complexity of gameplay to evoke a high level of curiosity.
- **Missing objects.** Game designers can introduce treasure hunts or add Easter eggs into a game to create hidden information, uncertainty, and conceptual curiosity.
- **Mysterious objects and opponents.** The game may provide players with mysterious objects, such as mysterious weapons or vehicles (e.g., the gravity-reversing machine in the game Limbo, Fig. 2), where their behavior and purpose are not fully explained. The players need to figure out how to use them to achieve goals. Mysterious objects create hidden information uncertainty that may trigger manipulatory curiosity, conceptual curiosity, adjust-react curiosity, and curiosity about complex and ambiguous [26]. The game may provide weapon lore or backstories to further heighten the mystery and curiosity. Similarly, game designers may also introduce mysterious opponents.
- **Puzzles.** Adding a puzzle to the game is like asking the player a question, which creates conceptual curiosity [14, 26].
- **Randomizers.** Randomizers, such as dice, card shuffling, and random pairing of players, make the game unpredictable, creating luck-based uncertainty [2] that triggers conceptual curiosity.

5.2   Discovery Motive

Game mechanics are very effective tools for evoking curiosity via discovery motive because there are many opportunities to introduce novel game mechanics. Novel game mechanics can trigger a variety of curiosities, such as manipulatory curiosity, conceptual curiosity, adjustive-reactive curiosity, and curiosity about complex or ambiguous. This is one of the reasons why many games are fun and addictive. Here are some common techniques for evoking curiosity.

- **Novel weapons or tools.** For example, the Portal Gun (Fig. 3) was a novel tool when the game Portal was released.

- **Novel tactics.** In the game Katamari Damacy, the players are tasked with rolling a sticky ball called Katamari to pick up various objects (Fig. 3), a very unusual tactic at the time. In Miegakure, players can use a fourth spatial dimension to go through a wall (Fig. 4).
- **Novel rules.** Some games introduce new and unusual rules to trigger player curiosity. Some games contain different game modes where the rules are different and sometimes reversed (e.g., Pacman). For example, in a sandbox mode, the normal rules of a game are abandoned or significantly changed. The changing of rules creates a sense of novelty that triggers curiosity.
- **Constructive play.** Constructive play allows players to build characters, objects, or environments. The freedom to create something new fits well with the discovery motive for curiosity.

(a)   (b)

Fig. 3. a. The portal gun in Portal; b. The sticky ball in Katamari Damacy

Repeated gameplay is often considered detrimental to curiosity and suspense because the players already know what is going to happen, and the uncertainty is reduced. However, if properly designed, repeated gameplay can be used to enhance curiosity. If the game has a pattern of introducing random and novel mechanics and objects, the repeated gameplay creates expectations for future novelties, just like foreshadowing in storytelling.

6   The Artifact Layer

The artifact layer includes all the graphical objects, sound, animation, and visual and environmental effects. Artifacts can be used to create both information gaps and novelties.

6.1 Deprivation Motive

Artifacts can be used to direct a player's attention to certain information, creating hidden information uncertainty and environmental uncertainty [2] that trigger conceptual curiosity [26].

- **Clues.** Game designers can use visual and audio clues to provide players with partial information. For example, in The Last Of Us, blood on a wall indicates something bad has happened, triggering conceptual curiosity. Lights and sounds can direct a player's attention to hidden things. In the game Journey, the light in the distance evokes the players' curiosity throughout the game (Fig. 4). Haptic feedback from game controllers can also indicate something is coming. Maps, posters, and signs in the game world can guide players toward an unknown place.
- **Camera control.** A camera can guide a player's attention toward certain intriguing information. The camera framing and angle can hide certain information, creating an information gap. For example, if the camera follows an unknown character from the back, it evokes curiosity about the character's identity. An extremely close shot of an object makes players wonder what the object is like as a whole.
- **Complex or intriguing objects.** Complex or intriguing objects, including intriguing sounds, may evoke curiosity about complex and ambiguous [26]. For example, in the game Control, the main character encounters complex machines and paranormal devices.

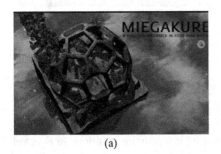 

(a)   (b)

Fig. 4. a. Miegakure includes a fourth spatial dimension. b. The mysterious light in Journey

6.2 Discovery Motive

The artifacts are particularly useful for creating perceptual curiosity, manipulatory curiosity, and adjustive-reactive curiosity [26].

- **Playable objects.** Create objects that players can pick up, manipulate, or play with. Create machines that players can operate. Playable objects, even those irrelevant to gameplay, evoke the players' manipulatory curiosity and adjustive-reactive curiosity.
- **Unusual visual or audio environment.** Create visual or audio novelties to evoke players' perceptual curiosity.
- **Extreme points in an environment.** Geomez-Maureira and Kniestedt [7,8] pointed out that players are often curious about extreme points in an environment, such as the peak of a mountain, and want to reach them.
- **Intriguing spatial connections.** Geomez-Maureira and Kniestedt [7,8] also pointed out that intriguing spatial connections may also evoke perceptual curiosity because of their novelty.

7   Secondary Game Design Techniques

Some game design techniques are not the primary triggers of curiosity, but they can help manage the intensity of curiosity. We call them secondary design techniques. Here are some examples.

- **Gradually reveal more information (hints) to narrow the information gap for the players.** Studies have shown that the intensity of curiosity is elevated when people feel they are close to finding the missing information [14,20].
- **Let players know that the resolution is coming soon.** Curiosity is generally short-lived [10]. The longer it takes to satisfy people's curiosity, the less pleasant they feel [20], and curiosity may turn into anxiety or even annoyance. Therefore, to maintain the intensity of curiosity, it is helpful to remind the players of their progress and promise a quick resolution.
- **Attach higher significance to the missing information.** Markey and Loewenstein [15] found that people displayed higher curiosity on topics of importance. In a game, a significant reward may be attached to finding the missing information, while failing to do so may result in punishment.
- **Show that other people are also interested.** For example, if you show that other people (NPCs) in the game are also pursuing a mystery object or person, the player will be more curious about it.
- **Apply time pressure.** Adding an expiration date to the missing information will elevate the curiosity because the players realize that they will be unable to close the information gap if the missing information (e.g., a document or a witness) is destroyed.
- **Make the missing information more exclusive.** People are generally more curious about information that is only accessible to a selected group. For example, giving access to a piece of information only to privileged members will increase the players' curiosity about it.
- **Insert famous people into the story.** Stein [25] indicated that readers are more curious about famous people than ordinary ones.

- **Insert famous places into the story.** People are curious about famous places that they have not visited. Video games provide an opportunity for people to experience that place in 3D.
- **Appeal to strong emotions.** For example, using an NPC to tell the player that a particular place is very scary or very beautiful will pique the player's curiosity.
- **Sudden change of stimuli.** A sudden change of stimuli can elevate the intensity of curiosity because it creates an additional layer of uncertainty. For example, footstep sounds may evoke curiosity regarding the person's identity. But if the footsteps suddenly stop, it creates additional uncertainty about what this person will do next. Similarly, a mysterious light may evoke curiosity. But when the light is suddenly turned off, the intensity of the curiosity is raised even higher.

8   Conclusion and Future Work

This paper provides a comprehensive review of game design techniques for evoking and managing curiosity, a key motivational factor for players. Despite the advances in curiosity theories for games, there is a missing link between theories and practices, and some game theorists have called for such a comprehensive review. This work is an attempt to address this issue. Our main contribution is to provide a framework for classifying game design techniques for evoking and managing curiosity. We have used this framework to analyze many game design techniques and link them with theoretical concepts. This work helps game designers understand the underlying psychological mechanism for triggering curiosity and provides a systematic approach for creating new curiosity-evoking game design techniques. It may also help people design serious games for learning or for studying human cognition.

Curiosity is an active subject in behavioral science. We will continue to study new theories and experiments and incorporate new findings into our theoretical framework. We also plan to conduct a genre-specific study on game design techniques for curiosity and carry out user studies to validate some of the claims in our theoretical framework.

References

1. Blom, J.N., Hansen, K.R.: Click bait: forward-reference as lure in online news headlines. J. Pragmatics **76**, 87–100 (2015). https://doi.org/10.1016/j.pragma.2014.11.010
2. Costikyan, G.: Uncertainty in Games. MIT Press (2015)
3. Dahabiyeh, L., Najjar, M.S., Agrawal, D.: When ignorance is bliss: the role of curiosity in online games adoption. Entertain. Comput. **37**, 10039 (2021). https://doi.org/10.1016/j.entcom.2020.100398
4. Dubey, R., Griffiths, T.L.: Reconciling novelty and complexity through a rational analysis of curiosity. Psychol. Rev. **127**, 455–476 (2020). https://doi.org/10.1037/rev0000175

5. FitzGibbon, L., Lau, J.K.L., Murayama, K.: The seductive lure of curiosity: information as a motivationally salient reward. Curr. Opin. Behav. Sci. **35**, 21–27 (2020). https://doi.org/10.1016/j.cobeha.2020.05.014

6. Galdieri, R., Haggis-Burridge, M., Buijtenweg, T., Carrozzino, M.: Exploring players' curiosity-driven behaviour in unknown videogame environments. In: De Paolis, L.T., Bourdot, P. (eds.) Augmented Reality, Virtual Reality, and Computer Graphics. LNCS, vol. 12242, pp. 177–185. Springer, Cham (2020). https://doi.org/10.1007/978-3-030-58465-8_13

7. Gómez Maureira, M.A., Kniestedt, I.: Games that make curious: an exploratory survey into digital games that invoke curiosity. In: Clua, E., Roque, L., Lugmayr, A., Tuomi, P. (eds.) ICEC 2018, WCC 2018, LNCS, vol. 11112, pp. 76–89. Springer, Cham (2018). https://doi.org/10.1007/978-3-319-99426-0_7

8. Gómez-Maureira, M.A., Kniestedt, I.: Exploring video games that invoke curiosity. Entertain. Comput. **32**, 100320 (2019). https://doi.org/10.1016/j.entcom.2019.100320

9. Gómez-Maureira, M.A., Kniestedt, I., Duijn, M.V., Rieffe, C., Plaat, A.: Level design patterns that invoke curiosity-driven exploration: an empirical study across multiple conditions. In: Proceedings of the ACM on Human-Computer Interaction, vol. 5. ACM (2021). https://doi.org/10.1145/3474698

10. Hidi, S.E., Renninger, K.A.: On educating, curiosity, and interest development. Curr. Opin. Behav. Sci. **35**, 99–103 (2020). https://doi.org/10.1016/j.cobeha.2020.08.002

11. Kashdan, T.B., et al.: The five-dimensional curiosity scale: capturing the bandwidth of curiosity and identifying four unique subgroups of curious people. J. Res. Personal. **73**, 130–149 (2018). https://doi.org/10.1016/j.jrp.2017.11.011

12. Knobloch, S., Patzig, G., Mende, A.M., Hastall, M.: Affective news: effects of discourse structure in narratives on suspense, curiosity, and enjoyment while reading news and novels. Commun. Res. **31**, 259–287 (2004). https://doi.org/10.1177/0093650203261517

13. van Lieshout, L.L., de Lange, F.P., Cools, R.: Why so curious? quantifying mechanisms of information seeking. Curr. Opin. Behav. Sci. **35**, 112–117 (2020). https://doi.org/10.1016/j.cobeha.2020.08.005

14. Loewenstein, G.: The psychology of curiosity: a review and reinterpretation. Psychol. Bull. **116**, 75–98 (1994). https://doi.org/10.1037/0033-2909.116.1.75

15. Markey, A., Loewenstein, G.: Curiosity. In: International Handbook of Emotions in Education. Routledge (2014)

16. Mekler, E.D., Bopp, J.A., Tuch, A.N., Opwis, K.: A systematic review of quantitative studies on the enjoyment of digital entertainment games. In: Proceedings of the SIGCHI Conference on Human Factors in Computing Systems, pp. 927–936. ACM (2014). https://doi.org/10.1145/2556288.2557078

17. Murayama, K., FitzGibbon, L., Sakaki, M.: Process account of curiosity and interest: a reward-learning perspective. Educ. Psychol. Rev. **31**, 875–895 (2019). https://doi.org/10.1007/s10648-019-09499-9

18. Muscat, A., Duckworth, J.: World4: designing ambiguity for first-person exploration games. In: Proceedings of the 2018 Annual Symposium on Computer-Human Interaction in Play, pp. 353–364. ACM (2018). https://doi.org/10.1145/3242671.3242705

19. Niehoff, E., Oosterwijk, S.: To know, to feel, to share? exploring the motives that drive curiosity for negative content. Curr. Opin. Behav. Sci. **35**, 56–61 (2020). https://doi.org/10.1016/j.cobeha.2020.07.012

20. Noordewier, M.K., van Dijk, E.: Deprivation and discovery motives determine how it feels to be curious. Curr. Opin. Behav. Sci. **35**, 71–76 (2020). https://doi.org/10.1016/j.cobeha.2020.07.017

21. Ortony, A., Clore, G.L., Collins, A.: The Cognitive Structure of Emotions. Cambridge University Press (1988)

22. Perron, B.: A cognitive psychological approach to gameplay emotions. In: Proceedings of the 2005 DiGRA International Conference: Changing Views: Worlds in Play (2005)

23. Power, C., Cairns, P., Denisova, A., Papaioannou, T., Gultom, R.: Lost at the edge of uncertainty: Measuring player uncertainty in digital games. Int. J. Hum. Comput. Interact. **35**, 1033–1045 (2019). https://doi.org/10.1080/10447318.2018.1507161

24. Ruan, B., Hsee, C.K., Lu, Z.Y.: The teasing effect: an underappreciated benefit of creating and resolving an uncertainty. J. Market. Res. **55**, 556–570 (2018). https://doi.org/10.1509/jmr.15.0346

25. Stein, S.: Stein on Writing. St. Martin's Press (1995)

26. To, A., Ali, S., Kaufman, G., Hammer, J.: Integrating curiosity and uncertainty in game design. In: Proceedings of the First International Joint Conference of DiGRA and FDG (2016)

27. Vorderer, P., Wulff, H.J., Friedrichsen, M. (eds.): Suspense: Conceptualizations, Theoretical Analyses, and Empirical Explorations. Lawrence Erlbaum (1996). https://doi.org/10.4324/9780203811252

28. Wojtowicz, Z., Loewenstein, G.: Curiosity and the economics of attention. Curr. Opin. Behav. Sci. **35**, 135–140 (2020). https://doi.org/10.1016/j.cobeha.2020.09.002

Aesthetics, Engagement, and Narration.
A Taxonomy of Temporal Constraints
for Ludo-Narrative Design

Cristian Parra Bravo[(✉)] [iD]

Universidad de Playa Ancha, Av. Guillermo González de Hontaneda, n° 855, Valparaíso, Chile
`cristian.parra@upla.cl`

Abstract. Harmony between narratives and game mechanics is essential for sustaining player interest; temporal structures assume critical importance in this regard. This paper introduces the notion of temporal constraints, a component that researchers and designers can use to quantify and assess the way videogame systems regulate the advancement of players within the game, utilizing empirical observation and a review of the relevant literature. Techno-aesthetic ludo-narrative elements, which merge the ludic and narrative components of the game, are how time limitations are represented in video games. Consequently, a taxonomy of these limitations is suggested after an examination of crucial facets of the novel notion.

Keywords: Engagement · ludo-narrative · Time

1 Introduction

VGs are technological aesthetic pieces. They are aesthetically pleasing, totally practical, and completed. Equally so, the medium and the object are mutually conditioned. Simondon states, "Techno-aesthetics is not primarily concerned with contemplation, but rather with application and action." Techno-aesthetics is an intercategorial fusion that is founded upon the fusing of experience (aesthetic delight) and a technical category (use, goal) [2].

A medium related with technological things is revealed via engagement with the topic. This connected medium, which Torres [3] defines as the collection of interaction alternatives offered to the technological item by its surroundings, is said to be the source from which the functioning of the technological being is conditioned. These potentialities engender discord with the user of a virtual game (VG) and impact the user's continued involvement in the game, which is referred to as "game engagement."

In the domains of software development and communication, therefore, it is essential to transcend idle thought and concentrate on implementation. In consideration of the above, this article presents temporal constraint as a valuable variable for striking the balance between reality and fantasy in any video game. This variable emphasizes temporal characteristics of player behaviors as permitted by the game, as opposed to agency,

A. L. Brooks (Ed.): ArtsIT 2023, LNICST 565, pp. 127–140, 2024.
https://doi.org/10.1007/978-3-031-55312-7_10

freedom, and originality, which are the primary narrative property components of video games. In addition to possessing a theoretically different temporal nature, every playing session is also unique. Speed runs, in which the player has unrestricted access to the (real) world, are manifestly extremely unlike to turn-based games, which impose temporal and geographical limits that speed runs ignore. By bearing this in mind, this endeavor proposes a previous taxonomy of this variable and advances the work of investigating empirical evidence of time limitations across all gaming genres.

2 Theoretical Framework

2.1 Ludo-Narrative Games

The popular assertion that Games Studies (GS) production unites broad groups within its domain is supported by empirical data. One is concerned with technological methods for comprehending and designing games (e.g., artificial intelligence, visualization, or computational modeling). Another utilizes a variety of methodologies from the humanities, arts, design, and social sciences to address non-technical elements of video games (e.g., user experience, virtuality, narrativity, communication, semiotics or hermeneutics). Despite their apparent opposition, interactions and synergies can be discovered between the two through ludo-narrativity [4]. This concept, similar to techno-aesthetics, aims to link the functional aspects of technology (rules, functions) with the imaginative aspects of the fictional realm e.g. [5–7].

In her work, Ryan [7] proposes a novel academic discipline that examines the interdependent nature of playfulness and narrative. This discipline consists of the following seven components: a open-ended research into the utilization of narrativity in video games; an examination of diverse functions and expressions of narrative in digital games; a delineation of the variety of narrative structures found in video games (including progression, emergence, and discovery narratives); and an investigation into the symbiotic relationship between playfulness and narrative. This conceptual notion does not attempt to distinguish the tale from the game rules from a literary standpoint; instead, it offers a functional viewpoint that integrates the two and recognizes the merits of both.

This approach has gained traction in scholarly circles and has been used to examine the realm of VGs in a more integrated fashion. The ludo-narrative perspective, as proposed by Ensslin [8], contributes to the analysis of video games (VG) by emphasizing the manner in which these digital games and its producers, through the integration of multimodal features and narrative, express meanings to players. At now, the term "ludo-narrative" have been studied addressed to the congruence or resonance of its constituent parts within the realm of video games. Ludo-narrative dissonance was characterized by Howe [9] as events o moments in games where something prohibits the user from perform an action designated originally as the proper. The concept of resonance -in ludo-narrative terms-, as defined by Watssman [10], pertains to ensuring a cohesive connection between the narrative and gameplay dynamics.

Furthermore, Pynenburg's work [11] presents ludo-narrative harmony. States that "the gameplay and narrative of an interactive story are harmoniously intertwined" (p. 24). To distinguish harmony from resonance in ludo-narrative, he stresses the importance of the game mechanics and story working together to enhance, rather than merely complement, the overall experience. However, by integrating ludological and narrative elements to conduct in-depth research on games, the fundamental notion of ludo-narrativity offers a lens through which the intricacies of video the exploration of ludic narration necessitates a nuanced analysis that transcends the mere consideration of the convergence or divergence of actions within the game, requiring a more intricate examination of the narrative fabric interwoven with the interactive elements.

Therefore, ludo-narrative would come to be linked with the most salient characteristics of video games. Nitsche [12] examines the ways in which game design elements such as game mechanics, places or aesthetics influence the headway of the fantasy and science fiction gaming experience. He recommends assessing the technological elements of VG media in consideration of their intrinsic relationship with story.

2.2 Narrative Time and Temporal Frames in Videogames

Temporal and geographical correlations are crucial for gaining a more comprehensive knowledge of tales than just noting the location and time of occurrences. Time and place are not just allusions in the narrative; they are intrinsic components of it. Their impact on the reader's understanding of the narrative and the cultivation of mental imagery in response to textual stimuli is profoundly significant. [13]. Moreover, player's decisions and actions are critical components in the enactment of time-related elements inside a VG. The gamer's time experience is also influenced by their actions inside the game as well as the behavior of the control software [14]. Numerous academics have delved into the examination of the intricate interplay between temporal dimensions and events within the context of gaming [15–18].

Time and space are both crucial components for a comprehensive understanding of design and experience in VGs. Digital games function as a transitory medium whereby the narrative progressively unveils itself as the player advances. Temporality encompasses not just deceleration and the clock, but also concepts like as Iteration, cadence, narrative trajectory, and culmination [13].

The design of time experience in VGs has been affected by filmmaking and other early platforms, however, because to the increasing processing capacity that enables dynamic, joyful play, time in VGs encompass attributes that concurrently facilitate and mediate the interplay between gameplay dynamics and the corresponding narrative elements. These games need more than passively read; they demand active agency in the unfolding occurrences of the tale and exert influence over the narrative [19].

In addition, these games serve as an illustration of the concept of "parallel linearity" described Sora [20]. Digital games function in a linear fashion responding to data indicative of the player's actions entails a complex and dynamic engagement with the information at hand. Likewise, Sora asserts that game actions means components of the game, thus, inaction aids the system's success; then, the player is obliged to engage in actions that serve as inputs for the generation of solutions.

The heterochrony [15] and interactions across many temporal frames provide a substantial component that distinguishes the temporality of video games. These frameworks provide a coherent theoretical approach to the examination of phenomena associated with various periods. Academic perspectives of time in the ludo narrative have surfaced as a result of the concept of chronotope, which examines the spatio-temporal underpinnings of all stories, language activity, and the canonical times of Bal [21]. A "fuzzy temporality" is a concept introduced by Herman [16]. Hitchens [17] further develops and introduces a novel model for playtime, incorporating elements such as world engine, and game progress time. Juul [22] suggests a binary framework consisting of playtime and fictional time. Nitsche [12] examines the intricate interplay and interdependence between temporal and spatial dimensions, linked to the player's experience and "positioning" within a game. Four temporal frames that are frequently applicable to the analysis of videogames were identified by Zagal and Mateas [23]: real-world, game world, coordination, and fidictive times. Tychsen and Hitchens [24] proposed seven time frames with a specific emphasis on role-playing games; Adams [25] posits the existence of a sequential or linear structural framework, a divergent and branching structural configuration, and a complex folding or intertwining structural arrangement. All above as a brief compilation of some studies concerning the exploration and examination of the multifaceted relationship between temporal aspects and narrative constructs within the realm of video games.

Additionally, the proposal of Federico Álvarez about time frames in VG is worth mentioning [26]. The temporality of video games is classified by Álvarez as "conditions," "space-time," and "change of state." In defining events, the term "change of state" incorporates all time-related nuances, including the apparent movement or blinking of a pixel inside the game's state. In the context of navigating from A to B, the time required to reach B increases proportionally with the distance between A and B. To access space-time barriers (barriers or looked entrances, for instance), the player is obligated to resume their quest for the item in question prior to revisiting and retracing the path they had previously taken and continue their path. In the last segment, "Conditions," an analysis is conducted on the impact of mechanics and temporal regulations. Time gauges serve to display the duration of time allotted for a player to finish a task or monitor the remaining chronometric tenure for a player to complete an activity, both of which have the potential to impact the final payout. Objectives that a player must accomplish may also impact the duration of the game, as the completion of said objectives may result in turns or the advancement of a player avatar within the game world.

In summary, several scholarly papers and books have explored time and space concepts in VGs. However, there is a dearth of research that identifies specific aesthetic phenomena associated with time, engagement, and play. Alvarez's contributions to this line of thinking are noteworthy; however, there remains scope for disagreement, particularly regarding the detection, evaluation, and utilization of temporal game features to augment the gaming experience and foster heightened levels of player engagement and immersion. A promissory alternative in this regard is to develop temporal constraints.

2.3 Temporal Constraints

Temporal constraints (TCs) mean the temporal parameters that dictate the order in which certain activities or occurrences must transpire or be fulfilled. Furthermore, they are capable of delineating temporal connections among occurrences, as in "event C must transpire within a certain timeframe subsequent to event D" or "event A must precede event B." In several fields, modeling, planning, and reasoning about time-related issues need temporal constraints because they give structure and direction to assure the accomplishment of desired goals and the efficient and effective use of resources within specified time bounds [27, 28]. Therefore, temporal constraints are time-related limits or restrictions that impact several facets of game design and gameplay within the domain of video games. As to Álvarez's categorizations, these limitations have the potential to impact game mechanics, design choices, narrative construction, and player experiences; therefore, it is plausible to classify them under the umbrella of ludo-narrative.

The act of successfully completing a computer game can be metaphorically likened to the process of narrating a captivating and immersive story, involving strategic decision-making, skillful execution, and progression through dynamic challenges within the digital realm, and TCs are the technical mediation through which computer games indirectly regulate playing time. As no terminology exist to indicate the events of the tale in computer games [29], "Progress Time" is a more suitable designation to describe the effects of TC. Progress Time (PT) is an integral element of the game time model proposed by Hitchens and Tychsen [24], which posits that game time is a nonlinear and dynamically generated characteristic of game play and gaming.

Tychsen and Hitchens considers PT as a conceptual and individualized perspective of the temporal dimension within the gaming experience that signifies the player's advancement inside the game. It is possible to represent several elements of advancement inside a single game. It utilizes "happens-before" and "happens-after" linkages to represent events. Since each player may progress at a distinct pace, it is inherently unique to each player. Timeline branching may result from player actions and/or reloads; hence, the game is nonlinear. While discrete branches may symbolize analogous temporal epochs within the gaming milieu, their impact is confined to augmenting the player's understanding of the revisited segment, rather than exerting influence on their subsequent progress.

In a similar vein, TC may be seen as anisochronies [30] that requires the repetition of actions to achieve a goal that allows for the continuation of the game's sequences. Anisochronies, as defined by Genette, pertains to a fluctuation in the pace of the narrative and suggests a non-linear transition among the four fundamental movements he delineates. These movements account for the narrative's velocity or the absence of equivalences between the fabula's duration and the envisioned temporal extent of the reading endeavor [30]. As a result, TC denotes the means by game designers use to regulate progress time. Game creators may use TCs to offer recurrent tasks that modify the flow of game time to maintain player interest in VGs.

From this standpoint, engaging in computer gaming might be likened to narrating a story. This procedure imparts discernible imprints on the discourse level, which establishes specific correlations with the narrative level, including temporal aspects. The unfolding of plot events in computer games transpires in tandem with the progression of the narrative discourse, and their instantiation is contingent upon the unfolding discourse itself. As Genette [30] explains, this is an interpolated narration rather than a direct contemporaneous narration; the narrative and the tale may become entangled to the extent that the latter impacts the former. Actual actions, narrative actions, and fake actions all transpire concurrently in computer games. The player is also aware of the fact that alternative routes and outcomes of the game have been established; the objective must be uncovered and achieved over the course of play [29].

Despite the fact that the game design work surpasses the limitations of narrative rhythm speed, prior writers have recognized these time speed control methods without providing enough information. Pérez-Latorre [31] labels this phenomenon "centric temporality," wherein the overall temporality of the game evolves as a consequence of the player's actions. These pertain to alterations in the length of the video-ludic story; The authors are compelled to engage in thoughtful deliberation and empirical testing of this artistic decision in collaboration with the player. This collaborative effort aims to ascertain the optimal equilibrium between the imposition of control and the provision of freedom, ensuring a delicate balance that sustains the player's interest throughout the gaming experience. Aarseth [14] proposes that the structure of the game is comprised of nodes, with players exerting impact on the narration via their narrative choices at each node. As previously stated, task progress is a component of progress time, according to Tychsen & Hitchens [24], the progression of tasks is contingent upon the successful completion of specific objectives, thereby facilitating advancement within the game. Furthermore, Aarseth [32] calls negotiation time to the necessary iteration of activities until a workable array is established. An additional exemplification may be found in Laurel's [33] metaphorical "flying wedge," which symbolizes the player's development from the conceivable to the probable to the necessary.

In summary, while the progress made in studies about heterochrony in VGs, techniques or methods for controlling time remain undefined as a design aspect. Time is a critical determinant of participation in a VG [34]. In the pursuit of player retention, one of the objectives incumbents upon game designers is to exert influence over and guide the temporal dimension within the gaming experience. Consequently, temporal manipulation in virtual games emerges as a pivotal design tool, affording the capacity for deliberate adjustments aimed at augmenting player engagement [24]. As a design feature that balances the rate of narrative on VGs, this represents the transcendence of TCs.

2.4 Videogames Engagement and Temporal Constraints

Ensuring equilibrium is the fundamental attribute of TCs. These TCs represent time in video games and have the potential to increase player involvement [34]. Player involvement will diminish, according to Przybylski et al. [35], if the degree of difficulty is uneven. Thus, through a meticulous examination of Temporal Constructs (TCs), it becomes viable to posit a balanced equilibrium that harmonizes the frequency of narrative rhythm alterations at the functional level with the intricate audiovisual narrative, often referred to as the "art," within the game. This aesthetic value is ensured regarding the player's interaction with the VG by means of this techno-aesthetic equilibrium.

In the realm of engagement, Boyle et al. [36] undertook an exhaustive assessment of literature on video game involvement with the aim of identifying the preeminent psychological theories of profound influence. Thus, in accordance with Boyle, theoretical framework address two engagement-relevant topics: the subjective sensations and pleasures of games and the motivations for playing games. Initially, flow theory is the prevailing perspective used to characterize subjective experiences during video game play [37]. The theories that are best suitable to describe video game involvement in the second example are the Self-determination theory [38], the Uses and gratifications theory [39], and the Technology Acceptance Model [40].

At its inception, the notion of flow pertained to a maximally joyful, subjective, and gratifying emotional state that emerged with the execution of actions that were viewed as beneficial. Activities must need a high degree of concentration, a feeling of individual agency, well-defined objectives, immediate and direct feedback, and intrinsic interest and satisfaction in order to promote flow. By using flow theory to video games, Sherry [41] described the level of player involvement that occurs during gameplay.

Ryan et al. [42] used SDT to scenarios including video games. The demonstration revealed that games can function as a wellspring of motivation when the psychological aspirations of connection, proficiency, and self-governance are duly met. Furthermore, according to the uses and gratifications theory, individuals possess unique preferences for amusement and leverage a spectrum of media to gratify their distinct desires. This exploration also delves into elucidating the motives behind individuals engaging in television viewing and music listening [39], this conceptual framework has further been employed to elucidate the rationale behind individuals engaging in the pursuit of computer gaming [43].

Given the incentives behind gaming and the subjective feelings and pleasures associated with video games, unquestionably, players necessitate captivating tasks, demanding scenarios, or worthwhile pursuits to perpetuate their engagement in the medium. Play designers establish time even when players are naturally driven to participate by assigning tasks that need a certain level of difficulty and rhythm in order to maintain players' interest throughout the session. As per this idea, the TCs are used for the purpose of quantifying this temporal design.

Thus, balance is the fundamental property of TCs. Moreover, the aesthetic orchestration of the real-time experience is inherently embedded in the regulation of temporal rhythms within this narrative genre. Players actively partake in an array of time-related activities meticulously curated and tuned by game designers. [34]. Consequently, TCs seem to be significant elements of the attractiveness of a video game.

2.5 Techno-Aesthetic of Gaming

A TC, which lies between the rules and the narrative, is a theoretical unit. A crucial aesthetic device that generates a "transductive" aesthetic placed between the observable and the playable, it permeates the "audiovisual layer" [2, p. 216]. Techno-aesthetics is founded upon an inter-categorical fusion of the aesthetic experience and a technical category (use, aim - aesthetic delight).

Given its distinctive vantage point on technology and culture, Simondon's argument is notably relevant in the context of video games. Simondon believes that individuation, the process by which creatures differentiate themselves from their surroundings, is central to his philosophy [44]. The technical work of Simondon focuses on the misalignment between industrial technology and culture and the way in which philosophy may assist in reestablishing harmony; aesthetic reasoning, thus, is necessary for this regulating role [45].

The design of video games is illuminated by the notion of "individuation." Individualization is promoted by the symbiotic interplay of player agency and immersive experiences within the intricate landscape of video games, which enable players to act and influence game narration and play [45]. Furthermore, this notion profoundly influences the process by which videogame designers fashion worlds and characters, enabling players to customize their experiences via branching narratives, open-world exploration, and character customization; furthermore, it strengthens the player's emotional bond with the virtual world; and ultimately, it increases player engagement [46].

The visual experience of play must take both the player and the game creator into account. Constantly imposing immovable limits on the player is mostly necessary for the game to generate resistance. A player's identity is constructed inside the game in accordance with their own interests and capabilities, the game's narrative, and its gameplay mechanics. The designer then creates the experience indirectly via the establishment of the player's restrictions and agency inside the world in which they reside, explore, and alter, as opposed to directly constructing the game experience. Aesthetics is the philosophical discipline concerned with elucidating the joys that individuals obtain from certain experiences and things; aesthetic experience is not reducible to the formal features of the object or the subject's experience alone [47].

In this instance, temporal manipulation is represented by temporal constraints that, depending on the design of the game, provide the player a greater or lower degree of freedom. Furthermore, TCs necessitates the completion of tasks repeatedly until a certain objective is achieved, hence sustaining the game's narrative. The author must evaluate the player's compatibility and make a technologically artistic decision [48–50].

3   An Effort to Classify Temporal Constraints

In addition to possessing a theoretically different temporal nature, every playing session is also unique. Speed runs, in which the player has unrestricted access to the (real) world, are manifestly extremely unlike to turn-based games, which impose temporal and geographical limits that speed runs ignore. Furthermore, it is essential for a theory of time in games to include either single-player or online games, or both. Can a universal temporal constraint foundation design be identified for all scenarios? Although factual information exists about TCs across all game genres, not all TCs are identical in every scenario.

Mobile, online PC, and console games are quite accessible. Every platform has a multitude of game genres and an extensive selection of titles within each category, with some titles being platform-specific and others being cross-platform. Although these genres are not mutually exclusive, all platforms share consistent game categories (Action,

Sports, Strategy, Role Playing, First Person, Simulation, or Third Person, just to mention a few). Even though each of these categories has distinctive qualities, TCs are used to affect the game time experience of the gamer.

3.1 Method

Electronic Arts, a global leader with more than 450 million registered players, is the most valuable video game corporation based on market capitalization. The EA platform provides a wide variety of games in the same categories than other platforms. This platform's games were utilized in a census.

EA's game platform, EA Play, serves as a template for this undertaking. Using the platform's categories as a reference point, a census research was done. EA Play is a membership service that offers a wide variety of video games. Given that Electronic Arts is a worldwide leader with more than 450 million registered users, it appears to be a fair representation of the interactive digital entertainment business. EA is now the largest video gaming company by market value [51]. Prior to 2019, EA was the third best Game Publisher [52].

It may be a trade-off to consider just titles accessible on EA Play. Given that just one developer's titles will be evaluated, it is plausible to examine the possibility of bias. On the other hand, the platform offers a vast selection of games in several categories (the same categories as a plethora of platforms), and many of the games can be played on PC or consoles (Xbox, PlayStation, Steam) with a single subscription (there is no need to buy 86 games to analyze). In light of the second point, the EA Play platform should be regarded as a reliable choice for convenience sampling.

This study is of a qualitative and analytic character. Using the methods given by Martins et al. [53], a gamer played and recorded 30 min of each game in the EA Play catalog using the "normal" difficulty option. Using the AQUAD CAQDAS software, these recordings have been directly and inductively coded in order to determine and identify the numerous thematic values of each TC. Thus, significant taxonomies are gained for future research.

The researcher coded the videos to differentiate the thematic values of each TC. Thus, taxonomies are obtained for future research. The proposed taxonomies to triangulate were addressed by informants who participated in a prior quantitative TC frequency analysis of the recordings. Through interviews, the content analysis of game recordings was "verified," enlarged, and refined.

3.2 Results

Each of the game recordings from the various EA Play catalog categories have been evaluated, resulting in 10 distinct temporal constraints. The above list covers the ludo-narrative techniques employed by videogame designers to restrict player mobility and impede uninterrupted linear play. It is a matter of rhythm, or the control of narrative time.

Table 1. Proposal of taxonomy of temporal constraints

TC-common	A common constraint satisfies all the previously described properties. They are interruptions to the natural flow of the game. It corresponds to spaces where you must eliminate a certain number of enemies (action, adventure, and simulation games), path changes (the appearance of sharp turns in a racing game, for example), complete a level within the allotted number of shots (puzzle games), or run out of resources to build, in a game like SimCity. Corresponds to situations in which the game system must be mastered through the application of elemental game mechanics
TC-advancement	This constraint on advancement happens when a player is unable to access a territory, resource, weapon, or other game variable owing to a need. It might be a lack of knowledge, resources, a product, a pending assignment, or anything else
TC-dialog	These restrictions extend to character interactions inside the game. Occasionally, these conversations are necessary to acquire extra information about the universe to enhance the player's immersion in the game environment. In addition, they produce new occupations inside the same game; for instance, identifying a certain character and conversing with him may result in the discovery of a new TC-quest, which may be compulsory or optional along the hypertextual path of the game
TC-cinematics	They are immersive cinematics, but they are merely reflecting, helping you connect with the game. It may give background, explanation, or narrative ellipsis
TC-battle-cinematics	It is TCs combination. The cinematic instructions or events before a fight. It is both reflection and notice of an approaching occurrence. There is a relationship between the animation or narrative and the tension that ensues
TC-action	The primary distinction between this TC and the standard TC is that the latter includes NPC opposition. The medium establishes criteria for TC-actions that need expertise or dexterity. In addition, action limits may vary in their degree of intricacy. At times, it may be as simple as performing a single motion, and at other times, it may require merging previously learned movements into a sequence of motions. In addition, a solved recurrent issue becomes an action TC since its solution is no longer a mystery; just the taught action must be executed (the same applies for TC-tutorial)
TC-extra	TCs are typical examples of time limitations. When they occur, the player has reached a location that presents a greater obstacle than normal advancement. However, this TC poses an even greater obstacle. Predicts failure owing to the intricacy of the situation. A TC-extra enhances the difficulty and likelihood of losing and having to respawn. It is a protracted stalemate that necessitates a shift in conduct. Either new methods, better equipment, or a higher skill level are required to break the knot

(continued)

Table 1. (continued)

TC-puzzle	The problem TC might consist just of an intervention in space, or it could also involve TC-action. This type of paired tension may be accompanied by other forms of pressure, such as the passage of time or the presence of NPCs seeking to stop the player. In this instance, an issue is selected since there is no prior guidance; only the TC impedes progress, hence the problem must be investigated. There are challenging, time-consuming puzzles with stacked TCs in adventure games. A puzzle is a short labyrinth impediment that requires a combination of actions to overcome. A unique hurdle that is not a tale element
TC-tutorial	A TC-tutorial implies the required completion of a gaming mechanism; it is a portion of the game in which the player is directed by graphics to complete a certain combination that allows them to overcome the provided obstacle. The required nature of this TC is a distinguishing feature; the player must repeat the task till achieving the desired outcome. Games having an RPG component, or a strong Narrative component should be addressed with caution when using immersion as a method. In some situations, any form of constraint is described, and its categorization as one type or another is not easily obvious. In this instance, an instructional limitation may be mistaken for a problem or an action; to differentiate between the three, the sequence must be examined until the game's stated objective is attained
TC-preparation	These limits correspond to situations requiring management in the game. The selection of an avatar's characteristics should only count as one TC, regardless of the number of versions. In games such as sports simulators, TC-preparation entails the selection of a lineup, the programming of a training plan (even if it is automatically filled out), reaching a certain point to manage the inventory, and the distribution of experience points

4   Conclusions and Discussion

This study contends that temporal constraints are a fundamental aspect of video-ludic narrative control and presents and analyzes temporal constraints as a valuable game design asset. Furthermore, by including Games Studies, aesthetics, and engagement, this article investigates the progression of time management system research from a ludic or narrative focus to an integrated ludo-narrative one.

The existence and importance of TCs have been clarified. An examination of the relevant scholarly works indicates that previous authors have noted the impact of ludological structures on time via the growth of games and the manner in which fundamental principles have been incorporated into game design and assessed within the field of Games Studies. Conversely, instances illustrate the prevalence of various forms of TCs that may vary between VG categories. Instead of perceiving TCs as a universally distinctive temporal design element in video games, this study advocates for a taxonomy wherein certain TCs assume greater prominence in specific game types, yet are not inherently exclusive to those genres.

In terms of the variety of TCs that may be discovered, this works propose ten types of this design elements. Beyond the mere categorization, the analysis performed showed the different thematic value that this ludo-narrative variable may acquire; sometimes the TC could be strategical management and another it may be narrative, ludical, competitive or challenging. Furthermore, TCs could follow an exact rhythm but game designers' creativity might stablish particular or unexpected order in TCs forms, this unequal time design make game designers' work distinctive.

Further comprehensive investigation is required in this domain since our efforts just provide an introductory assemblage of the diverse temporal constraints. However, it is now apparent that genres play a crucial role in offering a wide assortment of TCs.

Increasing is the analytical usefulness of this statement. TCs serve as a visual design resource that impacts player engagement in the game. The examination of whether the duration of a gaming session varies contingent upon the abundance of temporal constraints in a particular game holds inherent value. It is plausible to hypothesize that the player's time interest in a video game is inversely addressed to the quantity of TCs encountered.

Mobile games were excluded from the sample on purpose. TC need to implement a distinct mobile gaming strategy considering their current industry dominance in video games. Most mobile games are available for free with in-app purchases. In such games, it is critical to address player engagement via temporal design and the use of TCs to promote sales rather than produce money.

References

1. Déotte, J.L.: ¿Qué es un aparato estético? Santiago de Chile: Ediciones Metales Pesados (2012)
2. Simondon, G.: El modo de existencia de los objetos técnicos. Prometeo (2007)
3. Torres, A.: Máquinas con alma. Lo técnico y lo unano en Simondon y la cultura del anime. Astrolabio 10 (2013). https://doi.org/10.55441/1668.7515.n10.4500
4. Melcer, E., Dinh Nguyen, T.H., Chen, Z., Canossa, A., El-Nasr, M.S., Isbister, K.: Games research today: analyzing the academic landscape 2000–2014. In: Proceedings of the 10th International Conference on the Foundations of Digital Games, no. Fdg (2015)
5. Hocking, C.: Ludonarrative dissonance in Bioshock: The problem of what the game is about. In: Well played 1.0. VIdeogames, Value and Meaning, vol. 1, D. Davidson, Ed., ETC Press, pp. 255–260 (2009)
6. Kallio, O., Masoodian, M.: Featuring comedy through ludonarrative elements of video games. Entertain Comput. 31 (2019). https://doi.org/10.1016/j.entcom.2019.100304
7. Ryan, M.-L.: Avatars of Story. U of Minnesota Press, Minnesota (2006)
8. Ensslin, A.: Literary Gaming. MIT Press (2014)
9. Howe, L.A.: Ludonarrative dissonance and dominant narratives. J. Philos. Sport 44(1), 44–54 (2017). https://doi.org/10.1080/00948705.2016.1275972
10. Watssman, J.: Essay: Ludonarrative Dissonance Explained and Expanded, komentar na forumu objavljen, 23 (2012)
11. Pynenburg, T.: Games Worth a Thousand Words: Critical Approaches and Ludonarrative Harmony in Interactive Narratives, University of New Hampshire, [Online] (2012). https://scholars.unh.edu/honors/70

12. Nitsche, M.: From Faerie Tale to Adventure Game. Lublin: Maria Curie-Sklodowska Press (2007). https://s3.amazonaws.com/academia.edu.documents/30941360/Nitsche_F aerieToGame_07finalDraft.pdf?response-content-disposition=inline%253Bfilename%253 DFrom_Faerie_Tale_to_Adventure_Game.pdf&X-Amz-Algorithm=AWS4-HMAC-SHA 256&X-Amz-Credential=AKIAIWOWYYGZ2Y53UL3A%25

13. LeBlanc, M.: Tools for creating dramatic game dynamics. In: Salen, K., Zimmerman, E. (eds.) The game Design Reader: A Rules of Play Anthology, pp. 438–459. The MIT Press, Cambridge (2006)

14. Aarseth, E.: Cybertext: Perspectives on Ergodic Literature. JHU Press, Baltimore (1997)

15. Jayemanne, D.: Chronotypology: a comparative method for analyzing game time. Games Cult (2019). https://doi.org/10.1177/1555412019845593

16. Herman, D.: Story Logic: Problems and Possibilities of Narrative. U of Nebraska Press, Nebraska (2004)

17. Hitchens, M.: Time and computer games or 'no, that's not what happened. In: Proceedings of the 3rd Australasian Conference on Interactive Entertainment, pp. 44–51 (2006). https://doi.org/10.1155/2010/897217

18. Sora, C.: Temporalidades digitales. Editorial UOC (2016)

19. Wei, H., Bizzocchi, J., Calvert, T.: Time and space in digital game storytelling. Int. J. Comput. Games Technol. (2010). https://doi.org/10.1155/2010/897217

20. Sora, C.: Temporalidades digitales. Análisis del tiempo en los new media y las narrativas interactivas. Barcelona: UOC (2016)

21. Bal, M.: Teoría de la Narrativa. Ediciones Cátedra, Madrid (1990)

22. Juul, J.: Half-real: Video Games Between Real Rules and Fictional Worlds. Cambridge: MIT Press (2005). http://www.amazon.com/dp/0262101106

23. Zagal, J.P., Mateas, M.: Temporal frames: a unifying framework for the analysis of game temporality. In: 3rd Digital Games Research Association International Conference: "Situated Play", DiGRA 2007, December 2008, pp. 516–523 (2007)

24. Tychsen, A., Hitchens, M.: Game time: modeling and analyzing time in multiplayer and massively multiplayer games. Games Cult. **4**(2), 170–201 (2009)

25. Adams, E.: Fundamentals of Game Design. New Riders, Berkeley (2010)

26. Federico, A.I.: Structuring gametime a typology of the temporal constituents of video games. In: Abstract Proceedings of the 2019 DiGRA International Conference: Game, Play and the Emerging Ludo-Mix, DiGRA (2019). http://www.digra.org/wp-content/uploads/digital-library/DiGRA_2019_paper_320.pdf

27. Lundgaard, S.S., Kjeldskov, J., Skov, M.B.: Temporal constraints in human-building interaction. ACM Trans. Comput. Hum. Interact. **26**(2) (2019). https://doi.org/10.1145/330 1424

28. McBrien, P.: Temporal constraints in non-temporal data modelling languages. In: Conceptual Modeling-ER 2008: 27th International Conference on Conceptual Modeling, Barcelona: Springer Berlin Heidelberg, pp. 412–425 (2008)

29. Neitzel, B.: Narrativity in computer games. In: Handbook of Computer Game Studies, pp. 227–245 (2005)

30. Genette, G.: Figuras III. Lumen, Barcelona (1989)

31. Pérez-Latorre, Ó.: Análisis de la significación del videojuego. Fundamentos teóricos del juego, el mundo narrativo y la enunciación interactiva como perspectivas de estudio del discurso. Universitat Pompeu Fabra, Barcelona (2010)

32. Aarseth, E.: Aporia and epiphany in doom and the speaking clock: the temporality of ergodic art. Cyberspace Textuality Comput. Technol. Literary Theory, 31–41 (1999)

33. Laurel, B.: Toward the Design of a Computer-Based Interactive Fantasy System. Ohio State University (1986)

34. Rapp, A.: Time, engagement and video games: how game design elements shape the tempo-ralities of play in massively multiplayer online role-playing games. Inf. Syst. J. **32**(1), 5–32 (2022). https://doi.org/10.1111/ISJ.12328

35. Przybylski, A.K., Rigby, C.S., Ryan, R.M.: A motivational model of video game engagement. Rev. Gen. Psychol. **14**(2), 154–166 (2010). https://doi.org/10.1037/a0019440

36. Boyle, E.A., Connolly, T.M., Hainey, T., Boyle, J.M.: Engagement in digital entertainment games: a systematic review (2011). https://doi.org/10.1016/j.chb.2011.11.020

37. Csikszentmihalyi, M.: Flow: The Psychology of Optimal Experience. Harper & Row New York (1990)

38. Deci, E.L., Ryan, R.M.: Intrinsic Motivation and Self-Determination in Human Behavior. Springer Science & Business Media (2013)

39. Schramm, W., Lyle, J., Parker, E.B.: Television in the Lives of Our Children. Stanford University Press, Palo Alto (1961)

40. Davis, F.D.: Perceived usefulness, perceived ease of use, and user acceptance of information technology. MIS Q. 319–340 (1989)

41. Sherry, J.L.: Flow and media enjoyment. Commun. Theory **14**(4), 328–347 (2004). https://doi.org/10.1111/j.1468-2885.2004.tb00318.x

42. Ryan, R., Rigby, S., Przybylski, A.: The motivational pull of video games: a self-determination theory approach. Motiv. Emot. **30**(4), 344–360 (2006)

43. Lucas, K., Sherry, J.L.: Sex differences in video game play: a communication-based explanation. Commun. Res. **31**(5), 499–523 (2004)

44. Simondon, G.: Being and Technology. Edinburgh University Press (2012). http://www.jstor.org/stable/10.3366/j.ctv2f4vjrt

45. Mills, S.: Gilbert Simondon: Information, Technology and Media. Rowman & Littlefield (2016)

46. Rambo, D.: Building, Coding, Typing, Computational Culture, no. 8 (2021). http://computationalculture.net/building-coding-typing/. Accessed 07 Sep 2023

47. Terrasa-Torres, M.: Difficulty as aesthetic: an investigation of the expressiveness of challenge in digital games. Acta Ludologica **4**(1), 94–111 (2021)

48. Parra, C.: Narrative chrono-semiotics in video games: a state of the art in the study of ludo-narrative time. Signa **32**, 497–515 (2023). https://doi.org/10.5944/signa.vol32.2023.32747

49. Parra, C., Arros, A.: Semiotics of intertemporality in videoludic media. application of an analysis matrix: dragon age inquisition & the elder scrolls V: skyrim. Augusto Guzzo Rev. Acadêmica **1**(23), 41–58 (2019). https://doi.org/10.22287/ag.v1i23.879

50. Parra, C.: Semiótica de la intertemporalidad en medios videolúdicos. Propuesta de una matriz de análisis. ZER - Revista de Estudios de Comunicación **23**(45), 57–74 (2018). https://doi.org/10.1387/zer.19632

51. James, J.: Major Video Game Publishers and Developers That Still Operate Independently, GameRant. https://gamerant.com/microsoft-activision-independent-publishers-developers-ea-ubisoft-embracer/. Accessed 02 May 2022

52. Pickell, D.: The 13 Most Prominent Video Game Publishers in the World Learn Hub. https://learn.g2.com/video-game-publishers. Accessed 02 May 2022

53. Martins, N., Williams, D.C., Harrison, K., Ratan, R.A.: A content analysis of female body imagery in video games. Sex Roles **61**(11–12), 824–836 (2009). https://doi.org/10.1007/S11199-009-9682-9

Preliminary Study of the Artistic Potential of Video Games

Leonardo Porto Passos[1] (ID), Débora Mattos Peron[2] (ID), and Manuel Falleiros[3]([✉]) (ID)

[1] Interdisciplinary Nucleus of Sound Communication (NICS) – Postgraduate Program in Music (PPGM), Arts Institute (IA), State University of Campinas (Unicamp), Campinas, SP 13083-872, Brazil
leonardo.passos@nics.unicamp.br
[2] Postgraduate Program in Visual Arts (PPGAV), Arts Institute (IA), State University of Campinas (Unicamp), Campinas, SP 13083-854, Brazil
deboramperon@gmail.com
[3] Interdisciplinary Nucleus of Sound Communication (NICS) – State University of Campinas (Unicamp), Campinas, SP 13083-872, Brazil
mfall@unicamp.br

Abstract. Starting from consensual definitions about video games and arts, this article discusses the artistic potentialities of video games and their relationship with the gaming industry. Highlighting the importance of games for humanity in its cognitive aspect, power of socialization and the representation of society itself, this article discusses the possible artistic potential of video games, as well as the hypothesis that these potentialities may be totally or partially overshadowed by the emphasis on economic success that is given by the video game industry, mainly because it is a less established form of expression compared to cinema, music, literature and the visual arts, which do not need to prove anything as an artistic language, but which also experience dilemmas and counterpoints with their manifestations more focused on pop and media consumption, having as their first instance massive popularity and economic success, and not creativity and the full expression of human experience, subjectivity and sensibility.

Keywords: Video Games · Art Games · Gameart

1 Introduction

This article discusses the possible artistic potential of video games, as well as the hypothesis that these potentialities can be totally or partially overshadowed by the emphasis on economic success that is given by the video game industry, mainly because it is a less established form of expression compared to cinema, music, literature and the visual arts, which do not need to prove anything as an artistic language, but which also experience dilemmas and counterpoints with their manifestations more focused on pop and media consumption, with mass popularity and economic success as the first instance, rather than creativity and the full expression of human experience, subjectivity and sensitivity.

A. L. Brooks (Ed.): ArtsIT 2023, LNICST 565, pp. 141–153, 2024.
https://doi.org/10.1007/978-3-031-55312-7_11

In order to be able to establish the dichotomy "artistic manifestation vs. economic success" in video games, this article is segmented as follows: definition of game and video game: to outline the social importance of games and differentiate them from their digital version; definition of art: search for a modern concept that is compatible with the contemporary context; market, industry and commercial success: the focus on commercial success and the difference between indie and AAA games; benefits provided by video games: regarding cognitive, social and expressive aspects; video games as art: do video games have artistic potential or are they just entertainment?; finally, the conclusions, in which the lessons learned from the research for this article will be presented, as well as the answer to the hypothesis raised: can the artistic and transformative potential of video games be overshadowed by excessive concern with commercial success?

2   Definition of Game and Video Game

From the definitions presented by renowned game designers and ludologists, Passos [1] has reached the following definition of the game: "A game is an interactive, individual or collective, competitive or cooperative, intellectual and/or physical activity, which may or may not contain embedded narrative, but in any case, emergency narratives, consisting of pre-established rules and tools and objective-oriented, which usually involve conflict resolution, which lead to variable and quantifiable results.".[1] Such a definition fits any type of game, whether digital or not. But as video games have singularities, it is convenient to establish a definition for the digital version of games.

Video games emerged as a commercial product in the 1970s and are a specific type of videocentric electronic games [2], which became audiovisual and digital in the following decade. Currently, there are several types: consoles, PC games, mobile games, arcades, handheld consoles, web games, VR games (virtual reality games), social network video games, etc. Video games are also based on rules and follow the same premises as mechanical games [3], but the use of information and its rapid processing allows flexibility and many new possibilities, with more complex rules that do not need to be known in advance by the player, who learns them throughout the game [4]. Passos [1] also presents a succinct but comprehensive definition of video games, based on the definitions presented by important authors in the area: "Video games are electronic games with which the player(s) interact, in real time, with a computational simulation through a peripheral (user interface or input device) connected to the device (in the case of computers or consoles) or that is part of it (in the case of arcades, smartphones, tablets), such as a control (joystick or gamepad), keyboard, mouse or touchscreen, to

---

[1] "Um jogo é uma atividade interativa, individual ou coletiva, competitiva ou cooperativa, intelectual e/ou física, prazerosa e voluntária, que pode ou não conter narrativa embutida, mas, de todo modo, é propenso a narrativas emergentes, constituída de regras preestabelecidas, com espaço, tempo e ferramental próprios e orientada por objetivos, que geralmente envolvem resolução de conflitos, que levam a resultados variáveis e quantificáveis."

generate feedback to the player through output devices such as screen (screen), speakers and haptic sensors[2]".

After establishing definitions for games and video games, we are looking for a contemporary definition of art, one that is not dated and attached to traditionalisms that would prevent us from taking the proposed discussion forward.

3 Definition of Art

Before a specific definition of what art is, it is worth mentioning that authors such as Koslowski [5] and Solomos [6] differentiate two distinct types of artistic practice: pragmatic, functional or useful art and pure, fine art or artistic practice. "The first works are directed to an end, while the second are directed only to appreciation, delight and aesthetic experience."[3] [5]. At first sight, videogames may seem intrinsically belonging to functional art, since, even if they approach the concept of art, they are still games, therefore, and above all, they serve a purpose: to be played. As defended by Ebert [7]: "One obvious difference between art and games is that you can win a game. It has rules, points, objectives, and an outcome. [Someone] might cite an immersive game without points or rules, but I would say then it ceases to be a game and becomes a representation of a story, a novel, a play, dance, a film. Those are things you cannot win; you can only experience them.". In our view, the fact that video games have objectives and rules does not eliminate the possibility of being artistic objects, since, according to the interpretation, they can also have objectives (the exhibition of the work to the public), results (the finished work) and rules (which are generally defined by the artist himself, such as: a visual work painted with oil paint on fabric canvas only with "warm" colors; or an interactive installation mounted with video projections, ambient sounds, silk curtains and papers with texts randomly distributed on the floor). The score is just a way to quantify the results, as with the score assigned to films by IMDB, or awards such as the Oscar or the many lists of best literary works and music charts of a certain period. Such intrinsic characteristics of games (game mechanics) do not minimize their artistic possibilities, since they can be softened or adapted to artistic purposes, which, in fact, expands artistic expression to new frontiers, just as occurred with the emergence of new artistic languages, now established (literature, photography, cinema, comics, etc.).

Moreover, just like "pure" art, a video game, like any game, has an end in itself, that is, it is autotelic [8, 9], "which is not justified by any other end, other than existing for itself. Which has no purpose or objective outside or beyond itself"[4] [10]. Thus, playing video

---

[2] "Videogame é um jogo eletrônico com o qual jogador(es) interage(m), em tempo real, com uma simulação computacional por meio de um periférico (interface de usuário ou dispositivo de entrada) conectado ao aparelho (no caso dos computadores ou consoles) ou que faça parte dele (caso dos arcades, smartphones, tablets), como um controle (joystick ou gamepad), teclado, mouse ou tela sensível ao toque, para gerar feedback ao jogador por meio de dispositivos de saída, como tela (ecrã, ou écran), alto-falantes e sensores hápticos."

[3] "As primeiras obras são dirigidas a um fim, enquanto as segundas são dirigidas somente à apreciação, ao deleite e à experiência estética."

[4] "que não se justifica por nenhum outro fim, a não ser existir para si mesmo. Que não apresenta nenhuma finalidade ou objetivo fora ou além de si mesmo."

games, like appreciating art, is a self-motivated, self-rewarding, pleasurable, rewarding, and meaningful activity: autotelic; and video games are the quintessence of autotelic activity, as they enrich us with intrinsic rewards rather than fueling our appetite for extrinsic rewards [7]. Given this, we do not see video games and art as excluding, but as activities that may become complementary, if that is the purpose, according to the definition of art presented by Reicher [11]: "x is a work of art if, and only if, x is intended by an issuer (producer) as a means of an aesthetic experience.".[5]

That said, there are authors [12–14] who argue that there is not or is not possible a formal definition of art. However, we chose to resort to a definition that is comprehensive enough to account for the cultural and temporal diversities to which art definitions are subject, considering art as a broad concept, not reducible to a few homogeneous principles. Therefore, according to Dissanayake [15], art can be defined as a set of the following elements: demonstration of a remarkable ability; artificial creation (as opposed to "natural"); what evokes a sensation of beauty or pleasure, that is, what is sensorially attractive; fullness of sense experience, as opposed to ordinary experience; establish order, harmony, unity; orientation to originality, creativity, invention; willingness to beautify, decorate, adorn; self-expression, manifestation of a personal vision about something; communicate through a special language, symbolize; attribute importance, meaning, meaning; explore fantasy, make-believe, imagination; experience elevated, extraordinary existence: emotion, ecstasy, sublimation.

After a comprehensive definition of art, which brings together a series of concepts to encompass the subjectivity, scope and complexity of its meaning, we now seek to explore the motivations for the game, which goes beyond mere entertainment, which can explain the millennial ubiquity of games in the most varied human societies.

4   Why Do We Play Games?

Archaeological discoveries claim that ancient games, such as Senet (Egypt), Royal Game of Ur (Mesopotamia), Game of Twenty (Egypt, Babylon, Mesopotamia and Persia) and Fifty-Eight Holes, or Dogs and Jackals (Egypt), appeared about 5 thousand years ago [16], but it is possible that the games predate this period [17]. In any case, games have been part of our lives for long enough to be intimately rooted in human culture, to the point that many everyday expressions use the verb "to play" or the noun "game": open the game, hand over the game, hide the game, play fair, play dirty, game of life, game of scene, play around, word game, game of seduction, double game, political game, rules of the game, follow the game, turn the game, etc. Perhaps the reason for such ubiquity lies in the fact that games provide learning and training of physical, intellectual and social skills in a playful and safe way [1, 17] – with specific rules within the "magic circle", as defined by Huizinga [19] –, security that provides simulations of real situations that can be dangerous, unpleasant or boring [1]. Games require skill improvement,[6] which

---

[5] "x ist ein Kunstwerk genau dann, wenn x von einem Sender (Produzenten) als Medium einer ästhetischen Erfahrung intendiert ist."

[6] Such as agility, arithmetic, bargaining, communication, concentration, general knowledge, creativity, decision, dexterity, concealment, strategy, strength, leadership, logic, narrative, aim, precision, pattern recognition, reflex, social relations, resistance, rhythm, jumping, teamwork, speed, vigor, vocabulary, among others.

helps adapt to reality and cognitive behavior [20], and provide challenge, feedback, epic meaning, and autotelic pleasure [21].

Video games were launched commercially in the 1970s as an evolution of computer science experiments, based on mechanical games, carried out from the 1940s [1, 3]. It is a mostly audiovisual interactive media (although there are audio games consisting only of sounds and video games that are only visual, without sounds) that, since its inception, has become increasingly notorious by the public, the media, critics and also by academia, due to its aspects related to the development of brain capacities [22], the promotion of social transformations [23], entertainment, communication, cognition and artistic expression [24], which is of particular interest to this article.

Thus, in line with the interest of the related public (players, developers, critics, researchers, etc.), video games are expected to evolve in terms of technology, as has been the case since their genesis. With the evolution of digital technology, increasingly faster and with greater processing power in reduced-size components, many advances have been taking place with graphics, animations, audio, narrative, gameplay (game mechanics: procedures, interactions, and rules), hardware and tactile experiences [25], including in the field of arts, in which many experiments have been carried out.

In addition to having fun and training intellectual, motor and social skills, there are other motivations for people to feel attracted to video games, since, unlike the real world, they effectively motivate, maximize their potential and make them happy, in addition to being able to teach, inspire, involve and unite people in a way that society has not been able to meet [7], in full contrast to what is still disseminated by the media, politicians and religious, that video games are mere escapism, lead to addiction, idleness and unhealthy habits and encourage violent behavior [18]. Although excesses can cause problems, as in any other activity, video games are a ludic alternative to societies in which competitiveness, divergences, individualism, consumerism, manipulation, corruption, indifference, and violence – which destroy lives, relationships and families, not controlled and imaginary violence – make life increasingly distressing and frustrating [7].

Video games also demand hard work so that the player can master the required skills, overcome challenges, achieve goals and evolve, which leads to *fiero*,[7] the feeling of success in the face of an assignment or adversity [7] – which is minimized or nonexistent in passive (non-interactive) entertainment or activities that do not require deeper involvement –, and also relates to the flow, the holistic, satisfying, satisfying, stimulating, exciting feeling of creative achievement and heightened/enhanced functioning does, and playing games is a powerful Flow activity, it is the quintessential Flow experience [26]. A dynamic and holistic mental state achieved when one acts with full involvement, in which actions follow one another according to an internal logic that apparently does not require the person's conscious intervention, whose consciousness is harmoniously ordered to the activity to which that person surrenders for the mere satisfaction of living it, as a reward in itself, and therefore, flow experiences are generally autotelic, such as

---

[7] *Fiero* is a term of Italian origin, coined by psychologist Isabella Poggi [31], which refers to a specific type of pride experienced when one triumphs over a great challenge. Paul Ekman [32] subtly broadens the definition of *fiero*, describing it as the pleasure felt when an individual overcomes a challenge that enhances his capabilities, being a positive form of pride.

hobbies, sports, games and arts [27]. Thus, fun, skill training, hard work, *fiero* and flow are intimately interrelated in video games, and can find parallels in artistic practices.

Among the main benefits provided by video games and disseminated by the academic community, it can be highlighted: selective attention, improved self-monitoring, learning abilities, analytical and spatial skills, psychomotor and social skills, introduction of new concepts, short- and long-term memory, strategic thinking, planning, problem recognition and resolution, decision making, among others [29].

In addition, video games also promote confident and optimistic attitudes; division of major challenges into manageable steps; maintenance of interest; creative thinking; and, consequently, reduce the fear of failure, which inhibits innovative experimentation – very dear to the arts. Personalization of the experience for each participant, problem-solving, sense of control, and teamwork [30].

With such a ludic and cognitive potential, it is natural that video games have become one of the main forms of entertainment among entertainment habits in the country, with 70.1% of the Brazilian population playing some type of video game [28]. With such popularity, the power of influence and importance of video games today is noticeable. In view of this, it is to be expected that video games will begin to approach the field of the arts, as has happened with other languages, such as sounds, dance, writing, photography, cinema, etc.

5   Video Games as Art

The discussion about the possibility of video games being or not artistic manifestations is not new, and the subject has already been debated by several authors [33–41]. When we start from the assumption that an artistic object can be understood as any creation that proposes the reflective experience or aesthetic enjoyment [42], it is possible to understand that videogames can be perceived as works of art, according to the communicative and expressive potential that this language offers. Since video games are a multimedia medium whose objective is the interaction and exchange of experiences between the player and the game, by proposing video games as art, there is the idea of a rupture between artist and observer, in which both play a role in interpretation and experimentation, which generates an inter-relational experience, such as that observed in contemporary art.

According to Heinich [43], contemporary art is not limited to temporal issues, but is a paradigm shift and a new way of making and thinking about art. Classical art is the standard model of figuration (idealized or realistic), while modern art is the expression of interiority, which can go beyond the standards of classical figuration or even figuration itself (abstract art). Contemporary art, on the other hand, goes beyond the limits of common sense, not of classical figuration, but of the very notion of art, as well as the modern demand for a link between the work and the interiority of the artist. Contemporary art breaks with classical and modern arts, as it does not fit into their conceptions of what a work of art is; it does not demonstrate a link between the work of art and the interiority of the artist; irony and playfulness are more important than seriousness; technical or social mediations are necessary, as well as technology to effect the durability of the work; the work of art is not limited only to the object proposed by the artist, but to the whole set of operations, actions, interpretations, etc. provoked by his proposition, hence the importance of the context; the discourse about the work is part of the artistic proposal; and employs new types of materials or modes of presentation (installations, performances, land art, body art, video, photographs, multimedia, cybernetic art, etc.), breaking the boundaries between languages (visual, sound, body, etc.) [33].

For Archer [44], artists from the 1960s/70s sought to establish the political parameters of artistic creation, by adapting the marginality of avant-garde practices to the expression of the experience of cultural marginality, giving voice to those excluded from "high culture", which does not mean that the performance of the work of art in political terms is independent of its artistic merit, on the contrary, it is closely linked to the way in which art exerts an aesthetic influence on the observer and as a starting and convergence point for the investigation of meaning. Thus, contemporary art is configured and defined as such according to the needs of each artist and the time in which it is inserted, and the readings of this whole confer important paradigm shifts for the construction of new perspectives and knowledge about art and its definition. Thus, if videogames have been developed with artistic purposes, and nowadays we see that artists are increasingly appropriating digital technological resources for their productions, both end up offering the player-spectator an experience of immersion and interaction with their "works", not differentiating here what was proposed by Duchamp when he opens a huge paradigm shift by proposing direct questions to spectators in relation to their ready-mades, opposing established conventions and artistic values with ludic-conceptual conceptions to disturb the spectator and make him co-author [45].

Art is not defined by the production itself, but by its idealization, performance and realization [46], thus, the option for a certain technology or technique reflects, in addition to an aesthetic point of view, an ethical position of the artist [47]. Thus, "[...] the video game can be seen as an aesthetic sign – capable of offering interpretative developments that result in aesthetic experiences." [42]. This point of view offers us the understanding that many video games go beyond mere entertainment developed by an industry concerned with extravagant profits and the need to reach player-consumers from all corners of the world. It brings the possibility of thinking about the possibility of video games being in fact artistic objects, and in view of the discussions about it, we can understand, in the broad field of the definition of art, that many video games are lived and experienced by their players as an artistic manifestation of their time. By understanding that video games build dialogues and manifest themselves with their audience, they can be considered as an artistic object. For Stateri and Pfutzenreuter [48], "The technology that originated video games, as well as those that preceded it, raised the creative impulse of artists, inside and outside the video game industry. The result is that we see manifestations of this artistic intent within commercial video games and alternative circuits as well.".[8]

---

[8] "A tecnologia que originou os videogames, bem como as que a antecederam, suscitou o impulso criativo de artistas, dentro e fora da indústria dos videogames. O resultado é que vemos manifestações dessa intenção artística dentro de videogames comerciais e também circuitos alternativos."

Therefore, there are currently very concrete and theoretically elaborated definitions of what is defined by art games and game art, our intention here is to give video games status as art and defend that if we do it with the intention of being art we will possibly have a future where video games will be perceived and understood (not only in the academy, art galleries, museums, but by the whole society, including the most lay people) as an artistic expression and, just as in contemporary art, it will offer more and more reflection and transformation in society, modifying structural paradigms of society, as well as art has done and comes doing just like some video games are already being developed. Let's look at the Games for Change,[9] an NGO that is willing to boost entertainment and engagement aimed at social causes, proposing that game developers reflect and create games that can contribute to building a better world, in causes that positively impact their communities and promote social good.

There is a distinction between art games and artistic videogames. Art games are "[...] an interactive work, usually humorous, by a visual artist that does one or more of the following: challenges cultural stereotypes, offers meaningful social or historical critique, or tells a story in a novel manner. To be more specific art games contain at least two of the following: a defined way to win or experience success in a mental challenge, passage through a series of levels (that may or may not be hierarchical), or a central character or icon that represents the player." [49], and "Art games may be made in a variety of media, sometimes from scratch without the use of a prior existing game. They always comprise an entire, (to some degree) playable game... Art games are always interactive [...] art games explore the game format primarily as a new mode for structuring narrative, cultural critique. Challenges, levels and the central character are all employed as tools for exploring the game theme within the context of competition-based play."[10] [50].

Art games are oriented as artistic objects, which seems to us to take the focus away from the gameplay;[11] while artistic video games are focused on gameplay, but with artistic treatment in the aspects that make up the video game: narrative, graphics and sounds. The former are more related to the concept of pure/fine art/artistic practice, while the latter are more connected to the idea of pragmatic/functional/useful art. *The Path*[12] (2009), *The Cat and The Coup*[13] (2011), *Conglomeration*[14] (2013), *Rehearsals and Returns*[15] (2013) and *Mizmaze*[16] (2017) are examples of art games; while *Journey*[17]

---

[9] See: <https://www.gamesforchange.org/>.

[10] We do not agree with the final statement, "[...] within the context of competition-based play.", as not every video game is inherently competitive, as there are cooperative and even exploratory video games, which may not be competitive or cooperative, even though there are objectives, challenges, and rules, possibly supported by a narrative.

[11] The further away from gameplay and mechanics an art game is, the less relation there is with what is known as a game, approaching an artistic object that simply uses an aesthetic common to video games, such as pixel art, chiptune, 3D virtual world, etc.

[12] See: <https://tale-of-tales.com/ThePath/index.html>.

[13] See: <https://www.thecatandthecoup.com/>.

[14] See: <http://frederickostrenko.com/conglomeration/>.

[15] See: <https://www.rehearsalsandreturns.peterbrinson.com/>.

[16] See: <http://mizmazegame.com/>.

[17] See: <https://thatgamecompany.com/journey/>.

(2012), *Sound Shapes*[18] (2012), *Walden, A Game*[19] (2017), *Gorogoa*[20] (2017) and *Outer Wilds*[21] (2019) can be defined as artistic videogames. Many of these games, whether artistic video games or art games, offer the possibility for the player to freely explore the beauty and details of the fictional game world, which ends up guiding the game towards artistic expressions in terms of sounds, graphics and narrative [35].

Obviously, not every video game is oriented towards artistic intentions: "Although all video games should not be considered art, recent developments in the medium have been widely recognized as clear indications that some video games should be regarded as art works." [33]; and even when it is, the gameplay, that is, the game mechanics, is a crucial element for its design and development, since it is, above all, a videogame, and even if it is composed of different languages, since videogames are multilingual media, a videogame has its own characteristics: "I judge that though they have their own non-artistic historical and conceptual precedents, videogames sit in an appropriate conceptual relationship to uncontested artworks and count as art. [...] At the same time, videogames have their own distinctive features, meaning that as a form of art they should be treated on their own terms and not simply seen as derivative forms of pre-existing types. [...] Videogames are representational artifacts in the way that many other forms of art are, and though differing to traditional artworks in certain respects, they do have perceptual and formal structures that are the object of an aesthetic and interpretive engagement in much the same way as other artworks." [51].

Since gameplay is a distinct element of videogames, we believe that the tendency is for it to be treated more and more with a view to creativity and even expression, so that it comes closer aesthetically and conceptually to the languages that make up an artistic video game or art game, breaking with a possible utilitarian view of games, which Ebert [52] argues is intrinsic to video games: "I am prepared to believe that video games can be elegant, subtle, sophisticated, challenging and visually wonderful. But I believe the nature of the medium prevents it from moving beyond craftsmanship to the stature of art. To my knowledge, no one in or out of the field has ever been able to cite a game worthy of comparison with the great dramatists, poets, filmmakers, novelists and composers. That a game can aspire to artistic importance as a visual experience, I accept. But for most gamers, video games represent a loss of those precious hours we have available to make ourselves more cultured, civilized and empathetic.". In this sense, many indie

---

[18] See: <https://youtu.be/zfPbJE6XDxg> and <https://www.ign.com/games/sound-shapes-que asy-games>.

[19] See: <https://www.waldengame.com/>.

[20] See: <https://gorogoa.com/>.

[21] See: <https://www.mobiusdigitalgames.com/outer-wilds.html>.

games[22] have offered creative solutions, seeking to offer an artistic experience added to the "meaningful ludic interaction", resulting from the interaction between players, the game system and the context in which the game is played, that is, the player's actions result in the creation of new meanings in the system, and thus the newly established relationships originate new sets of meanings created by the players' actions [24].

According to the above, video games can present, if not all, at least a large part of the elements present in the definition of Dissanayake [15] presented in Sect. 3. *Definition of Art* of this article, provided that the game is conceived with artistic intentions, in contrast to the marketing intentions of the big studios, whose games are more oriented to popularity and obtaining the maximum profit, which ends up leading games to commercial success formulas, clichés and populism to the detriment of expression. Creative, questioning, and elevated artistic.

6   Conclusion

In the same way that writing began to receive artistic treatment until it became literature, or mimesis evolved into theater, storytelling became narrative, rock art mutated into graphic arts, noises turned into music and photography gave rise to cinema, games, as ingrained in human life as these languages and even prior to some of them, are being transfigured into art games or artistic video games, which seems to us a natural course in view of the relevance of games and, more currently, video games. It is up to those involved (graphic designers, concept artists, screenwriters, programmers, music composers, sound designers, animators, etc.) to seek creative solutions for video game mechanics to bring them closer to the proposed experience and artistic concept, as well as to the other aspects that make up a game. And if they do so, time will take care of protecting a place in the pantheon of the arts for these videogames, just as it happened with the other languages that are now considered artistic.

---

[22] Indie games can be considered as the opposite of mainstream video games: "[…] the mainstream is one whose goal of popularity and profit overrides creativity, self-expression, and art. So attitude is what defines independence, not just a billing and popularity metric. […] 'indies' are not characterized by their underground nature, but by a matter of conceptual and ideological positioning, where cultural and creative production finds its centrality." ("[…] o mainstream é aquele cujo objetivo de popularidade e lucro sobrepõe a criatividade, a autoexpressão e a arte. Portanto a atitude é que define independência, não somente uma métrica de faturamento e popularidade. […] os 'indies' não se caracterizam por sua natureza underground, mas por uma questão de posicionamento conceitual e ideológico, onde a produção cultural e criativa encontra sua centralidade.") [53]. That is, indices are games with their own proposals and wide creative freedom, which do not follow the corporate standards of the industry and the big studios, focused on popularity and profit, with large teams and millionaire investments. Indie games usually have an elaborate graphic look in a retro or unconventional 2D style, are developed by small teams, bring experimental mechanics and often affronting or intimate themes [54, 55]. Indies can be subdivided into professionalized, who adopt professional work practices, similar to the rest of the industry; and non-professional, linked to amateur and hobbyist practices [55].

References

1. Passos, L.J.P.: Um estudo interdisciplinar sobre a comunicação sonora em games narrativos. Campinas. 103 f. Dissertação (Mestrado em Música) – Instituto de Artes, Universidade Estadual de Campinas (2022). https://hdl.handle.net/20.500.12733/7830. Accessed 10 Jul 2023
2. Passos, L.P.: O visocentrismo nos videogames: um assunto a ser debatido. Blog C4. https://uni campc4.blogspot.com/2022/05/o-visocentrismo-nos-videogames.html. Accessed 10 Jul 2023
3. Passos, L.P.: "Jogos mecânicos": uma proposta taxonômica. Blog C4. https://unicampc4.blo gspot.com/2022/08/jogos-mecanicos-uma-proposta-taxonomica.html. Accessed 10 Jul 2023
4. Juul, J.: The game, the player, the world: looking for a heart of gameness. In: Copier, M., Raessens, J. (eds.) Level Up: Digital Games Research Conference 2003, pp. 30–45, Utrecht University, Utrecht (2003). http://www.digra.org/wp-content/uploads/digital-library/05163.50560.pdf. Accessed 10 Jul 2023
5. Koslowski, A.: Acerca do problema da definição de arte. Revista Húmus: Educação e Cultura na Pós-modernidade, São Luís: Universidade Federal do Maranhão, May-Ago 2013, vol. 3, no. 8, pp. 1–9 (2013). https://periodicoseletronicos.ufma.br/index.php/revistahumus/article/view/1675. Accessed 06 Jul 2023
6. Solomos, M.: From Music to Sound: The Emergence of Sound in 20th- and 21st-Century Music. Routledge, London (2019). https://doi.org/10.4324/9780429201110
7. Ebert, R.: Video games can never be art. Roger Ebert's J. (2010). https://web.archive.org/web/20111010001841/http://blogs.suntimes.com/ebert/2010/04/video_games_can_never_be_art.html. Accessed 17 Jul 2023
8. McGonigal, J.: A realidade em jogo, Tradução Eduardo Rieche. Best Seller, Rio de Janeiro (2012)
9. Luz, A.R.: Gamificação, motivação e a essência do jogo. In: Santaella, L., Nesteriuk, S., Fava, F. (Orgs.). Gamificação em debate, pp. 39–50. Blucher, São Paulo (2018)
10. Autotélico. In: Michaelis, Dicionário Brasileiro da Língua Portuguesa Online. Melhoramentos, São Paulo (2015). http://michaelis.uol.com.br/. Accessed 06 Jul 2023
11. Reicher, M.E.: Einführung in die philosophische Ästhetik, 3rd edn., p. 164. Wissenschaftliche Buchgesellschaft, Darmstadt (2015)
12. Coli, J.: O que é Arte, 15th edn. Brasiliense, São Paulo (1995)
13. Warburton, N.: O que é a arte? Bizâncio, Lisboa (2007)
14. Weitz, M.: The role of theory of art. J. Aesthetics Art Criticism **15**, 27–35 (1956)
15. Dissanayake, E.: What is art for?, 5th edn. University of Washington Press, Seattle (2002)
16. Whitehill, B.: Toward a classification of non-electronic table games. In: Conference on 11th Board Game Studies Colloquium 2008, pp. 53–63. Associação Ludus, Lisboa (2009). http://jnsilva.ludicum.org/PBGS08.pdf. Accessed 12 Jul 2023
17. Masukawa, K.: The origins of board games and ancient game boards. In: Kaneda, T., Kanegae, H., Toyoda, Y., Rizzi, P. (eds.) Simulation and Gaming in the Network Society. TSS, vol. 9, pp. 3–11. Springer, Singapore (2016). https://doi.org/10.1007/978-981-10-0575-6_1
18. Peron, D.M., Passos, L.P., Camargo, F.E., Fornari, J.: Videogames: transcendendo vício, violência e escapismo. In: 22th SBGAMES 2022, pp. 448–457. Sociedade Brasileira de Computação, Porto Alegre (2022). https://doi.org/10.5753/sbgames_estendido.2022.226088
19. Huizinga, J.: Homo ludens: o jogo como elemento da cultura, 7th edn. Trad. João Paulo Monteiro. Perspectiva, São Paulo (2012). (Col. Estudos)
20. Bystřina, I.: Tópicos de semiótica da cultura: aulas do professor Ivan Bystřina, maio 1995, PUC-SP. Trad. Norval Baitello Jr. e Sônia B. Castino. Cisc, São Paulo (2009)
21. Luz, A.R.: Gamificação, motivação e a essência do jogo. In: Santaella, L., Nesteriuk, S., Fava, F. (eds.). Gamificação em debate, pp. 39–50. Blucher, São Paulo (2018)

22. Bavelier, D.: Your brain on video games. TED Talks, June 2012. https://www.ted.com/talks/daphne_bavelier_your_brain_on_video_games. Accessed 13 Jul 2023
23. McGonigal, J.: Gaming can make a better world. TED Talks, February 2010. https://www.ted.com/talks/jane_mcgonigal_gaming_can_make_a_better_world. Accessed 13 Jul 2023
24. Salen, K., Zimmerman, E.: Regras do jogo: fundamentos do design de jogos: principais conceitos, vol. 1. Blucher, São Paulo (2012)
25. Schell, J.: A arte de game design: o livro original. Trad. Edson Furmankiewicz. Elsevier, Rio de Janeiro (2011)
26. Csikszentmihalyi, M.: Beyond Boredom and Anxiety: The Experience of Play in Work and Games, 4th edn. Jossey-Bass Publishers, San Francisco (1985)
27. Csikszentmihalyi, M.: Flow: a psicologia do alto desempenho e da felicidade. Trad. Cássio de Arantes Leite. Objetiva, Rio de Janeiro (2020)
28. Go Gamers: Pesquisa Game Brasil: edição gratuita 2023. SX Group, São Paulo (2023). https://www.pesquisagamebrasil.com.br/pt/edicao-gratuita/. Accessed 13 Jul 2023
29. Passos, L.J.P., Novo, J.E.F., Jr.: Proposta de um role-playing audiogame acusmático para educação musical. In: Nassif, S.C., Mendes, A.N.A., Cruz, F.V., Gilberti, F.P., Weidner, K.E. (eds.) 2021 14th Encontro de Educação Musical da Unicamp (EEMU), IA-Unicamp, Campinas, pp. 208–216 (2021). http://sites.google.com/dac.unicamp.br/eemu/anais/2021. Accessed 13 Jul 2023
30. Werbach, K., Hunter, D.: For the Win: How Game Thinking can Revolutionize Your Business. Wharton Digital Press, Philadelphia (2012)
31. Poggi, I., D'Errico, F.: Types of pride and their expression. In: Esposito, A., Vinciarelli, A., Vicsi, K., Pelachaud, C., Nijholt, A. (eds.) Analysis of Verbal and Nonverbal Communication and Enactment. The Processing Issues. LNCS, vol. 6800, pp. 434–448. Springer, Heidelberg (2011). https://doi.org/10.1007/978-3-642-25775-9_39
32. Ekman, P.: Emotions Revealed: Recognizing Faces and Feelings to Improve Communication and Emotional Life, 2nd edn. Owl Books, New York (2003)
33. Smuts, A.: Are video games art? Contemp. Aesthet. 3(3) (2005). https://digitalcommons.risd.edu/liberalarts_contempaesthetics/vol3/iss1/6. Accessed 17 Jul 2023
34. Stalker, P.J.: Gaming in art: a case study of two examples of the artistic appropriation of computer games and the mapping of historical trajectories of "art games" versus mainstream computer games. Monography (Master of Fine Arts), University of the Witwatersrand, Johannesburg (2005). http://hdl.handle.net/10539/1749. Accessed 17 Jul 2023
35. Lanchester, J.: Is it art?. In: London Review of Books, vol. 31, no. 1 (2009). https://www.lrb.co.uk/the-paper/v31/n01/john-lanchester/is-it-art. Accessed 17 Jul 2023
36. GAME Developer: The art history... of games? Games as art may be a lost cause (2010). https://www.gamedeveloper.com/pc/the-art-history-of-games-games-as-art-may-be-a-lost-cause. Accessed 17 Jul 2023
37. Samyn, M.: Almost art. The Escapist (2011). https://www.escapistmagazine.com/almost-art/. Accessed 17 Jul 2023
38. Tavinor, G.: Video games as mass art. Contemp. Aesthetics 9(9) (2011). https://digitalcommons.risd.edu/liberalarts_contempaesthetics/vol9/iss1/9/. Accessed 17 Jul 2023
39. Jones, J.: Sorry MoMA, video games are not art. The Guardian (2012). https://www.theguardian.com/artanddesign/jonathanjonesblog/2012/nov/30/moma-video-games-art. Accessed 17 Jul 2023
40. Parker, F.: An art world for artgames. Loading 7(11), 41–60 (2012). https://journals.sfu.ca/loading/index.php/loading/article/view/119. Accessed 17 Jul 2023
41. Smith, E.: Why games matter blog: indie games aren't art games. International Business Times (2013). https://www.ibtimes.co.uk/why-games-matter-indie-movie-fez-meat-427544. Accessed 17 Jul 2023

42. Stateri, J.: O signo estético no videogame: paralelos entre a criação de jogos e a arte como processo. In: Proceedings of the 12th SBGames 2013, pp. 23–29. Sociedade Brasileira de Computação, Porto Alegre (2013). https://www.sbgames.org/sbgames2013/proceedings/cd/_Julia%20Stateri_CD_2013.pdf. Accessed 13 Jul 2023

43. Heinich, N.: Practices of contemporary art: a pragmatic approach to a new artistic paradigm. In: Zembylas, T. (ed.) Artistic Practices: Social Interactions and Cultural Dynamics, pp. 32–43. Routledge, London (2014). https://doi.org/10.4324/9781315863092

44. Archer, M.: Arte contemporânea: uma história concisa. Trad. Alexandre Krug e Valter Lellis Siqueira. Martins Fontes, São Paulo (2001)

45. O'Doherty, B.: No interior do cubo branco: a ideologia do espaço da arte. Trad. Carlos S. Mendes Rosa. Martins Fontes, São Paulo (2002)

46. Pareyson, L.: Os problemas da estética. Martins Fontes, São Paulo (1989)

47. Vargas, A., Bahia, A. B., Born, R.: Da arte ao game: processo de criação artística para mobile game. In: Proceedings of the 12th SBGames 2013, pp. 106–114. Sociedade Brasileira de Computação, Porto Alegre (2013). https://www.sbgames.org/sbgames2013/proceedings/artedesign/14-dt-paper.pdf. Accessed 12 Jul 2023

48. Stateri, J., Pfutzenreuter, E.P.: A complexidade sensível: um paralelo entre videogames e arte. Curso online (MOOC). Unicamp, Campinas; Coursera, Mountain View (2016). https://www.coursera.org/learn/videogames. Accessed 12 Jul 2023

49. Holmes, T.: Arcade classics span art? Current Trends in the Art Game Genre Archived 2013-04-20 at the Wayback Machine. Melbourne DAC, pp. 46–52 (2003). https://web.archive.org/web/20130420092835/http://hypertext.rmit.edu.au/dac/papers/Holmes.pdf. Accessed 18 Jul 2023

50. Cannon, R.: Introduction to artistic computer game modification. PlayThing Conference, Sydney, Australia, pp. 7–14 (2003) apud Bittanti, M.: Game art: (this is not) a manifesto, (this is) a disclaimer. Bittanti, M., Quaranta, D.: Gamescenes: art in the age of videogames. Johan & Levi, Milano (2003)

51. Tavinor, G.: The Art of Videogames. Wiley-Blackwell, Malden (2009)

52. Ebert, R.: Why did the chicken cross the genders. RogerEbert.com (2005). https://www.rogerebert.com/answer-man/why-did-the-chicken-cross-the-genders. Accessed 18 Jul 2023

53. Zambon, P.S., Chagas, C.J.R.: Produção independente de jogos digitais: o desenvolvedor "Lone Wolf". In: SBGAMES, 17., out./nov. 2018, Foz do Iguaçu. Anais eletrônicos…: Industry Track – Full Papers. Porto Alegre: Sociedade Brasileira de Computação (2018). Disponível em http://www.sbgames.org/sbgames2018/files/papers/IndustriaFull/189970.pdf. Accessed 10 Dec 2022

54. Prieto, D.T., Nesteriuk, S.: Indie Games BR: estado da arte das pesquisas sobre jogos independentes no Brasil. In: SBGAMES, 20., out. 2021, Gramado. Anais eletrônicos…: Industry Track – Full Papers. Sociedade Brasileira de Computação, Porto Alegre (2021). Disponível em http://www.sbgames.org/proceedings2021/IndustriaFull/218217.pdf. Accessed 10 Dec 2022

55. Pereira, L.S.: A independência dos jogos: um estudo sobre a percepção do jogador brasileiro. In: SBGAMES, 17., out./nov. 2018, Foz do Iguaçu. Anais eletrônicos…: Culture Track – Short Papers. Porto Alegre: Sociedade Brasileira de Computação (2018). Disponível em http://www.sbgames.org/sbgames2018/files/papers/CulturaShort/186779.pdf. Accessed 10 Dec 2022

Interactive Technologies, Multimedia, and Musical Art

Singing Code

Jasmina Maric[1]([✉]) and Lekshmi Murali Rani[2]

[1] Department of Computer Science and Engineering, Interaction Design Division,
Chalmers University of Technology, Gothenburg, Sweden
`jasmina@chalmers.se`
[2] Department of Computer Science and Engineering, Software Engineering Division,
Chalmers University of Technology, Gothenburg, Sweden
`lekshmi@chalmers.se`

Abstract. This paper provides a comprehensive examination of the Creative Coding project's impact on female empowerment. The project addresses the gender disparity in STEAM fields by integrating coding and art for girls aged 10–15, using the Strudel tool that combines real-time coding and music composition. Through working with female teachers and mentors from the Computer Science and Engineering Department as relatable role models, the Creative Coding project generated surprising effects on learning due to increased curiosity. The integration of art, music, and audio-visual elements, supported by user-friendly software, created an engaging learning experience. Collaboration and peer learning are emphasized to foster social capital and teamwork skills. Positive outcomes included increased interest, self-confidence, creativity, and aspirations in STEAM subjects among the girls. However, a more comprehensive and supportive environment is recognized as necessary to engage a broader group of girls in STEAM activities. The findings highlight the importance of cultivating curiosity, providing mentorship, promoting inclusivity and collaboration, and integrating creativity in education to inspire and empower girls in coding and STEAM disciplines. The paper concludes by emphasizing the potential of Creative Coding practices and the audio-visual perception of music to enhance girls' empowerment, creativity, and interest in coding and STEAM fields.

Keywords: STEAM · Art & Tech · Creative Learning · Live Coding · Social Capital · Mentorship

1 Introduction

In today's job market, which is heavily influenced by technology, where constant change is the norm, and technology is gradually replacing manual labor, future employees need a diverse set of skills to succeed. It's increasingly important to nurture creativity and innovation in individuals, as they play a crucial role in navigating these evolving dynamics of the modern world. To empower students to realize their full potential as adults and equip them with knowledge of worth

A. L. Brooks (Ed.): ArtsIT 2023, LNICST 565, pp. 157–174, 2024.
https://doi.org/10.1007/978-3-031-55312-7_12

to become future leaders, productive workers, and responsible citizens [11,24] modern education systems are under increasing pressure to cultivate essential skills in students, including problem-solving, critical thinking, communication, collaboration, and self-management. Contemporary academia is acutely aware of this pressure and strives to employ effective teaching methods where traditional educational models are being replaced by innovative approaches that prioritize the development of critical thinking, problem-solving, and innovation skills among the younger generation.

This needs for educational transformation that should cater to increased creativity has led to the widespread adoption of STEM education (Science, Technology, Engineering, and Mathematics) and, more recently, the integration of arts into the STEM disciplines through the STEAM model [29]. While arts and STEM may initially seem incompatible, they actually complement each other by fostering the generation of fresh, creative ideas and facilitating new thought processes [14]. The STEAM framework promotes holistic learning by combining systematic thinking skills from both scientists and artists [3]. It recognizes the innovative ideas that arts contribute to artistic, scientific, and societal domains [15,26]. Moreover, the diverse nature of arts enables students to explore human nature, understand complex world dynamics, and develop empathy, which is considered one of the crucial skills in the 21st-century skills framework [5,18].

However, gender disparity persists in STEAM fields, particularly in technical areas like engineering and computer science, where men outnumber women [25]. This disparity is even present in Sweden, a country renowned for its commitment to gender equality, where women are underrepresented in Swedish universities [23]. To address this significant issue and counteract the effects of stereotype threat [25], we have developed a Creative Coding project at the Chalmers University of Technology, Department of Computer Science and Engineering (CSE), Sweden.

In this paper, we present a deep analysis of the Creative Coding project and its effect on female empowerment. Creative Coding is an initiative that integrates coding and art across interdisciplinary areas specifically designed for girls aged 10–15. The project's objective is to equip these girls with knowledge, creativity, and social capital-essential skills that will empower them and boost their confidence in pursuing their future endeavors [8].

2   The Creative Coding Project

The Creative Coding project developed at Chalmers University in Sweden, Computer Science and Engineering Department (CSE), was offered in the spring term, from March to June of 2023. It was delivered as a series of workshops spanning over 10 weeks, where girls aged 10–15 are introduced to coding using the innovative Strudel tool [17]. Strudel is a fascinating and unique tool that provides fast audio and visual feedback, allowing participants to compose music and engage in live coding sessions as depicted in Fig. 1. Live coding music is a growing international phenomenon where programmers communicate their musical intentions to a computer and receive real-time visual and auditory output

[22]. In this context, live coders write code that generates sound in real time, bridging the gap between coding and artistic music creation.

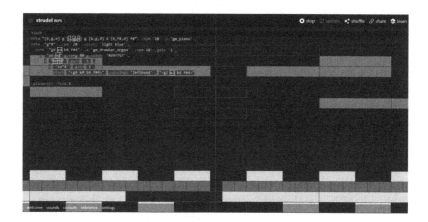

Fig. 1. Strudel Interface

In spite of the conventional division between art and coding, we have purpose-fully decided to prioritize art, in this case, digital art as a central element in our endeavors to tackle the issue of boosting girls' interest in STEAM fields. Drawing from scientific research and our collaborative experiences with the Gothenburg Opera House [18], we have merged these interdisciplinary domains to create an engaging and enriching learning experience for young individuals. Our hope was and is, that by providing a positive learning and coding experience we can help girls overcome stereotypes associated with STEM fields and inspire them to pursue careers in STEAM.

In order to enhance girls' learning in Creative Coding sessions, we made a conscious effort to choose female teachers and mentors primarily from the CSE department. Knowing that mentorship is clearly recognized as mutually bene-ficial for both mentors and mentees, going from personal satisfaction to career advancement we focused on recent findings that good mentors need a combina-tion of soft and technical skills [27]. Among the selected teachers and mentors, the majority were women, while there was only one male teacher. By including female role models in the sessions, we aimed to create an environment that fos-tered positive learning experiences for the girls. In order to provide more relatable mentors [21], we invited a mix of female members of the CSE department - some of them were professional software developers while some of them do not know to code also as mentees. To boost girls' confidence and also encourage them to explore coding and technology with enthusiasm, the presence of female teachers and mentors was crucial for breaking down gender stereotypes and promoting inclusivity within the field of computer science.

Thinking of the best ways to support and improve girls' learning experience we have put special importance on close collaboration between mentors, teachers, and their peers. By working together, we aimed for the creation of a supportive network where the girls could learn from each other and build social capital [16,19,21]. This collaboration would not only facilitate their technical skills development but also cultivate teamwork and communication abilities, which are essential not only in the field of coding but as one of the most important 21st-century skills. Overall, our goal was to positively influence girls' learning experiences and empower them through the integration of female role models, collaborative learning, and the acquisition of social capital in Creative Coding sessions.

Finally, the Creative Coding project has been designed to respond to the social demand for more girls in STEAM fields and thus, help them become active solution designers of their social integration and mobility. To achieve that and to contribute to addressing the under-representation of women in STEAM due to societal, stereotypical, and other factors, we have opted for a process that involves early intervention in schools to encourage more girls from dis-empowered neighborhoods of Gothenburg to pursue STEAM education and careers.

The Creative Coding Project is supported by three pillars. The first pillar is learning to code through the informal learning practice of making music. The second pillar is social capital acquisition through the mentorship of positive role models and meeting peers. The third pillar is being present at Chalmers University building Kuggen through which we implicitly say that girls are very welcome.

3   Ethical Considerations

This initiative prioritized the equitable treatment of all girls, irrespective of their backgrounds, ethnicity, socioeconomic status, or abilities. A comprehensive endeavor was undertaken to ensure that every girl had an equal opportunity to participate. To achieve this, the project's information was diligently disseminated through diverse and viable means, ensuring its broad reach and maximizing the chances of engagement from a large pool of potential candidates. In order to conduct the project's research, working both as researchers insiders and outsiders [2], which was primarily focused on females between the ages of 10 and 15, parental consent was absolutely necessary. Therefore, it was made sure that the parents' informed agreement was gained before the research began and that they were given a thorough explanation of the research and interview process. The technique of handling the data acquired and how the personal data is kept securely was explained to the parents.

The participants' psychological well-being was also given a lot of consideration by creating a friendly and encouraging environment. This entailed attending to any emotional or psychological difficulties that could emerge throughout the project and offering suitable support and direction. The teachers took the initiative to inform the candidates about the project and the steps involved because the majority of the participants were unfamiliar with the collaborative live coding experience. Communication with the participants and their families was kept open and transparent throughout all phases of the research. This required outlining the project's goals, actions, and any potential dangers or advantages in explicit terms. At the conclusion of the session, feedback from the participants was collected in order to determine what needs to be improved in the sessions that will follow.

4 Methodology

The Creative Coding sessions were organized as two-hour sessions from 5–7 pm each Thursday and two four-hour sessions from 12–4 pm on two Saturdays. Workshops were hosted by two PhD students of the CSE department, and the venue was the Department of Interaction Design at Kuggen, the Chalmers University of Technology Sweden. The series of workshops culminated in a concert where girls performed their coded music in Visual Arena, Lindholmen.

Through the workshops, the participants were iteratively acquainted with the fundamentals of the Strudel tool. Taking good care of a friendly positive and safe atmosphere, girls were granted the opportunity to explore the logic of coding, meet the basics of live coding, and the basic concepts associated with Strudel commands. Subsequently, through their collaboration with PhD students and close mentorship which is based on old Bloom's study [4] that revealed a significant enhancement in the grades of students who received intense tutoring, participants advanced towards coding music using these rudiments. Gradually, they learned to improvise their code based on their preferences and investigated working collaboratively in pairs and teams.

The Creative Coding project started with 16 girls of different cultural backgrounds. Since the whole process is novel and unusual, we wanted to learn from the girls [18] and hear in their own words how they describe their experience in this process. Bearing in mind their very young age (10–15), we sent hard copy requests to all parents for consent to interview the girls. After obtaining parental consent for running a research study in parallel with a coding session we chose two complementary methods for analysis of Creative Coding sessions impact - observation and interview. During the sessions we conducted a close observation of the workshop participants, studying their interactions and approaches towards teachers, mentors, and peers. A weekly observation diary was created to carefully monitor their transitions and document participants' progress. Once Creative Coding workshop sessions were finished, an open-ended focus group interview was conducted to evaluate girls' progress, motivation, the feelings of belonging, and well-being. This method was chosen to make girls feel comfortable, safe, and more engaged suggested in the scientific literature [1]. Their identity was completely anonymized in the data, and the data was saved just for the purpose of the study. The data collected were noted and subjected to rigorous analysis using Braun and Clarke's (2006) six-phase framework for thematic Analysis [6].

5   Analysis and Results

In this section, we present the analysis of the collected data and its results. As mentioned above we analyze the Creative Coding project through observation and interview.

5.1   Observation

Through detailed thematic analysis of weekly diary data, we gained a more comprehensive understanding of the various aspects and experiences within the Creative Coding project. Based on the observations, the following 8 themes and sub-themes were identified:

Theme 1: Initial Excitement and Exploration.

- **Sub-theme 1.1: Unboxing and Equipment Exploration**
  As for the full participation in the Creative Coding program composing music played a central role, a lot of the audio equipment was necessary to support coding on computers. We have bought sound cards and headphones for each participant. To start with coding music girls needed to get acquainted with the equipment and learn how to connect the devices. The girls exhibited eagerness and exhilaration when unboxing the coding equipment, indicating a sense of creativity, engagement, and joy when introduced to novelty.
- **Sub-theme 1.2: Interest in Coding**
  The participants actively engaged in exploring the new tools and features, demonstrating curiosity about the coding process and its potential for creativity. The concepts of Strudel and live coding were entirely novel to the participant, but their desire to delve into the software indicated an eagerness to learn something new. Strudel represents a novel live coding platform that enables the composition of dynamic music pieces within the confines of a web browser. This platform has been designed to be easily accessible to individuals of varying levels of expertise, ranging from novices to professionals. As Strudel requires quite a low threshold of knowledge of coding for sending interesting audio-visual feedback to the user, the experiments with code, in other words making changes and listening, and watching it, were easy and ensured creative flow.

Theme 2: Learning and Retention.

- **Sub-theme 2.1: Understanding Coding Concepts**
  Due to Strudel's repressiveness participants were able to grasp the fundamental syntax of the code with relative ease, as it proved to be quite straightforward and well-suited for beginners. A subset of the participants exhibited prior experience with Scratch (coding language with a simple interface suitable for young learners) and Block Coding (coding using visual methods),

which facilitated their comprehension of the coding principles in Strudel. The participants exhibited an aptitude for understanding and applying the coding concepts conveyed during the sessions, thus attesting to their progress along the learning continuum. They were able to establish logical connections between the musical notes and the corresponding codes, which is indicative of their progress in their learning journey. Participants coding their own music pieces can be observed in Fig. 2.

Fig. 2. Participants coding music

– **Sub-theme 2.2: Challenging to Remember Commands**
The Strudel system offered a wide range of commands and functions for generating music. However, certain participants encountered difficulties in recalling all the coding commands, indicating the necessity for supplementary reinforcement and practice. The teachers stressed the importance of regular practice in achieving greater familiarity with the commands, rather than attempting to memorize them all at once. As a helpful measure, the participants were introduced to the well-defined documentation in strudel to support their coding works.

– **Sub-theme 2.3: Recall and Discussion**
During Creative Coding sessions, participants had to recall their previous knowledge from memory to effectively engage with new concepts in the Strudel programming language. The initial portion of the Creative Coding session was consistently dedicated to reviewing the concepts and commands acquired during the preceding lesson, thereby ensuring that participants possess a sound understanding of the material. Notably, participants were able to recall and actively discuss previous session topics, revealing a strong retention of acquired knowledge. Interestingly, we observed that younger participants below 12 years encountered greater difficulty in recalling commands when compared to their elder counterparts, however, this is a common occurrence.

Theme 3: Collaboration and Peer Learning

- **Sub-theme 3.1: Group Work and Sharing**

  The participants of the project were girls between the age group of 10–15 from different schools and areas in Gothenburg. Initially, it was observed that the participants exhibited a reluctance to engage in communication, owing to their unfamiliarity with one another. But in due course of time, they started observing each other, discussing their works and doubts with their classmates, and started editing the codes of their peers as part of group work. The project fostered a collaborative environment where girls worked in groups, shared their work, and exchanged feedback.

- **Sub-theme 3.2: Peer Learning and Feedback**

  During the sessions, the educators consistently highlighted the significance of effective communication and productive discussion within the classroom environment to enhance comprehension and collaborative efforts. Consequently, the attendees attentively listened to and provided feedback on each other's musical creations, promoting mutual learning and a harmonious community. The instructors ensured that all participants' work was presented to the entire class during each session, enabling peers to evaluate and enhance their own presentations by incorporating the received feedback.

- **Sub-theme 3.3: Encouragement and Motivation**

  The project created a supportive community where girls encouraged and motivated each other, contributing to a positive learning environment. The teachers, mentors, and other members of the Creative Coding team also encouraged the participants by giving positive reinforcements and comments on their works in Strudel. Although there were instances when girls felt discouraged by the perceived superiority of their peers' work, the teachers offered relevant guidance and positive reinforcement to lift their spirits and encourage them to continue.

Theme 4: Mentoring and Guidance

- **Sub-theme 4.1: Teacher's Role**

  Teachers played a crucial role in providing guidance, explaining coding concepts, and offering feedback to help girls overcome challenges. They encouraged the participants to try and experiment with code and to explore the software to unleash its potential to create music. The teachers were able to give individual attention to the candidates and ensured the inclusion of all participants in the session.

- **Sub-theme 4.2: Mentor's role**

  In addition to the presence of teachers, mentors were selected from the pool of Ph.D. students and lecturers affiliated with the CSE department. These mentorships were specifically designated to serve as exemplars for the participating girls. Our primary objective was to furnish participants with guidance and positive role models, with the intent of inspiring them through the mentorship relationship. Nonetheless, the practical realization of the mentors' role proved to be hindered by various constraints, including restricted time allocations for sessions, inflexible scheduling on the part of the mentors, and limited availability of mentors. As a result, the effective implementation of the mentors' involvement in the initial phase of the project was compromised due to the absence of a comprehensive strategy and methodological support for leveraging their support.

- **Sub-theme 4.3: Engaging Discussions**
  Teachers were engaged in discussions with the girls about their project progress, offering strategies and support to enhance their learning experience. The participants actively participated in the discussions and collaboratively worked while creating music. The teachers gave the participants the freedom to raise their queries and concerns regarding their work and progress.

Theme 5: Overcoming Shyness and Fostering Inclusion

- **Sub-theme 5.1: Encouragement to Participate**
  Some girls were initially shy and required encouragement to ask questions and actively participate in the sessions. With the support of teachers and peers, they were able to overcome this challenge and thereby fostering inclusivity and creating a safe space for everyone to engage. It was also apparent that parents played a significant role in motivating the participants to overcome their shyness. This was evident when some participants mentioned during a discussion that their parents encouraged them to practice for the concert at home and urged them to ask questions and clarify doubts during training sessions to achieve the best results.

Theme 6: Iterative and Creative Process

- **Sub-theme 6.1: Experimentation with Sounds and Colors**
  The project aimed to inspire girls to explore various sounds, colors, and coding techniques, thus promoting a creative and iterative mindset. Throughout the learning process, and due to the amazing audio-visual capacity of the Strudel tool, the participants were encouraged to iterate and learn from their mistakes. Notably, the girls utilized the Freesound platform to identify and recreate sounds within Strudel and used online collaborative tools like Google Docs to share and work collaboratively. This proactive involvement showcased their eagerness to experiment and work creatively, indicating their explicit ability to learn and experience more technical aspects.
- **Sub-theme 6.2: Editing and Improvisation**
  Participants were given the opportunity to edit and improvise existing code, promoting their individuality and creativity. They also edited and improvised using the code of their group members and peers, which gave them further opportunities to explore the coding strategies. This allowed them to broaden their knowledge while discussing with their peers.

Theme 7: Concert Preparation

- **Sub-theme 7.1: Motivation for Performance**
  The participants were clearly introduced to the idea of having a concert at the end of the Creative Coding project. The clear objective of preparing for a concert motivated girls to work towards their public performances individually and as a team. From the beginning of the project, the participants were aware that they can perform on the stage and in front of the audience. Those who did not feel good about being on the stage were given the opportunity to perform in some other way, for example, just playing the code. Still, all girls embraced this challenge in an impressive, brave way.
- **Sub-theme 7.2: Planning and Roles**
  Planning for solo, duo, and trio acts provided girls with the opportunity to take on different roles, promoting personal growth and skill development. It was admirable to see the girls taking the initiative to do a solo, duo, and trio. The participants were given the freedom to choose their partners for the performance and this gave them the freedom to explore and communicate further with their team members. Participants actively selecting their partners for their performance is observable in 3.

Fig. 3. Participants creating groups

Theme 8: The Performance

- **Sub-theme 8.1: Transformation and Collaboration**
  The candidates' performance demonstrated the significant transformation they underwent throughout the project, from initial hesitation to becoming a cohesive team, promoting cooperation and synergy. We observed that the participants were able to collaborate harmoniously with their fellow team members on stage during live coding.
- **Sub-theme 8.2: Self-Confidence and Expression**
  Regardless of their very young age, the girls displayed impressive and unexpected self-confidence during their Creative Coding presentation. They proudly showcased their individual and collective aptitude for task completion and their passion for creatively expressing themselves through music. (A sample music from one of the participants can be experienced through the following link: https://strudel.tidalcycles.org/?fYMWmFziVVnh). Their enthusiastic and well-prepared self-introduction and impressive live coding were indicative of this aspect. Participants engaging in live coding can be observed in Fig. 4.

In conclusion, through the careful and detailed implementation of Braun and Clarke's (2006) six-phase framework [6], we managed to reveal crucial insights into the girls' experience with Creative Coding sessions. The identified themes and sub-themes highlighted the girls' enthusiasm, learning, collaboration, mentoring, inclusivity, iterative and creative process, and final performance.

Fig. 4. Performing on stage - Live Coding in Strudel

5.2   Results of Observation

By giving a deep, holistic look at these eight topics a number of conclusions can be drawn. The initial theme of excitement and exploration displayed by the girls indicates a strong curiosity to explore new tools and features, which is a crucial skill in motivation and learning [12]. Furthermore, the girls demonstrated a surprising ability to comprehend coding concepts, even when faced with challenges in recalling specific commands. With more practice, they were able to improve their retention of these concepts. When exploring in a supportive and safe environment, girls' group work, sharing, and feedback sessions demonstrate their great capacity for collaboration and peer learning.

It also became apparent that mentoring and guidance can promote inclusivity and encourage introverted participants to engage actively, ensuring an empowering and inclusive environment for all. We agree with Millar and colleagues (2022) that to be more successful in supporting a diverse group of participants more carefully developed strategies should be planned to ensure closer mentor-participant relations. It is important to note, that the Creative Coding iterative and creative process enhanced experimentation, improvisation, and individual expression through coding techniques. Together with the concert preparation, and performance, this stimulated a goal-oriented approach that advances personal growth, self-confidence, and skill development.

5.3   Interviews

Girls' first-hand testimony takes us to an even deeper understanding of the effects and capacities of the Creative Coding project. Here we present the analysis and results of the data collected through direct communication with the girls. In this communication, we used open-ended questions and a quantitative Likert scale [28]. Again, by using Braun and Clarke's (2006), we identified 5 themes. They read as follows.

Theme 1: Interest and Engagement. There were participants who lacked prior experience in coding, while others had a basic understanding of block coding and Scratch, which served as their foundational knowledge. A few girls had previously engaged in coding using game platforms. Integration of music/art with coding made the learning experience more engaging. Girls found the combination of creative elements more interesting than traditional coding methods or games. One of the girls said *"I was interested in coding, but I like this combination better"* making a clear appreciation and preference for coding together with art.

Participants expressed a high level of interest in coding and art. The majority of participants expressed a strong passion for art, and when asked how many stars (from 1 to 5) they would give to art, all of them gave a rating of 5/5. Some girls mentioned attending guitar and piano classes, indicating their active engagement in artistic pursuits. The following quote clearly illustrates the relationship between coding and art. *"I would give a 4/5 for coding and 5/5 for art"*.

Girls found Creative Coding more enjoyable and stimulating compared to conventional coding techniques or game-based learning approaches. The project's emphasis on creative expression through coding captured their attention and sustained their interest. A participant said, *"This combination of music and coding is better than coding with games"*.

Theme 2: Positive Experience and Confidence. All girls reported a positive and enjoyable experience during the Creative Coding project. They expressed enthusiasm for attending the sessions and actively participating in coding activities. *"I joined Creative Coding as I like music and I thought it would be fun to learn to code music."* said one of the girls. The Creative Code project was described as a fun and enjoyable activity that some participants considered pursuing as a hobby. "I feel like I have a new hobby now" were exact words. The girls felt positive about their overall experience, indicating a sense of satisfaction and enjoyment. Participation in the Creative Code project contributed to an increase in participants' confidence in their coding abilities, leading to feelings of empowerment about their future in the science and technology fields. This is how they describe it. *" Strudel is fun, I feel like I have learned something interesting."* *"Here I had the freedom to explore my own interests."* Girls' testimonies clearly show that participation in Creative Coding workshops fostered a sense of empowerment and self-confidence.

Theme 3: Importance of Female Mentors and Role Models. Participants highlighted the significance of having female mentors who could relate to their experiences. Female mentors provided guidance, support, and encouragement, which boosted their confidence and motivation. A participant said,*"I feel it is good to have female mentors/teachers. The teacher was approachable, and she patiently heard and solved my doubts"*. Girls felt supported and believed in by their mentors, creating a positive learning environment. Having mentors who understood their challenges and aspirations helped them overcome barriers and doubts.

Theme 4: Building Relationships Social Capital Acquisition and Feeling of Well-Being. Participants formed strong bonds with their mentors, built on trust, respect, and mutual understanding. Positive relationships with peers fostered collaboration, shared learning, and a sense of camaraderie. The girls expressed a preference for mentors who provided positive feedback and conveyed information in a positive manner. Some girls mentioned that the decision of whether to have a mentor should be left to the individual, allowing them to decide if they need one. Teachers and mentors created a supportive environment that encouraged open communication and collaboration. Participants felt comfortable asking questions, seeking help, and expressing their ideas. One of the interviewees said, *"I felt secure and included in the group."* Collaborative activities and group projects allowed participants to bond with their peers. Sharing experiences and collaborating with other girls enhanced their sense of community and support. Participants desired more time for breaks and interactions with classmates, indicating the importance of socializing and connecting with peers. This is how they describe it - *"Working together on a piece of code was fun, we made a horror theme for the concert"*.

Theme 5: Time Constraints and Attrition. Some participants mentioned that time constraints and other activities might have contributed to a few participants leaving the course. The Creative Code project finished 8 girls out of 16. Limited availability and conflicts with other commitments impacted their ability to fully engage in the project. The girls noted that flexibility in scheduling sessions and accommodating individual time constraints could help address attrition and increase retention rates. Adapting the project to fit participants' busy schedules would allow more girls to fully participate.

5.4   Results of Interview

Based on the findings derived from the interviews of the participants in Creative Coding workshops, and the analysis of the collected data, several conclusions can be drawn. As already pointed out by scientific literature [7,9,13] the integration of art, into the teaching and learning processes, has the potential to enhance engagement and attract girls to STEAM fields. The integration of music, or art as a final result of coding practice, and audio-visual perception of music holds great potential in enhancing the interest of girls in the fields of coding and STEAM. Music has been found to positively affect physiological factors such as heart rate, blood pressure, and hormonal levels, as well as psychological experiences like restlessness, anxiety, and nervousness [10]. Therefore, the incorporation of music to support interdisciplinary learning can lead to immersive and emotionally engaging learning experiences for girls.

Additionally, coding visual elements (enabled through the Strudel software tool) to correspond with various aspects of music such as tempo, pitch, and intensity can further elevate the audio-visual perception, resulting in the creation of captivating experiences. With that in mind, we argue that the adoption of art as a final destination increases plausible learning experiences, feeling of belonging, and well-being. The participants that undergo such learning and experience performance of their creation are more confident, and empowered to explore further within science and technology fields.

The support and encouragement delivered through collaboration with their teachers, emphasizes the significance of representation and the need for more diverse role models to inspire girls in STEAM. We should not forget that girls stressed the importance of having opportunities to bond and collaborate with their peers to foster a sense of community, belonging, and support. Finally, when thinking of quite a high level of dropouts (almost 50%), one could argue that better methodological approaches and creativity in teaching, together with scheduling flexibility should be considered to accommodate more girls and increase retention rates.

The insights from the results of interviews and observations are summarized in Table 1.

Table 1. Key Insights from Combined Interview and Observation Results

Insight	Description
1. Passion for Creative Coding	Participants displayed a strong passion for combining music and coding, finding it engaging and enjoyable
2. Positive Learning Experience	The project fostered a positive learning experience, leading to increased confidence and empowerment
3. Importance of Female Mentors	Female mentors played a crucial role in providing guidance, support, and motivation, enhancing participants' confidence
4. Collaborative and Supportive Environment	Collaboration and positive relationships with mentors and peers created a supportive learning environment
5. Encouragement and Inclusivity	Shy participants required encouragement to participate actively, contributing to inclusivity in the learning process
6. Exploration and Creativity	Participants actively experimented with coding techniques, promoting creativity, experimentation, and individuality
7. Motivation through Performance	The prospect of a concert motivated girls to work individually and in teams, fostering goal-oriented efforts
8. Transformation and Self-Confidence	Participants transformed from hesitant individuals to a cohesive team on stage, displaying self-confidence and creativity

6   Conclusion

The research findings demonstrate that Creative Coding workshops have proven to be highly beneficial for girls' engagement in STEAM fields. The Creative Coding project had a positive impact on various aspects of the girls' involvement, including their interests, self-confidence and well-being, creativity, explorations, and aspirations in STEAM subjects. The girls exhibited great enthusiasm and passion for exploring new tools and concepts, and their dedication to practice allowed them to enhance their learning and retention. Notably, they demonstrated remarkable stamina and bravery in preparing for their public performance and surprised everyone with their newfound self-confidence. It is evident that the girls are eager to showcase their abilities and assert their power in STEAM domains.

The integration of art, music, and audio-visual elements in the teaching and learning processes proved to be highly and surprisingly effective in creating a positive learning flow and attracting girls to STEAM subjects. The use of user-friendly software, such as Strudel, that provided interesting audio-visual feedback played a crucial role in facilitating this positive learning experience. Additionally, the preparation for the public performance added a final touch to the girls' personal empowerment and well-being, as they felt proud and enthusiastic about presenting their musical creations. However, it is important to emphasize the need for more efforts to create an inclusive and supportive environment where a wider group of girls feel comfortable and empowered to fully engage in STEAM activities.

Overall, the findings highlight the significance of nurturing curiosity, providing mentorship and guidance, promoting inclusivity and representation, fostering collaboration, and integrating creative elements in education to inspire and empower girls in coding and STEAM disciplines. The report argues that the Creative Coding project has the potential to enable innovative learning experiences by simulating synesthetic experiences, where musical elements are linked to visual elements, thus allowing the creation of multimedia art that engages multiple senses simultaneously. The report highlights the potential of Creative Coding practices and the audio-visual perception of music in boosting girls' personal empowerment, creativity, and interest in coding and STEAM disciplines. It emphasizes the significant role of music creation in this process, as it contributes to increased curiosity. The music is the key.

References

1. Adler, K., Salanterä, S., Zumstein-Shaha, M.: Focus group interviews in child, youth, and parent research: an integrative literature review. Int J Qual Methods **18**, 1609406919887274 (2019)
2. Alderson, P., Morrow, V.: The Ethics of Research with Children and Young People: A Practical Handbook. Sage (2020)
3. Bazler, J., Van Sickle, M. L.: Cases on STEAM education in practice. IGI Global. https://doi.org/10.4018/978-1-5225-2334-5 (2017)
4. Bloom, B.S.: The 2 sigma problem: the search for methods of group instruction as effective as one-to-one tutoring. Educ. Res. **13**(6), 4–16 (1984)
5. Catterall, L.G.: A brief history of STEM and STEAM from an inadvertent insider. The STEAM J. **3**(1), 5 (2017)
6. Clarke, V., Braun, V., Hayfield, N.: Thematic Analysis. In: Qualitative Psychology: A Practical Guide to Research Methods, 3rd edn., pp. 222–248. Springer (2015)

7. Cook, K., Bush, S., Cox, R.: Engineering Encounters: From STEM to STEAM. Sci. Child. **54**(6), 86–93 (2017)

8. Chalmers University of Technology: Creative Coding - Chalmers University of Technology. https://www.chalmers.se/en/collaborate-with-us/activities-for-schools/creative-coding/ (2023)

9. Crayton, J.: Designing for immersive technology: Integrating art and STEM learning (2015)

10. De Witte, M., Spruit, A., van Hooren, S., Moonen, X., Stams, G.J.: Effects of music interventions on stress-related outcomes: a systematic review and two meta-analyses. Health Psychol. Rev. **14**(2), 294–324 (2020)

11. Ge, X., Ifenthaler, D., Spector, J.: Moving forward with STEAM education research. In: X. Ge, D. Ifenthaler, J. Spector (Eds.), Emerging technologies for STEAM education. Educational communications and technology: Issues and innovations, 383–396. Springer. https://doi.org/10.1007/978-3-319-02573-5-20

12. Gruber, M.J., Valji, A., Ranganath, C.: Curiosity and learning: a neuroscientific perspective. In: Proceedings of the Conference on Curiosity in Science and Beyond, pp. 397–417 (2019)

13. Land, M.H.: Full STEAM ahead: the benefits of integrating the arts into STEM. Proc. Comput. Sci. **20**, 547–552 (2013)

14. Leavy, A., Dick, L., Meletiou-Mavrotheris, M., Paparistodemou, E., Stylianou, E.: The prevalence and use of emerging technologies in STEAM education: a systematic review of the literature. J. Comput.-Assist. Learn. (2023)

15. Liao, C.: From interdisciplinary to transdisciplinary: an arts-integrated approach to STEAM education. Art Educ. **69**(6), 44–49 (2016)

16. Lluch, A.M., Lluch, C., Arregui, M., Jiménez, E., Giner-Tarrida, L.: Peer mentoring as a tool for developing soft skills in clinical practice: a 3-year study. Dentistry J. **9**(5), 57 (2021)

17. Maric, J., Rani, L.M.: Coding Music For No Stress Learning. Proc. SMM23. Workshop Speech, Music Mind **2023**, 6–10 (2023). https://doi.org/10.21437/SMM.2023-2

18. Marić, J.: Digital storytelling in interdisciplinary and inter-institutional collaboration-lessons from our youngest. Cult. Manag. Sci. Educ **4**, 129–144 (2020)

19. Marić, J.: Web communities, immigration, and social capital (2014)

20. Maric, J.: Who wants to grow old in Welfare Sweden?. In European Conference on Social Media (Vol. 9, No. 1, pp. 130–136) (2022 April)

21. Millar, V., Hobbs, L., Speldewinde, C., van Driel, J.: Stakeholder perceptions of mentoring in developing girls' STEM identities: "You do not have to be the textbook scientist with a white coat". Int. J. Mentor. Coach. Educ. **11**(4), 398–413 (2022)

22. Ruthmann, S.A.: The Routledge companion to music, technology, and education. A. King, E. Himonides, A. Ruthmann (Eds.). New York and Abingdon: Routledge (2017)

23. Saline, M., Sheeran, M., Wittung-Stafshede, P.: A large 'discovery' experiment: gender initiative for excellence (genie) at chalmers university of technology. QRB Disc. **2**, E5 (2021). https://doi.org/10.1017/qrd.2021.3

24. Salmon, G.: May the fourth be with you: creating education 4.0. J. Learn. Development. **6**(1), 95–115 (2019)

25. Sullivan, A., Bers, M.U.: Investigating the use of robotics to increase girls' interest in engineering during early elementary school. Int. J. Technol. Des. Educ. **29**, 1033–1051 (2019)

26. Swaminathan, S., Schellenberg, E.G.: Arts education, academic achievement, and cognitive ability. In: P. P. Tinio J. K. Smith (Eds.), The Cambridge Handbook of the Psychology of Aesthetics and the Arts, 364–384. Cambridge University Press (2015)
27. Torres-Ramos, S., et al.: Mentors as female role models in STEM disciplines and their benefits. Sustainability **13**(23), 12938 (2021)
28. Taherdoost, H.: What is the best response scale for survey and questionnaire design; review of different lengths of rating scale/attitude scale/Likert scale. Hamed Taherdoost (2019)
29. Yakman, G.: STEAM education: an overview of creating a model of integrative education. Pupils' attitudes towards technology (PATT-19) conference: Research on technology, innovation, Design and Engineering Teaching, Salt Lake City, Utah, USA (2008)

Glitch Art Generation and Performance Using Musical Live Coding

Noriki Amano[✉]

Mukogawa Women's University, Nishinomiya 6338558, Hyogo, Japan
amnrk@mukogawa-u.ac.jp

Abstract. A glitch means an error in image, video, or sound, and glitch art is art in which glitches are intentionally generated by destroying digital data or physically manipulating electronic devices for artistic purposes. The effects of such glitches are often unpredictable and aesthetically interesting. As part of our exploration of glitching methods and performance, we have attempted to generate glitch art images using musical live coding. Specifically, we generated glitches in PNG image data in conjunction with sound effects using the Sonic Pi: musical live coding system. Furthermore, such glitch art images are generated in real-time. We have confirmed that our method can generate glitches with regularity assuming the filtering of PNGs. This research proposes a method of art expression across sound and image and is an attempt to explore the performance of glitch art.

Keywords: Glitch Art Images · Live Coding · PNG

1 Introduction

A glitch means an error in images, video, or sound. Glitch art is art in which glitches are intentionally generated by destroying digital data or physically manipulating electronic devices for artistic purposes. The effects of such glitches are often unpredictable and aesthetically interesting.

Live coding means writing and executing a program on the spot and has been a common practice in programming workshops. However, live coding as a performing art that considers the improvisational act of coding as a means of expression has been attracting attention in recent years.

We have been exploring glitching methods and their performance, and as part of this exploration, we have attempted to generate glitch art images using musical live coding. Specifically, we generated glitches in PNG image data in conjunction with sound effects using the Sonic Pi: musical live coding system. Furthermore, such glitch art images are generated in real-time. We have confirmed that our method can generate glitches with regularity assuming the filtering of PNGs. This research proposes a method of art expression across sound and image and is an attempt to explore the performance of glitch art.

A. L. Brooks (Ed.): ArtsIT 2023, LNICST 565, pp. 175–185, 2024.
https://doi.org/10.1007/978-3-031-55312-7_13

This paper is organized as follows: Sect. 2 describes the background of this research on glitch art and live coding as performing art, and related works. Section 3 describes the basic approach of this research. Section 4 outlines the prototype system and its implementation. Finally, Sect. 5 concludes the paper.

2   Research Background and Related Works

2.1   Glitch Art

A glitch means an error in images, video, or sound. Glitch art is art in which glitches are intentionally generated by destroying digital data or physically manipulating electronic devices for artistic purposes.

Such glitch art is classified as follows [1].

- **Data manipulation**: changing the data in a file and causing glitches
- **Misalignment**: opening a file from one app in another app
- **Hardware malfunction**: causes a machine to malfunction, producing sound or video
- **Misregistration**: physical noise in analog media
- **Distortion**: creating physical distortion with magnets, etc.

The effects of such glitches are often unpredictable and aesthetically interesting. In addition to the publication of a book on art using glitches [2], an international conference on glitch art such as GLI.TC/H [3] has been held, and glitching is already recognized as a form of expression in art.

Theoretical research has also been conducted on this kind of glitch art. Rosa Menkman uses information theory to propose an understanding of glitch art as a specific genre of contemporary art [4]. In 2010, an international conference on glitches, GLI.T/CH, was held by Roasa Menkman and others. Glitch art research has been also presented at the international conference evomusart, which has been held every year since 2011 [5].

2.2   Live Coding

Live coding as a performing art has been taking place worldwide since 2000. Events such as Algorave [6] have been held in various locations, led by the live coding community TOPLAP [7], and academically, international conferences such as ICLC [8] have been held annually since 2015.

Live coding as a performing art is dominated by coding performances that generate music and video improvisationally. In particular, music-based live coding is popular because it feels similar to live music performance.

Many tools have been developed for live coding as a performing art. Specifically, there are tools for live music coding such as Chunk [9], OverTune [10], TidalCycles [11], and Sonic Pi [12], as well as tools for live video coding such as Fluxus [13], LiveCodeLab [14], and Hydra [15].

2.3 Related Works

The use of glitches in artworks is not new, as seen in Nam June Paik's work "Magnet TV" [16], and as described in Sect. 2.1, glitches are already recognized as a form of expression in art. Some artists have published works on glitching in PNG, and we have referred to them in this study [17]. There is also a technique called data moshing, which causes errors during video playback, and research has been conducted on its effects and methods [18]. An installation work of glitches using live coding [19] has also been produced. This work is very interesting because it includes not only sound and images but also glitching by physical devices. Although not competing with this research, there is an attempt to use glitch art in contemporary dance [20]. This is very interesting because it uses glitch art based on sensor error to obtain dance motions as images for mixed reality.

The differences between these glitch artworks and this research are as follows.

- Manipulate (control) image glitches based on audio processing
- Intentionally cause glitches that match sound effects

In other words, the trigger for the glitch is audio processing (sound effects), which generates glitches in the image in real-time while enjoying the effect applied to the music. Live coding that generates video in real-time is not uncommon. Live coding systems that generate music have also been realized. Some live coding systems, such as Gibber [21], can generate music and video simultaneously. However, existing live-coding systems basically do not generate glitches, but rather generate video and music from scratch, and can be said to belong to the category of generative art [22]. Generative art and glitch art are distinctly different.

Another unique aspect of this research is the generation of regular glitches. This is related to the generation of glitches that are mapped to audio processing. That is, it is intended to generate glitches that match specific audio processing, not to generate glitches that are unpredictable. Although this point is debatable, one of the goals of this research is to use glitches to visualize audio processing. In other words, this study aims to enjoy music and glitch art at the same time, with the assumption that the two are interrelated.

3 Research Goals and Basic Approaches

3.1 Research Goals

In this study, our goals are as follows.

- Propose a method of expressing art across sound and image
- Create glitch art in an elegant method
- Performance of glitch art

The first goal is to propose a method of art expression in which sound and image processing are linked. Then, we will use an elegant method for creating glitch art. It's not difficult to create glitches. However, we are also concerned with the beauty of the method itself. We are not attracted to methods that are unreproducible, or just random. Nor do we feel sympathy for using image processing software to create glitch art. Glitches are inherent errors, and glitches are distinctly different from image processing. In addition to the above, this research aims to make the creation of glitch art itself a kind of performance.

3.2 Basic Approaches

The basic approach of this study is as follows.

- Linking sound and image processing
- Image format should be PNG [23]
- Mapping audio processing to glitching
- Generate real-time glitch art images with live coding

First, the image is glitched by audio processing to generate a glitch art image. We assume that the format of this image is PNG. This requires a mapping between the audio processing and the glitching. The above is achieved in real-time using live coding. This is the basic approach taken in this research. We regard this approach as an elegant method to generating glitch art through meaningful audio processing, rather than generating glitch art through random destruction of data. In addition, by linking it with live coding, we will be able to see the generation of glitch art in real-time, aiming to turn glitch art into a performance. In order to generate glitch art through audio processing, we use a music-based live coding system in this research.

There is a reason why we chose PNG as the image format. Before that, the characteristics of PNG are listed below.

- File format that compresses and records image data
- Compressible without degradation
- Compression encoding of bitmap images
- Lossless compression and decompression
- Pre-filtering to improve compression efficiency by using image characteristics

In the PNG feature, we focused on filtering. We focused on the filtering process because we thought it would be possible to achieve regular glitches corresponding to the processing of each filter. The processing of each filter is as follows.

- **None**: Do nothing.
- **Sub**: Subtracts the byte to the left.
- **Up**: Differentiate from the byte directly above.
- **Average**: Differs from the average of the bytes directly above and to the left.
- **Paeth**: Calculates the Paeth value from the left neighbor, directly above, and above the left byte, and subtracts the difference.

Figure 1 shows the basic mechanism of glitching in this study.

In this study, glitching is used to display PNG images, based on the basic mechanism of PNG. This is based on the basic mechanism of PNG, which uses the characteristics of an image to perform filtering to improve compression efficiency before the ZIP compression process. This process is described above. When displaying a PNG image, the reverse process (reverse filtering) is performed according to each filter.

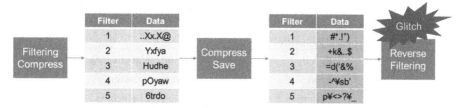

Fig. 1. Basic Mechanism of Glitch

Care must be taken when creating glitches in PNG images. That is, the PNG format must not be destroyed. In other words, if a PNG file is corrupted in a random way, the image itself cannot be displayed at all. Therefore, in this study, PNG images are loaded, the format is not broken, only the data portion is compressed simply, and the image is saved again. The key point is to skip the filtering process. This causes glitches when inverse filtering is performed. This mechanism generates glitches by inverse filtering even though the filtering process is skipped. In other words, the glitch art in this study is based on the data manipulation and misalignment described in Sect. 2.1.

The above glitching mechanism, which focuses on the filtering process of PNG, makes it possible to generate glitches with regularity as follows.

- **Sub filter action**: causes horizontal glitching
- **Up filter action**: causes vertical glitching
- **Average filter action**: causes diagonal glitching
- **Paeth filter action**: causes complex glitches

The next step is to map these glitches to audio processing. The purpose of this research includes making glitch art performances. Instead of viewing glitch art in a frame, we envision a performance in which glitches are continuously generated in the original image while the song is being played through live coding of music, and the glitches are shown in real-time. Sound and images seem to have nothing to do with each other, but if we consider images as patterns of color recognition based on the reflection of light, they have something in common. This is because both sound and light have the characteristics of waves, and both can be displayed as waveforms.

Based on the above, this study focuses on typical audio processing and links it to the glitching mechanism. We will employ the following audio processing.

- **Reverb**: creates a sense of spatial depth and spaciousness
- **Echo**: creates atmosphere and broadens the range of expression
- **Distortion**: creates a powerful, thick sound
- **Compressor**: makes overly loud sounds more consistent and easier to hear.

The above audio processing is not uncommon and is implemented in many music production software.

In this study, these audio processes are linked to the PNG filtering process, and glitches are generated at the time of audio processing. Table 1 shows the correspondence between PNG filtering and audio processing.

Table 1. Audio processing and PNG filtering corresponding and glitching

Filter	Audio Process	Glitch
Sub	Reverb	Horizontal glitching
Up	Echo	Vertical glitching
Average	Distortion	Diagonal glitching
Paeth	Compressor	Complex glitches

4   Implementation

In this study, a prototype system that generates glitches in PNG images in conjunction with Sonic Pi, a musical live cording system, was developed. The system configuration is shown below (Fig. 2).

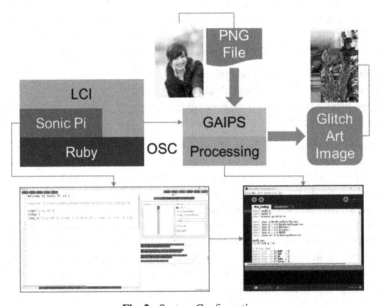

Fig. 2. System Configuration

Processing is used to generate glitch art images. The commands for generating glitch art images are implemented on Sonic Pi, a musical live coding system that is linked to, and Ruby, the language in which Sonic Pi is implemented, is also used for some of the commands. OSC communication is used to control glitching through live coding.

Live coding is performed on the Sonic Pi by writing a program in the Sonic Pi workspace similar to the Sonic Pi program and pressing the Run button (Fig. 2). To perform live coding for glitch control, the definition file (live_coding.rb) is first loaded with the require function of the Ruby language.

The definition file contains the instructions necessary for live coding to manipulate glitches. The instructions for controlling glitches are as follows (Table. 2).

Table 2. Glitch Control Commands and Usage Examples

Commands	Syntax	Example
revrb	revrb(cmd)	revrb("play 60")
revrb_o	rever_o(cmd, op)	revrb_o("play 60 \n play 62", "amp: 2, room: 0.5")
ech	ech(cmd)	ech("play 60")
ech_o	ech_o(cmd, op)	ech_o("play 60 \n play 62", "amp: 2, decay: 2")
dist	dist(cmd)	dist("play 60")
dist_o	dist_o(cmd, op)	dist_o("play 60 \n play 62", "amp: 2, distort: 0.5")
comp	comp(cmd)	comp("play 60")
comp_o	comp_o(cmd, op)	comp_o("play 60 \n play 62", "amp: 2, mix: 0.5")

These commands are written in the Ruby language and are read by the require function when Sonic Pi is run. Each command is a wrapper function for a Sonic Pi effect; Sonic Pi has a with_fx command for effects, which can generate sound effects.

The following is a simple program for Sonic Pi.

```
with_fx :reverb do    # reverb effect
    play 60           # make C sound
end
```

This program applies a reverb effect to midi note number 60, the C sound in the middle of the keyboard. The implementation of the glitch control instruction above is a function for wrapping such a Sonic Pi effect. In other words, the above code is equivalent to the following code.

```
revrb("play 60")
```

The effects by Sonic Pi have many options; the revrb command is a simple command with default options, but if you want to specify detailed options, use the revrb_o command.

```
revrb_o("play 60 \n play 62", "amp: 2, room: 0.5")
```

The above code plays C sound and D sound at the same time, applying reverb with specified volume (amp option) and room size (room option), which is a measure of reverb. The above code is equivalent to the following Sonic Pi code.

```
with_fx :reverb, amp: 2, room: 0.5  do      # reverb with options
    play 60              # make C sound
    play 62              # make D sound
end
```

We show examples of glitch art generated in real-time using the above glitch control instructions below.

Fig. 3. Original image   **Fig. 4.** Capture screen

Figure 3 shows a PNG image of the sample used in the experiment, and Fig. 4 shows a captured image of the execution screen of the system implemented in this study.

Figure 5 shows glitch art generated in real-time in conjunction with sound effects by the implemented system. The effects of the effects are clearly shown.

① Generation by **revrb** command: Horizontal glitching by **Sub** filter
② Generated by **ech** command: Vertical glitching by **Up** filter
③ Generated by **dist** command: Diagonal glitching by **Average** filter
④ Generated by **comp** command: Complex glitching by **Paeth** filter

The numbers ① through ④ are for convenience only.

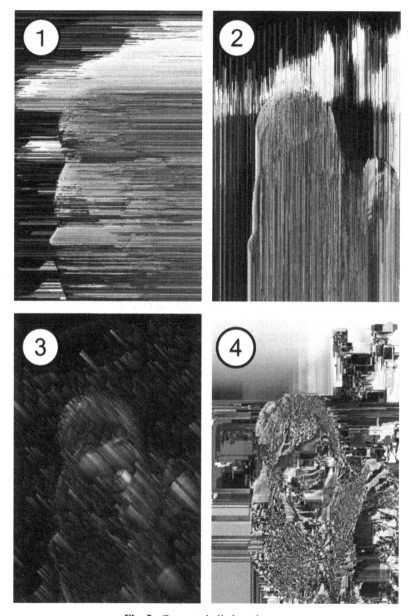

Fig. 5. Generated glitch art images

5 Conclusion

In this study, we explored the generation and performance of glitch art images using musical live coding. Specifically, we implemented and experimented with a prototype system that generates glitched art images in real-time by generating glitches in PNG image data in conjunction with sound effects using the Sonic Pi: musical live coding system. In the experiments, we confirmed that our method can generate glitches with regularity, assuming the filtering of PNGs. This regular glitch generation is based on audio processing and is a result of the basic approach of this research. This research proposes a method of art expression across sound and image and is an attempt to explore the performance of glitch art.

This research has only just begun, and the following are planned for future development.

- implementation of glitches corresponding to audio processing options
- implementation of audio processing and corresponding glitching
- Practice and evaluation of glitch art performance

The prototype implemented in this study does not implement glitches corresponding to audio processing options completely. For example, Sonic Pi's reverb has an option to specify the size of the space, which allows the reverb to change, but the prototype does not implement the corresponding glitching. A possible effect would be to change the size of the glitch in the horizontal direction, but we will consider appropriate glitches for such audio processing options in the future.

Moreover, the audio processing implemented this time and the corresponding glitches are at most 4 types, and the performance of glitch art lacks impact. Sonic Pi has various sound effects, and we plan to design and implement glitches corresponding to them.

In addition to the above, we would also like to conduct a live-coded glitch art performance in front of a large number of people and evaluate the results. Based on the evaluation results, we will consider whether the way of enjoying glitch art proposed in this study can be established as performance art.

References

1. Betancourt, M.: Glitch Art in Theory and Practice: Critical Failures and Post-Digital Aesthetics. Routledge, New York and London (2016)
2. Moradi, I.: Glitch: Designing Imperfection. Mark Batty Publisher (2009)
3. McCormack, T.: Code Eroded: At GLI.TC/H (2020). https://rhizome.org/editorial/2010/oct/13/code-eroded-at-glitch/
4. Menkman, R.: The Glitch Moment(um), Institute of Network Cultures, No. 04. networkcultures.org (2011)
5. Heijer, E.: Evolving glitch art. In: Machado, P., McDermott, J., Carballal, A. (eds.) EvoMUSART 2013. LNCS, vol. 7834, pp. 109–120. Springer, Heidelberg (2013). https://doi.org/10.1007/978-3-642-36955-1_10
6. Algorave (2012). https://algorave.com/
7. TOPLAP (2004). https://toplap.org/
8. International Conference on Live Coding (2015). https://iclc.toplap.org/
9. Chunk (2003). http://chuck.stanford.edu/
10. OverTone (2018). https://overtone.github.io/
11. TidalCycles (2009). https://tidalcycles.org/
12. Sonic Pi (2012). https://sonic-pi.net/
13. Fluxus (2005). https://monoskop.org/Fluxus_(software)
14. LiveCodeLab (2014). https://livecodelab.net/

15. Hydra (2019). https://hydra.ojack.xyz/
16. Magnet TV (1965). https://whitney.org/collection/works/6139
17. The Art of PNG Glitch (2015). https://ucnv.github.io/pnglitch/
18. Yuichi, I., Carl, S., Masashi, Y., Shinya, M.: Datamoshing technique for video art production. J. Soc. Art Sci. **13**(3), 154–168 (2014)
19. Witchcraft at Livecodera (2022). https://www.youtube.com/watch?v=XyUGDKd3lME
20. Stephan, J., Nuno, N.C., Raul, M.: Designing glitch procedures and visualisation workflows for markerless live motion capture of contemporary dance. In: Proceedings of the 7th International Conference on Movement and Computing, MOCO 2020, pp.1–8 (2020)
21. Roberts, C., Kuchera-Morin, J.: GIBBER: LIVE CODING AUDIO IN THE BROWSER. In: Proceedings of the International Conference on Mathematics and Computing, ICMC 2012, pp.64–69 (2012)
22. Matt, P.: Generative Art: A Practical Guide Using Processing. Manning Publications (2011)
23. Roelofs, G.: PNG: The Definitive Guide, 2nd edn. O'Reilly & Associates (2003)

Psychological Evaluation of Media Art Focusing on Movement

Go Kazawa[1](✉), Naoko Tosa[1], Manae Miyata[2], and Ryohei Nakatsu[1]

[1] Kyoto University, Kyoto, Japan
kazawa.gou.62x@st.kyoto-u.ac.jp
[2] Seiko Epson Corporation, Suwa, Japan

Abstract. This study investigates the psychological impact of the movement elements of media art on viewers. As a new art form with contemporary media often exhibited in public places, media art plays a crucial role in stimulating thoughts, raising social issues, and designing various experiential spaces. Therefore, understanding how each element within media art affects human psychology is indispensable.

We took the abstract and diverse characteristics of Tosa Art as a subject and categorized the movement elements of the artwork into five categories: vertical, horizontal, scattered, rotational, and symmetrical. We then had 30 participants evaluate the impact of each category on four psychological factors: impression, relaxation, motivation, and creativity. As a result, while the "rotational" category significantly influenced impression, motivation, and creativity, the impact of the "vertical" movement was minimal.

Our findings align with existing research that visual movements draw attention, reinforce memory, and induce emotions and behaviors. This study adds a new perspective to understanding the psychological effects of art and emphasizes the importance of evaluating movement as an element of the artwork.

However, as a limitation, this study primarily focuses on movement, not considering other potential influencing factors such as other visual elements, the context of the work, and individual personality traits. Given that the research is specifically centered on Tosa art, there is also a need to experiment with a broader range of artists. Future research needs to explore these factors to fully grasp the overall psychological impact of media art.

Keywords: Media art · movement in media art · psychological evaluation

1 Introduction

1.1 Research Background

Determining which elements of an art piece influence people's psychology and how this connects to the evaluation of the work is an essential yet challenging question. It can be meaningless to extract specific elements of an art piece, such as color, movement, or shape, and investigate whether these elements leave impressions on individuals.

© ICST Institute for Computer Sciences, Social Informatics and Telecommunications Engineering 2024
Published by Springer Nature Switzerland AG 2024. All Rights Reserved
A. L. Brooks (Ed.): ArtsIT 2023, LNICST 565, pp. 186–197, 2024.
https://doi.org/10.1007/978-3-031-55312-7_14

Assessing art pieces created by artists by breaking them down into their components is a valuation of those components themselves and, as numerous previous studies have recorded, cannot always be considered an evaluation of the work. In other words, when evaluating a piece, it is critical to look at and assess the elements within the work. In other words, if we can gain a fundamental understanding of how elements such as color, shape, and in the case of media art, movement within an art piece influence human psychology, this would be beneficial in understanding how to appreciate art and what types of art pieces are suitable for specific situations.

Media art is a new form expected to be viewed by many people in public places using the latest media, such as projectors and LED displays. As seen in New York's the 9/11 Memorial Museum and Times Square, media art is anticipated to raise social issues through visual expression, stimulate thought, and create various experiential spaces for visitors. Therefore, its exhibition in various situations and locations will advance.

Given recent trends, it is essential to conduct evaluative research on the impact of various elements within media art on human psychology.

1.2 Tosa Art

Naoko Tosa, one of the authors, has been creating media art where new technologies play a significant role. She has used high-speed cameras to uncover the beauty hidden within various natural and physical phenomena. In particular, she has been interested in the behavior of fluids and has attempted to create media art by capturing the dynamics of fluids using a high-speed camera [1]. This field is known as "fluid dynamics," and a variety of research has been conducted in this area [2, 3]. The beautiful movement of fluids captivates many people, and another field of study is "visualization of fluid motion" [4]. However, most visualizations only show the behavior of stable fluids, with unstable and unpredictable behaviors being rare. Naoko Tosa, a new media artist, has created many media art pieces using high-speed cameras to express such unstable and unpredictable behaviors in the form of art.

This study uses her media art as a subject to analyze how the components within media art psychologically impact the audience, contributing to the future application and evaluation methods of media art.

2 Related Research

The idea that evaluating art based on its constituent parts can lead to different interpretations from viewing the art as a whole that can be explained by Gestalt psychology, a concept developed in Max Wertheimer's "Experimental Studies on Seeing Motion (1912)" [5]. Through a series of experiments using frame-switching in film projectors, he conducted experiments on the perception of motion through vision.

Wertheimer discovered that when individual static images are displayed in succession at a constant speed, they are perceived visually as a continuous motion. He analyzed this phenomenon, concluding that this type of perception is not simply due to the quick succession of individual images but rather a tendency of the human brain to perceive continuous motion as a "whole."

This study was instrumental in establishing the foundations of Gestalt psychology and supporting the theory that the process of visual perception is more than just the simple sum of individual visual stimuli. In other words, it laid the groundwork for advocating the main principle of Gestalt psychology, "the whole is greater than the sum of its parts."

The central idea of Gestalt theory, "the whole is greater than the sum of its parts," has deeply influenced the visual understanding and practice of art and design. It was vital in discussions about how visual organization, spatial arrangement, balance, and symmetry influence the interpretation of a piece. This is illustrated in Roy R. Behrens' "Art, Design and Gestalt Theory (1998)" [6]. In the paper, Behrens investigates in detail how the principles of Gestalt psychology influenced the art and design movements of the early 20th century, particularly the Bauhaus, and discusses how Gestalt theory has been integrated into the visual understanding and practice of art and design. For instance, Behrens highlights the foundational courses in Bauhaus education, where the exploration of essential visual elements like color, shape, and material aimed to understand how these elements shape a piece's overall visual impression and meaning. This exploration is deeply rooted in the principles of visual organization and perception of Gestalt theory.

Furthermore, Behrens provides specific examples of how Gestalt theory has influenced artists and designers' creations and understanding of their work. The "Dada Cafe" logo design demonstrates how visual elements are grouped to form a pattern. The logo is designed based on the principle of similarity in Gestalt theory, where similar visual elements are perceived as belonging.

Thus, the elements have no meaning when evaluated separately for art and design. Instead, what matters is how these elements impact the viewer. At the same time, crucial previous research has been presented from the viewpoint of evaluating works, which will be introduced.

On the other hand, research on how people evaluate art forms such as paintings began with Fechner's initiation of experimental aesthetics in the late 19th century. This started with quantitatively measuring human emotions, such as comfort and discomfort. Since then, various psychological studies have been conducted on the beauty and style of art.

For instance, Farkas [7] conducted a study using Surrealist paintings to investigate the types of art that people prefer and discovered that famous artworks were favored. Winston and Cupchik [8] experimented on whether experts or novices prefer fine art paintings or secular paintings. The result showed that experts tend to prefer fine art, while novices tend to prefer secular paintings.

Remarkably, numerous studies have compared abstract paintings with figurative ones [9–11]. For example, Friedman compared the preferences for the abstractness and complexity of paintings among three adult groups: students receiving elementary education, students receiving art education, and professors and graduate students. She found that the first group preferred paintings with the lowest levels of abstractness and complexity, while the third group preferred paintings with high abstractness and complexity [9]. Pihko et al. varied the degree of abstraction in paintings and examined the preferences for abstract art among amateurs and professionals. They found that abstraction influences the aesthetic judgments of amateurs but not professionals [10]. Belke et al. investigated whether evaluations change when style information is provided and discovered that such information positively shifts the subjects' opinions [11].

While evaluations of fine art, such as paintings, have been widely studied in prior research, temporal art forms like media art are scarce. With the proliferation of high-luminance projectors and large LED displays, the frequency of media art appreciation is increasing, and evaluating media art is a critical issue. This study is a challenging endeavor that attempts to evaluate media art, drawing on the evaluation methods of fine art established so far.

3 Media Art to Be Evaluated

Media art introduces an additional element of time or motion compared to traditional fine arts like painting. As discussed in Sect. 2, while evaluations are plentiful for traditional art forms, focusing on this unique aspect - the temporal axis - is essential when evaluating media art. Hence, it is necessary to define criteria for aspects beyond the time dimension, including only those that meet these standards, to streamline the evaluation of media art. Here, we employ the following criteria:

(a) We focus on works that do not incorporate human subjects. When artworks use human subjects, they resemble art forms such as film, leading audiences to evaluate narrative and other aspects. We can better understand how viewers value media art by excluding humans as subjects.
(b) Works that engage with natural objects fit this category. When artworks involve artificial artifacts, viewers, similar to the situation (a), often judge these based on their interpretation of what they represent. To prevent this, we consider it best to include works that primarily deal with natural phenomena or objects.
(c) Consequently, works that do not use computer graphics (CG) fit the criteria.

The art created by Tosa aligns well with these criteria for the following reasons:

(1) The primary focus is on natural phenomena. Tosa art is characterized by capturing and portraying the hidden beauty in nature rather than human-made creations. In this way, it satisfies conditions (a) and (b).
(2) Given that CG is not used, thus fulfilling condition (c).

4 Media Art Content Used in the Experiment

4.1 Targeted Media Art

Table 1 presents a selection of 24 representative works from Tosa art [12–15], also detailing their distinct features concerning the following aspects:

(1) What are the predominant types of movements included?
(2) Is the expression abstract or figurative?
(3) What primary colors are used?

In this psychological experiment, we decided to select artworks for evaluation by focusing on movement among these factors.

Table 1. Representative media artworks by Naoko Tosa

No.	Title	Movement	Figurative /Abstract	Color
1	Sound of Ikebana:Spring	Hor.+Ver.	Abstract	Red
2	Sound of Ikebana:Summer	Hor.+Ver.	Abstract	Blue
3	Sound of Ikebana:Autumn	Hor.+Ver.	Abstract	Purple
4	Sound of Ikebana:Winter	Hori.+Ver.	Abstract	Dark
5	Genesis : Blue	Horizontal	Abstract	Blue
6	Genesis : Red	Horizontal	Abstract	Red
7	Genesis : Green	Horizontal	Abstract	Green
8	Genesis : Purple	Horizontal	Abstract	Purple
9	Foud Gods : Turtle	Vertical	Abstract	Blue
10	Four Gods : Red Phoenix	Vertical	Abstract	Red
11	Foug Gods:Blue dragon	Vertical	Abstract	Blud
12	Four Gods:White tiger	Vertical	Abstract	White
13	Volcano	Vertical	Abstract	Black
14	Shan-Shui	Vertical	Abstract	White
15	Moon Flower	Scatter	Fig.+Abs.	Flower Color
16	Space Flower	Scatter	Fig.+Abs.	Flower Color
18	Space Jungle	Scatter	Fig.+Abs.	Flower Color
17	Oiran	Scatter	Fig.+Abs.	Flower Color
19	Utsuroi	Scatter	Figurarive	Flower Color
20	Miyabi	Rotation	Fig.+Abs.	Flower Color
21	Twin Lions	Symmetric	Fig.+Abs.	Blue, Red
22	Organic Geometry	Symmetric	Abstract	White, Red
23	Wind god	Horizontal	Abstract	Blue, Red
24	Thunder god	Horizontal	Abstract	Blue, purple

Hor.: Horizontal, Ver.:Vertical

4.2 Preparation for the Experiment

Artwork Classification: The artworks were categorized into five distinct classes according to the type of movement they exhibited: vertical, horizontal, scatter, rotation, and symmetry.

Artwork Selection: Two artworks were chosen for each type of movement, specifically:

Vertical movement: Artworks No. 10 and No. 12
Horizontal movement: Artworks No. 23 and No. 24
Scattering: Artworks No. 15 and No. 19
Rotation: Artwork No. 20
Symmetry: Artworks No. 21 and No. 22

Artwork Grouping: Artworks were then arranged into two separate groups:

Group α: Artworks No. 10, No. 23, No. 15, No. 20, and No. 21
Group β: Artworks No. 12, No. 14, No. 19, No. 20, and No. 22

Creation of Movement Sets: Five sets were established, each sorted by "movement" in a different order:

Set 1: vertical, horizontal, scatter, rotation, symmetry.
Set 2: scatter, rotation, vertical, horizontal, symmetry.
Set 3: rotation, vertical, symmetry, scatter, horizontal.
Set 4: horizontal, rotation, scatter, vertical, symmetry.
Set 5: symmetry, vertical, rotation, horizontal, scatter.

Participant Division: The subjects were distributed into ten separate groups.

Assignment of Group and Set: Based on the experimental design method, each group was allotted an artwork group and a movement set.

5 Experimental Conditions

5.1 Subjects

Thirty Kyoto University students were used as subjects. The breakdown was 13 males and 17 females. They were randomly assigned to one of 10 groups of 3 students.

5.2 Experimental Equipment

The following experimental equipment was used.

Projector model (brightness): Epson EB-2155W (5000 lumens)
Screen size: 70 in.
Distance between subject and screen: 2 m

After viewing each artwork, the subjects were asked to answer the questionnaire using the subjects' smartphones using Google Forms.

5.3 Experimental Environment

The laboratory used for this experiment was equipped with the abovementioned devices, as illustrated in Fig. 1. Moreover, viewing the media art pieces took place under dimmed lighting conditions.

5.4 Evaluation Items

The evaluation was centered on two main aspects: the impression elicited by each image (Impression factor) and the psychological impact each image had on the subjects (Effect factor). The Effect factor consisted of assessing whether participants felt relaxed (Relaxation factor), motivated (Motivation factor), and inspired to think increased creatively (Creativity factor). The questions that formed the Impression, Relaxation, Motivation, and Creativity factors are listed below. A psychologist, one of the authors, established these questionnaire items and validated them in other art evaluation experiments [16].

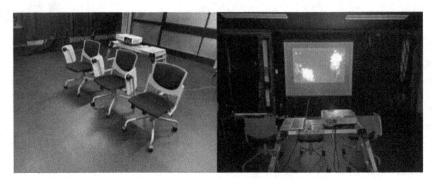

Fig. 1. Experimental environment

Impression Factor involves questions such as: Comfortable or Uncomfortable, Familiar or Unfamiliar, Beautiful or Not beautiful, Calm or Restless, Interesting or Boring, Warm or Cold, Changeable or Not changeable, Luxury or Sober, Individual or Ordinary.

Relaxation Factor involves questions such as: At ease or Not at ease, Secure or Not secure, Pleasant or Not pleasant, Relaxed or Not relaxed, Healed or Not healed.

Motivation Factor involves questions such as Enthusiastic or Not enthusiastic, Immersed or Not immersed, Curious or Not curious, Motivated or Not motivated, Aroused or Not aroused.

Creativity Factor involves questions such as Associate or Do not associate, Immersive or Not immersive, Activated or Not activated, Inspired or Not Inspired, and In the zone or Not in the zone.

6 Experimental Results and Discussion

6.1 Experimental Results

The original data obtained from the evaluations were first averaged for each of the five types of movements and then further averaged for each factor, including the impression, relaxation, motivation, and creativity factors. Subsequently, to enable straightforward comparisons of the results, these were standardized with an average of 50 and a standard deviation of 10. The resultant values are presented as bar graphs in Figs. 2, 3, 4 and 5. Figure 2 presents the evaluation results for the Impression Factor. The vertical movement received a score of 39.66, horizontal movement scored 48.22, scatter scored 51.71, rotation scored 65.99, and symmetry scored 44.42. Subsequently, Fig. 3 displays the evaluation results for the Relaxation Factor, with vertical movement scoring 52.53, horizontal movement scoring 58.02, scatter scoring 53.46, rotation scoring 53.46, and symmetry scoring 32.53.

Furthermore, Fig. 4 reveals the evaluation results for the Motivation Factor: vertical movement was rated 32.21, horizontal movement scored 50.47, scatter scored 52.81, rotation scored 59.83, and symmetry scored 52.68. Lastly, Fig. 5 outlines the evaluation results for the Creativity Factor, with vertical movement receiving a rating of 43.75, horizontal movement scoring 53.26, scatter scoring 35.93, rotation scoring 60.4, and symmetry scoring 56.66. Through the analysis of these data, it became evident that there is a notable difference in the psychological impact elicited by each movement element in Tosa art.

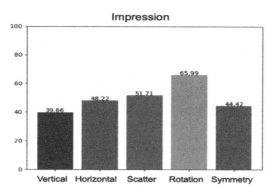

Fig. 2. Results related to the impression factor.

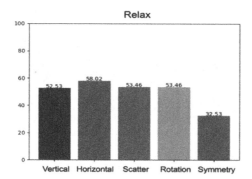

Fig. 3. Results related to the relaxation factor.

6.2 Considerations

This study was undertaken to provide answers on how elements of motion in media art impact viewers' psychology. Our proposed inquiry centered on how artworks influence individuals' psychology and, consequently, how these effects shape evaluations of the pieces. To deduce answers, our research primarily focused on new forms of media art, assessing how their motion elements affect viewers' psychology.

Our results revealed significant variability in the psychological impact exerted by the motion characteristics of artworks. Specifically, rotating motions had a profound influence on viewers' impressions, motivation, and creativity. In contrast, vertical motions were found to have minimal impact. These findings suggest that rotating motions capture viewers' attention, stimulate spontaneous actions, and activate imagination through visual emphasis, energy, vitality, and the introduction of new visual patterns and forms.

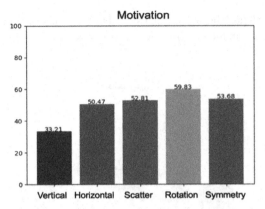

Fig. 4. Results related to the motivation factor.

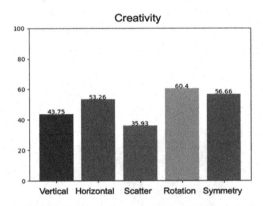

Fig. 5. Results related to the creativity factor.

These findings align with existing research. For instance, prior studies have reported that visual motions capture attention and enhance memory and recognition [17]. Research on the capacity of motion to elicit emotions and behaviors [18] has also been conducted, and our results support these investigations.

Moreover, this research offers a fresh perspective on understanding the psychological impacts artworks have on individuals. Traditionally, artworks have been evaluated based on elements such as color and shape. However, this study unveiled the impact of an artwork's motion elements on viewer psychology. This constitutes an essential step towards a deeper comprehension of art evaluation, providing new insights into how art appreciation influences individual psychological experiences.

Our research contributes to understanding the specific impacts when evaluating media art, especially when anticipated for public display. By considering the effects on viewers' impressions, emotions, behaviors, and creativity, it becomes feasible to contemplate more appropriate artwork selections and exhibition methods.

However, our study primarily focused on motion elements, not giving due consideration to other visual elements, the context of the work, and other potential influencing factors. Resolving this requires analyzing factors that are both conscious and unconscious. To uncover conscious factors, introducing qualitative investigations like semistructured interviews to deeply explore viewers' experiences, feelings, interpretations, and reactions to the artwork would be essential. Additionally, integrating psychological data, such as heart rate, for analyzing unconscious factors is deemed necessary.

Furthermore, the evaluation of viewers' experiences primarily relied on self-reports. While self-reports are valuable sources of information, they might only capture a partial account of viewers' experiences. Impressions of beauty, color, and movement in artworks are heavily influenced by an individual's past experiences, cultural background, and environment, as demonstrated in many preceding studies [19–21]. Hence, future research needs to determine whether the impressions derived from the artwork's motion are universally intuitive or reflect past experiences.

Additionally, our experiments were conducted focusing on works by Tosa. Based solely on these results, drawing conclusions for artworks by other artists would be presumptuous. Moving forward, experiments on various artists' works are essential to validate if these findings are universally applicable. Nonetheless, our research is significantly meaningful as a precursor to evaluating media art pieces appropriately in the future.

In conclusion, our research offers new insights into the effects of motion elements in media art on viewer psychology. These findings present a new framework for understanding the relationship between art appreciation and viewer psychology, providing valuable insights for future research. We anticipate that the results of our study will find further applications in understanding the creation, appreciation, and evaluation of art and deepening comprehension of human psychology.

7   Conclusion

This study elucidated the impact of motion elements in media art on the psychology of viewers. Specifically, it revealed that different movement characteristics of art pieces, such as vertical, horizontal, scattered, rotational, and symmetrical movements, exert varied effects on viewers' impressions, relaxation, motivation, and creativity. Notably, rotational movements were found to have the most significant impact on viewers' psychological responses. This suggests that rotational movements, by presenting visual emphasis, energy, vigor, and introducing new visual patterns and shapes, capture the attention of viewers, stimulate spontaneous actions, and enhance imagination.

These findings deepen our understanding of how motion, as a visual element, is involved in the evaluation of art pieces and the psychological experiences of viewers. They offer new insights into how the appreciation of media art affects individual psychological experiences. Moreover, these insights contribute to a more concrete understanding of the impact on evaluations of media art intended for public exhibitions, providing guidelines for the selection of more appropriate artworks and exhibition methods.

For future research directions, it is essential to consider other visual elements of art pieces, the context of the work, cultural backgrounds, and individual experiences as potential influencing factors. The incorporation of qualitative research and psychological data, such as heart rate measurements, should also be considered. Furthermore, by studying works from different artists, there's a need to verify the extent to which our results can be generalized across a broader spectrum.

In conclusion, this research provides new insights into the relationship between motion elements in media art and viewers' psychology, confirming its significance. These findings propose a novel framework for the creation, appreciation, and evaluation of art, potentially deepening our understanding of human psychology. We anticipate that this research points to new directions that can enrich the evaluation and experiences of viewers in media art.

References

1. Chen, F., Sawada, T., Tosa, N.: Sound based scenery painting. In: 2013 International Conference on Culture and Computing, IEEE Press (2013)
2. Munson, B.R., Rothmayer, A.P., Okiishi, T.H., Huebsch, W.W.: Fundamentals of Fluid Mechanics (2012)
3. Bernard, P.S.: Fluid Dynamics. Cambridge University Press, Cambridge (2015)
4. Lim, T.T., Smiths, A.J. (eds.): Flow Visualization: Techniques and Examples. Imperial College Press, London (2012)
5. Wertheimer, M.: Experimental studies on motion vision. Zeitschrift für Psychologie **61**(1), 161–265 (1912)
6. Behrens, R.R.: Art, design and gestalt theory. Leonardo **31**(4), 299–303 (1998)
7. Farkas, A.: Prototypicality-effect in surrealist paintings. Empir. Stud. Arts **20**(2), 127–136 (2002)
8. Winston, A.S., Cupchik, G.C.: The evaluation of high art and popular art by naïve and experienced viewers. Vis. Arts Res. **18**, 1–14 (1992)
9. Freedman, K.: Judgements of painting abstraction, complexity, preference, and recognition by three adult educational groups. Vis. Arts Res. **14**(2), 68–78 (1988)
10. Pihko, E., et al.: Experiencing art: the influence of expertise and painting abstraction level. Front. Hum. Neurosci. **5**(5), 1–10 (2001)
11. Belke, B., Leder, H., Augustin, M.D.: Mastering style - effects of explicit style-related information, art knowledge and affective state on appreciation of abstract paintings. Psychol. Sci. **48**, 115–134 (2006)
12. Tosa, N., Nakatsu, R., Yunian, P.: Projection mapping celebrating RIMPA 400th anniversary. In: 2015 International Conference on Culture and Computing, pp. 18–24 (2015)
13. Tosa, N., Yunian, P., Zhao, L., Nakatsu, R.: Genesis: new media art created as a visualization of fluid dynamics. In: Munekata, N., Kunita, I., Hoshino, J. (eds.) ICEC 2017. LNCS, vol. 10507, pp. 3–13. Springer, Cham (2017). https://doi.org/10.1007/978-3-319-66715-7_1
14. Tosa, N., Pang, Y., Yang, Q., Nakatsu, R.: Pursuit and expression of Japanese beauty using technology. Arts J. MDPI **8**(1), 38 (2019)
15. Pan, Y., Tamai, H., Tosa, N., Nakatsu, R.: Sound of Ikebana: creation of media art based on fluid dynamics. Int. J. Humanit. Soc. Sci. Educ. **8**(3), 90–102 (2021)
16. Nakatsu, R., et al.: Construction of immersive art space using mirror display and its preliminary evaluation. In: WMSCI 2023 (2023). (accepted)
17. Franconeri, S.L., Simons, D.J., Junge, J.A.: Searching for stimulus-driven shifts of attention. Psychon. Bull. Rev. **11**, 876–881 (2004)
18. Aronoff, J., Barclay, A.M., Stevenson, L.A.: The recognition of threatening facial stimuli. J. Pers. Soc. Psychol. **54**(4), 647–655 (1988)

19. Masuda, T., Gonzalez, R., Kwan, L., Nisbett, R.E.: Culture and aesthetic preference: comparing the attention to context of East Asians and Americans. Pers. Soc. Psychol. Bull. **34**, 1260–1275 (2008). https://doi.org/10.1177/0146167208320555
20. Vessel, E.A., Starr, G.G., Rubin, N.: Art reaches within: aesthetic experience, the self and the default mode network. Front. Neurosci. **7**, 258 (2013). Published online 30 December 2013. https://doi.org/10.3389/fnins.2013.00258. PMCID: PMC3874727. PMID: 24415994
21. Roberson, D., Davies, I., Davidoff, J.: Color categories are not universal: replications and new evidence from a stone-age culture. J. Exp. Psychol. Gen. **129**(3), 369–398 (2000). https://doi.org/10.1037/0096-3445.129.3.369

Describing and Comparing Co-located Interaction in Interactive Art Using a Relational Model

Dan Xu$^{(\boxtimes)}$, Maarten H. Lamers, and Edwin van der Heide

LIACS, Leiden University, Leiden, The Netherlands
{d.xu,m.h.lamers,e.f.van.der.heide}@liacs.leidenuniv.nl

Abstract. Co-located interaction in interactive art takes place among two or more co-located audience members and the technical system of an artwork. In this paper, we aim to assess the descriptive and comparative qualities of our previously developed relational model for describing and analysing such forms of interaction. The model focuses on specifying the actions of the interacting elements, such as the audience and art system, and the various forms of communication between them. To assess its significance, we first develop selection criteria and classification dimensions to select eight artworks that are representative of diverse forms of co-located interaction. The relational model is shown to be suitable for describing the selected artworks and comparing their similarities and differences. As outcome, it reveals different types of relationships between the actions of interacting elements that would otherwise not be highlighted. As such, it provides a context for analysing and discussing strategies for co-located interaction and points to opportunities for research and creation in this field.

Keywords: Co-located interaction · interactive art · interaction model · mediated communication · audience-artwork interaction

1 Introduction

Co-located interaction in interactive art takes place when two or more audience members are physically present at the same location, participating simultaneously in an interactive experience. Previously, we developed a relational model with the aim of describing co-located interaction in interactive art [1]. While the interaction between the audience and art system is often viewed as a dialogue [2], in the context of co-located interaction the dialogue model falls short in capturing the diverse relationships between the interacting elements and the unique roles each element plays in the overall interaction. In response to this limitation, our relational model focuses on describing the various forms of communication occurring among the interacting elements and how these elements relate to and influence one another [1].

We believe that the relational model provides a systematic approach to examine and compare different forms of co-located interaction, further enabling us to discover patterns

A. L. Brooks (Ed.): ArtsIT 2023, LNICST 565, pp. 198–217, 2024.
https://doi.org/10.1007/978-3-031-55312-7_15

and identify new research areas. Therefore, there are two main objectives of this study: firstly, to assess the descriptive and comparative capabilities of the relational model with regard to diverse forms of co-located interaction; and secondly, to gain insights into the characteristics and features of such interaction using the relational model.

In the following section, we very shortly introduce the relational model and discuss its features and scope. In Sect. 3, we outline the selection criteria, the classification dimensions and the process of selecting a total of eight artworks that represent the range and diversity of co-located interaction. In Sect. 4, we apply the relational model to describe the co-located interaction in each artwork and compare their similarities and differences. Finally, in Sect. 5, we reflect on the effectiveness of the relational model and discuss our insights into the various forms of co-located interaction.

2   A Relational Model of Interaction

The detailed motivation for, description of, and background of the relational model can be found elsewhere [1]. The model starts by identifying the individual actors participating in the interaction, each of which is described as an element. The type of an element can be 'audience', referring to the participating audience, or 'art system', referring to the technical system(s) of an artwork. To describe an interaction, the model examines the actions performed by the interacting elements, such as a movement, an update of a display, or pressing of a button. Following an action, a form of communication is created and directed at (an)other element(s). Communications can be either direct, only taking place between two communicating elements, or mediated, taking place between two communicating elements via a third element. And they can be either private, which can only be perceived by the communicating elements, or public, which can also be perceived by other elements. An action can be performed by one element or multiple elements together and it can be directed at one or multiple elements. The same action can also be directed at different elements thus creating different forms of communication. For each action, the model examines what role(s) it plays in the context of the interaction and how it influences other elements.

Although the model was developed for describing co-located interaction, it does not set any requirements for the number and types of elements or the location of interactions. The concepts involved can potentially describe interaction in general and artworks involving non-human actors such as animals or plants. It is important to note that the model focuses on describing the relational exchange between and among the elements. It does not specify the internal processes of the elements such as data processing and cognitive processes. Besides, as the model mainly accounts for actions that are directed at other elements, it does not explicitly indicate the receptive actions, such as sensing and observing. Therefore, it also does not include the audience members who are only spectators of the interaction.

3   Artwork Selection and Classification

To evaluate the descriptive and comparative capabilities of the relational model for analysing different forms of co-located interaction, we develop three selection criteria and seven classification dimensions to select different artworks that are representative of different forms of co-located interaction. The selection criteria help us to specify and narrow down the type of artworks of our interest. The classification dimensions aim to systematically delineate forms of co-located interaction that share common attributes and show the breadth of the landscape of co-located interaction in interactive art, which can in turn help us position the selected artworks within this landscape.

3.1 Selection Criteria

Our first criterion is that the artworks should involve active audience participation and at least one autonomous art system. Guljajeva distinguishes interactive art from other types of participatory art by highlighting the active engagement of the audience and the presence of an interactive system that is responsive to the audience's input [3] (p. 66). We agree that active, conscious participation of the audience is central to the interactive experience. However, we believe that an interactive system, or what we refer to as an art system in the relational model, does not only respond to the audience but can also trigger or initiate interaction and even mediate interaction between audience members.

Criterion two is that the artworks we consider use computational technologies at the core of the art system. In such cases, the computers are often used for processing data captured by sensors or input devices and programming the responses and behaviour of the art systems [4].

The third and last criterion specifies that specifies that active participation of multiple co-located audience members is required. The interaction takes place not only between the audience and art system, but also between audience members. Therefore, we do not consider interactive artworks that are designed for one audience member, despite the fact that such artworks can sometimes accommodate multiple audience members to observe the interaction.

3.2 Classification Dimensions

Before introducing our classification dimensions, we first review previous works in this area and discuss their similarities and differences compared to our approach. Mubarak proposed a taxonomy to classify co-located interaction in art installations based on factors influencing the audience experience [5]. The first factor *scale* describes the number of participating audience members and is classified as small ($\leq$ 10), medium (11 - 100), and large (> 100). The *interaction modality* refers to the method by which the audience interacts with the artwork, which can be through direct physical manipulation (direct), through individual remote input devices (facilitated), and captured by non-invasive sensing technologies (ambient). The *input and output distribution* indicates the distribution of input and output devices of the art system and is classified as centralized, partially distributed or fully distributed. The ability of the audience to recognize the effect of their actions in the artwork is described as *feedback attributability* ranging from low to medium to high. *Activity type* describes the audience activity solicited by the artwork, which can be collaborative when the audience must work together, competitive when the audience must challenge each other, and solitary when each audience member acts independently. The last factor proposed by Mubarak, *participation symmetry,* is defined by Bell as the "distribution of actions and contributions between participants" [6], which can be either symmetrical when the audience participates equally, or asymmetrical when audience members play different roles.

Taking Mubarak's taxonomy as a reference, we propose seven classification dimensions that focus on differentiating the various forms of organization and participation, instead of the factors influencing the audience experience discussed by Mubarak. To start, the levels of *scale* are delineated by arbitrary numbers and can hardly be considered a defining factor of an artwork. Instead, we propose *participation style* (PS) to describe the arrangement of the audience. It can be classed as duo, which is for two audience members; group, which takes place among one or more groups of audience members with limited numbers (>2); crowd, which takes place among one or more crowds of audience without any practically imposed limits on audience numbers. Additionally, the audience may have varying levels of commitment, either participating consistently throughout the interaction or having the freedom to join and leave as they wish. To distinguish between these two types, we propose *audience constellation* (AC) with fixed or fluid levels respectively.

Mubarak's *interaction modality* and *input and output distribution* pertain to aspects of the input and output set-up of an art system. For *interaction modality*, we find that it does not significantly impact the ways of participation. For instance, the audience can directly manipulate a virtual object via camera-tracking technologies, which blurs the distinction between direct and ambient modalities. For *input and output distribution*, we find that the technical configuration of an art system cannot sufficiently describe the access the audience has to it. A set of distributed devices can still be accessed by the audience equally and publicly. Instead, we propose the dimension *input and output access* (IA & OA) and distinguish private (accessible to one audience member only), partially public (accessible to more than one but not all audience members), and public (accessible to all audience members). An art system can have multiple private, partially public, public or mixed varieties of inputs and outputs.

As Mubarak's *feedback attributability* of the audience's actions can be highly subjective and contingent on the number of audience members, we do not consider it a defining factor for delineating forms of co-located interaction. Here we adopt the dimensions of *activity type* (AT) and *participation symmetry* (PSy). In many interactive artworks, the audience engages in open, hybrid and/or exploratory social activities that can neither be defined as collaborative nor as competitive. Sometimes the audience can also invent and transition between different types of activities. Therefore, we classify *activity type* as individual, when audience members participate individually and independently; social, when the activities require more than one audience member to perform together; mixed, when both types are supported. For *participation symmetry*, besides symmetrical and asymmetrical, we include varied to indicate when the audience can transition between the two.

Lastly, as mentioned before, the art system can not only respond to the audience but also initiate interaction. We use *initiator* (Ini) to indicate which element initiates the interaction, and it can be the audience, art system, other, and varied when the elements have equal chance to initiate the interaction. All the dimensions and their scale levels are summarized in Table 1.

Table 1. Classification dimensions and their scale levels for co-located interactive artworks.

Dimensions		Scale levels			
PS	Participation style	Duo	Group	Crowd	
AC	Audience constellation	Fixed	Fluid		
IA	Input access	Private	Partially public	Public	Mixed
OA	Output access	Private	Partially public	Public	Mixed
AT	Activity type	Individual	Social	Mixed	
PSy	Participation symmetry	Symmetrical	Asymmetrical	Varied	
In	Initiator	Art system	Audience	Other	Varied

3.3 Selecting Artworks

Based on the selection criteria, we can identify a wide variety of artworks. Our seven classification dimensions allow over 3,000 unique combinations, each representing a potential form of co-located interaction. However, these dimensions are not entirely independent from each other, for instance, the *input and output access* cannot be partially public if the *participation style* is duo. It is also reasonable to assume that not each unique combination has been realized. Nonetheless, it is impractical to cover all existing varieties of co-located interaction within the scope of this paper.

To tackle this dilemma, we suggest prioritizing some dimensions and scale levels to narrow down the artwork selection while preserving their potential to show the diversity of co-located interaction. Considering that not all combinations of the dimensions can be realized for each *participation style* and that the group and crowd share quite some similarities, we do not consider each of them independently but aim to cover all the variations in the final selection. The different types of *input and output access* indicate how much freedom each audience member has in the interaction and what kind of information is available to them. They affect the organization of the audience but are less significant than *participation symmetry* when it comes to considering the effects of the audience's action on the art system and the interaction. Therefore, we do not consider the *input and output access* independently when selecting the artworks.

Moreover, for *participation symmetry*, we propose to take the scale level 'varied' as an alternative to either 'symmetrical' or 'asymmetrical'. Given the definition of co-located interaction, the audience is likely to engage in social or mixed instead of individual activities. As co-located interaction mainly concerns the involvement of the audience and the art system, we focus on the audience or the art system as *initiator* of the interaction.

3.4 Selected Artworks

We searched for artworks in online archives such as the Ars Electronica Archives [7] and the Archives of Digital Art [8], and from personal knowledge of existing artworks. We selected eight artworks that satisfy the selection criteria and represent a meaningful distribution within the classification. They encompass a range of interactive experiences, including participatory performances, games, and interactive installations, and were created between 1997 and 2016. These eight artworks and their corresponding classification scale levels are summarized in Table 2. Below we introduce them individually based on their positions in the classification dimensions.

Table 2. Eight selected artworks and their classification scale levels.

Artwork	PS	AC	IA	OA	PSy	AT	In
Brainball (2003)	Duo	Fixed	Private	Public	Symmetrical	Social	Audience
Randomly Generated Social Interactions (2016)	Duo	Fixed	Private	Private	Symmetrical	Social	Art system
World Skin (1997)	Group	Fixed	Private	Public	Asymmetrical	Social	Audience
Zoom Pavilion (2015)	Crowd	Fluid	Public	Public	Asymmetrical	Mixed	Art system
Boundary Functions (1998)	Group	Fluid	Public	Public	Symmetrical	Social	Audience
Spatial Sounds (100dB at 100km/h) (2000)	Crowd	Fluid	Public	Public	Varied	Mixed	Art system
Lights Contacts (2009)	Crowd	Fluid	Partially public	Public	Asymmetrical	Social	Audience
Body Movies (2001)	Crowd	Fluid	Public	Public	Symmetrical	Mixed	Audience

Brainball **(2003) by Smart Studio** is a game between two audience members based on brain-computer interface technology [9]. The art system consists of EEG sensors, computers, software programs, a steel ball, a constructed table, and a display. Two players interact via the EEG sensors. Seated on both ends of the table, either one of the players can press a button to start the game. Their brainwaves are measured by the EEG sensors and used to control the movement of the ball on the table. The more relaxed player, as assessed via the brainwaves, scores a point over the other player. Meanwhile, both the brainwave data and game status are shown on the display next to the table.

Brainball requires two audience members to participate in a game, therefore it can be classed as a duo *participation style* in a fixed *audience constellation*. The *input access* to the art system is private as each player uses the sensor individually while the *output access* is public as the ball movement and the display can be seen by all present audience. The *participation symmetry* is symmetrical as both players contribute equally to the interaction. The interaction takes place as a competitive game, therefore the *activity type* is social. An audience member (*initiator*) can initiate the interaction by pressing a button. Brainball demonstrates a form of co-located interaction with a fixed and symmetrical participation style in a social activity initiated by the audience.

***Randomly Generated Social Interactions* (2016) by Anastasis Germanidis** is an interactive performance in which the audience is asked to interact with each other following instructions generated by the art system [10]. The art system consists of headphones, mobile phones, and software programs. Each audience member accesses the art system via a mobile phone and headphones. The art system assigns a fictional identity to each participant and randomly matches them together in pairs. Subsequently, the participants are given computer-generated instructions via the headphones for what to say and do during their interaction with each other.

Although Randomly Generated Social Interactions enables a group of audience members to participate, the interaction is intended for pairs of participants; thus we classify it as duo *participation style* and fixed *audience constellation*. The artwork provides both private *input and output access* to each participant via the use of mobile phones and headphones. Although each participant is assigned a different fictional identity, the *participation symmetry* is symmetrical as their roles in the interaction are equal. The *activity type* is social as the participants are required to interact with each other. The art system initiates the interaction by instructing the audience (*initiator*). Based on the classification scale levels, Randomly Generated Social Interactions is largely similar to Brainball, the key difference being that its art system initiates the interaction instead of the audience. It demonstrates a form of co-located interaction with a fixed and symmetrical participation style in a social activity initiated by the art system.

***World Skin* (1997) by Maurice Benayoun** is an interactive artwork in the form of a photo safari inside an immersive projection of a war zone landscape [11]. The art system consists of a projection room, projectors, a joystick for navigation, sensor-fitted cameras, computers, software programs, and speakers. The audience interacts via the input devices and plays different roles in the interaction. The artwork requires at least two audience members to participate. The exact number of participants depends on the number of input devices used in the exhibition set-up. One audience member plays the role of a 'driver' and navigates in the virtual landscape with the joystick while others play the role of 'photographers' who can take photos of the landscape with sensor-fitted cameras. The positions and orientations of the cameras are tracked and once a shot is taken, the corresponding surface of the virtual landscape is removed.

Similar to the previous two artworks, in World Skin the *audience constellation* is fixed and the *activity type* is social as the audience must work together to navigate and take photos. The main difference is that more than two audience members participate in the interaction. As the number of participants is limited by the number of input devices, we class it as a group *participation style*. Each participant has private *input access* via the input devices and public *output access* as the projection can be perceived by everyone present. The audience plays different roles, making the *participation symmetry* asymmetrical, and initiates the interaction (*initiator*). The work demonstrates a form of co-located interaction with a fixed and asymmetrical participation style in a social activity initiated by the audience.

***Zoom Pavilion* (2015) by Rafael Lozano Hemmer and Krzysztof Wodiczko** is an interactive installation with an immersive projection of live images of the audience inside the exhibition space [12]. The art system consists of projectors, infrared cameras, infrared illuminators, computers, software programs and speakers. The audience interacts mainly with their bodies. The camera views of the audience are displayed on the left, front, and right walls. The back wall, or 'archive wall', shows a grid of small, coupled face images of the audience. Several recognition algorithms detect the faces and bodies of the audience. When no one is detected, the cameras will zoom-out and show the maximum field of view. When one person is detected, a white rectangle is drawn around their face or body, and the camera will zoom-in until this rectangle fills the screen. When two people are detected, a line is drawn between their bodies and a word describing their relation based on their spatial distance is shown. When two people's faces are shown simultaneously, they are displayed together as a 'couple' and stored on the archive wall.

Unlike the previously mentioned artworks, although with constraints of the physical capacity of the exhibition space, Zoom Pavilion allows for a variable number of audience members with no clear demarcation between those who are observing and those who are participating. We consider its *participation style* as crowd and *audience constellation* fluid. The *input and output access* are both public, as they can be shared and perceived by all audience equally. As the art system selects the faces of some audience members to zoom-in or draw connections between their bodies, there is an asymmetrical *participation symmetry*. The *activity type* is mixed because the audience can participate both individually and collectively. For instance, two audience members can control the line drawn between them together. The art system initiates the interaction by actively selecting and drawing connections between audience members (*initiator*). This artwork demonstrates a form of co-located interaction with a fluid and asymmetrical participation style in mixed types of activities initiated by the art system.

***Boundary Functions* (1998) by Scott Snibbe** is an interactive installation that requires the active participation of more than one audience member [13]. The art system consists of a camera, a computer, software programs, a projector, and a retroreflective floor. The audience interacts mainly with their bodies. When there is one person on the floor, nothing happens. With two people present, a single line cuts between them bisecting the floor and dynamically changes as they move. With more than two people participating in the interaction, the floor divides into cellular regions that demarcate the space closest to the person inside—a pattern known as a Voronoi diagram. If two or more audience members touch each other, they are registered as one unit by the art system impacting the Voronoi diagram.

The *participation style* of Boundary Functions is group as the number of participants is limited by the physical set-up of the installation. Like Zoom Pavilion, the *audience constellation* is fluid as people can join and leave the interaction freely. The *input and output access* are both public, as they can be accessed and perceived by all participants.

Since the art system responds to each participant equally, its *participation symmetry* is symmetrical. The art system requires multiple audience members to act together, and therefore the *activity type* is social and the audience is the *initiator* of the interaction. It demonstrates a form of co-located interaction with fluid and symmetrical participation style in a social activity initiated by the audience.

Spatial Sounds (100dB at 100km/h) **(2000) by Marnix de Nijs and Edwin van der Heide** is an interactive installation that consists of a speaker mounted on a robotic arm, custom developed software for the interaction and real time sound generation, ultrasonic sensors, and an angle sensor [14]. The audience mainly interacts with the position of their bodies in relation to the installation. The robotic arm rotates and scans the space with ultrasonic sensors mounted on the speaker. The artwork has four interaction modes. In mode 1, the arm rotates slowly and scans the space with a low humming sound that changes briefly when a person is detected. Once someone is detected, the art system continues scanning for a while and switches to mode 2, in which the arm makes one full rotation and stores at which angles it detects people. Then it randomly selects one person and attempts to follow them. The sound also changes depending on the arm's rotation speed. If the art system keeps detecting people, it will change to mode 3. If no one was detected, it switches back to mode 1. In mode 3, the arm has a fixed rotation speed but changes direction when someone is detected. The actual speed of the arm is the main parameter for the sound generation and the detection of audience is expressed by a pulse train sound. When the art system detects too many people nearby it moves to mode 4. If this does not happen, it moves back to mode 2. In mode 4, the more people there are and the closer they stand, the faster the arm rotates, while it will slow down when they move away from the arm. The sound is powerful and influenced by the rotational speed. After a fixed duration it will switch back to mode 3.

Similar to Zoom Pavilion, the *participation style* of Spatial Sounds (100dB at 100km/h) is 'crowd' as it allows for a variable number of audience members only limited by the exhibition space. Like the previous two artworks, the audience can join and leave the interaction freely in a fluid *audience constellation*. Unlike all previous artworks, here the art system has different modes of behaviour, which results in a varied *participation symmetry* and a mixed *activity type*. For instance, in mode 2 only one audience member is selected by the art system while in mode 3 all members have equal chance to engage in a social activity by 'passing' the art system to each other. The *input and output access* to the art system are both public and the art system initiates the interaction (*initiator*). It demonstrates several forms of co-located interaction with a fluid and varied participation style in mixed types of activities initiated by the art system. It also shows that different forms of co-located interaction can be combined to enrich the interactive experience.

Lights Contacts **(2009) by Scenocosme** is an interactive installation that responds to physical contact among the audience with varying sounds and lights in a public setting [15]. The art system consists of a sensor ball, a computer, software programs, LED lights, and a speaker. The audience mainly interacts with their bodies and has different functions in the interaction. A first individual is required to put their hand on the sensor ball. In doing so, the audience activates the art system and becomes an extension of its sensing unit. If that person remains alone, nothing else happens. Due to conductivity of the human body, when another person touches their skin, the art system detects changes in the electrostatic charge, then generates sounds and alters the light accordingly. The more people touch each other the more sound sources are generated.

The *participation style* in Lights Contacts is crowd as there can be an unlimited number of participants. Like the previous three artworks, people can join and leave the interaction at any time in a fluid *audience constellation*. As the sensor ball is only available to a limited number of people, the *input access* is partially public, while the *output access* is public as the generated lights and sounds are displayed publicly. Similar to World Skin, the *participation symmetry* is asymmetrical as some people must touch the sensor ball. The participants rely on each other in a social *activity type* and initiate the interaction (*initiator*). It demonstrates a form of co-located interaction with fluid and asymmetrical participation style in a social activity initiated by the audience.

Body Movies (2001) by Rafael Lozano-Hemmer is an interactive projection installation for public spaces [16]. The art system consists of projectors, robotic controllers, portraits, a public surface, Xenon lights, a screen, a camera, computers, software programs, and speakers. The audience mainly interacts with their bodies. A set of portraits is projected on a surface and washed out by lights positioned at a distance on the floor. The audience enter the space and their shadows are thrown onto the projection surface so that portraits are revealed. The artwork tracks the edges of the shadows and once they overlap with a portrait, a hotspot is activated for a few seconds and an audio track is played. When all portraits are revealed, the artwork blacks out the projection and displays a new set of portraits at different locations on the projection surface. Meanwhile, the tracking interface is displayed in real-time on a display next to the projection.

Similar to Lights Contacts and Zoom Pavilion, the *participation style* of Body Movies is crowd as there is no practically imposed limit on the number of participants, and the audience has the freedom to join and leave the interaction in a fluid *audience constellation*. The *input and output access* are both public, as they can be accessed and perceived by all audience members equally. The *participation symmetry* is symmetrical as all people participate equally. The audience can interact both individually and with each other to reveal portraits or perform together in a mixed *activity type*. The audience initiates the interaction by casting shadows on the projection surface (*initiator*). It demonstrates a form of co-located interaction with fluid and symmetrical participation style in mixed types of activities initiated by the audience.

4   Application of the Relational Model

In this section, we apply the relational model to describe the forms of co-located interaction in the selected artworks[1]. Based on these descriptions, we further compare them and identify their similarities and differences.

---

[1] Visualisations of co-located interaction in the selected artworks are available online: https://git hub.com/danxxxu/co-located_interaction.

4.1 Describing the Selected Artworks Using the Relational Model

Brainball (2003) by Smart Studio

Here the interaction takes place among one art system and two audience members. There are two forms of direct private one-to-one communication from the audience to the art system. Firstly, either audience member can press the button to start the game and initiate the interaction. Secondly, both audience members share their brainwaves with the art system to compete in the game and participate in the interaction.

There are two forms of direct public one-to-many communication from the art system to the audience. The art system calculates the difference between the audience's states of relaxation based on their brainwave data and uses this information to move the steel ball as a response. This also informs the audience of their relative states of relaxation and provides a game for them to compete in. Meanwhile, the brainwave data and game status are displayed on a screen, which informs and provides feedback about the audience's states of relaxation. However, such information can also distract them and disrupt their performance.

As the ball movement and screen display reveal the mental states of the audience, it creates a form of mediated public one-to-one communication between them via the art system. The audience can also converse with each other through speech and body language in a form of direct public one-to-one communication, allowing them to exchange information, influence and disrupt each other's performance.

Randomly Generated Social Interactions (2016) by Anastasis Germanidis

Here the interaction takes place among one art system and two audience members and multiple instances of the interaction can take place simultaneously. There are two forms of direct private one-to-one communication from the art system to the audience. Firstly, the art system generates a fictional identity for each audience member to provide background for the interaction. Secondly, it instructs the audience to initiate the interaction and directs their performance. As an audience member enacts the instructions, they also communicate the information from the art system to the receiving audience member. This creates a form of mediated public one-to-one communication from the art system to the receiving audience member via the enacting audience member.

The audience can also converse with each other through speech and body language in a form of direct public one-to-one communication, allowing them to respond to the instructions and participate in interaction. Meanwhile, this form of communication provides channels for them to exchange information and coordinate actions that are not prescribed by the art system.

World Skin (1997) by Maurice Benayoun

Here the interaction takes place among one art system and at least two audience members. There are two forms of direct public one-to-one communication from the audience to the art system that depend on the role of the audience. If an audience member is the 'driver', they navigate the landscape to initiate and participate in the interaction and facilitate the 'photographers'. If an audience member is one of the 'photographers', they take photos of the landscape to express themselves, initiate and participate in the interaction. Multiple instances of such communications can take place.

There are two forms of direct public one-to-one communication from the art system to the audience. Firstly, the art system updates the virtual landscape as a response to the 'driver' and provides content for interaction. Secondly, the art system removes the pixels of taken shots in the landscape as a response to the 'photographers' and provides means for expression. Multiple instances of such communication can take place. Meanwhile, the art system plays audio to enhance the interactive experience in a form of direct public one-to-many communication to the audience.

Among the audience members, there are two forms of public communication and multiple instances of such communications can take place. Firstly, they can express themselves by changing the landscape either through navigation or taking photos in a form of mediated one-to-many communication via the art system. Secondly, they can converse with each other through speech and body language in a form of direct many-to-many communication to exchange information and coordinate actions.

Zoom Pavilion (2015) by Rafael Lozano-Hemmer and Krzysztof Wodiczko

Here the interaction takes place among one art system and one or more audience members. There are two forms of direct public communication from the audience to the art system and multiple instances of both forms of communication can take place. Firstly, the audience members enter the exhibition space to participate and perform in the interaction in a form of many-to-one communication. Secondly, once selected as a pair, the audience members can move to vary the line drawn between them to participate in the interaction and express themselves in a form of many-to-one communication.

There are five forms of direct public communication from the art system to the audience and multiple instances of such communications can take place. Firstly, the art system displays live camera images on the walls to provide content for interaction and a means for expression in a form of one-to-many communication. Secondly, the art system can zoom-in on the face of a chosen audience member to initiate interaction, isolate them, and provide a means for expression in a form of one-to-one communication. Thirdly, the art system can draw a line between two chosen audience members and label their spatial relations to initiate interaction, provide shared control of its visual response and a means for expression in a form of one-to-many communication. Fourthly, the art system stores and displays paired images of the audience to draw connections between them and provide a visual history of the interaction in a form of one-to-many communication. Lastly, the art system plays audio to enhance the perception of its actions in a form one-to-many communication.

Among the audience members, there are three forms of public communication and multiple instances of such communications can take place. Firstly, once an audience member's face is magnified by the art system, they can make facial expressions to participate in the interaction and express themselves in a form of mediate one-to-many communication via the art system. Secondly, pairs of audience members can move to vary the line between them to communicate and perform with each other in a form of mediated one-to-one communication via the art system. Lastly, the audience can also converse with each other through speech and body language in a form of direct many-to-many communication, which allows them to exchange information and coordinate actions.

Boundary Functions (1998) by Scott Snibbe

Here the interaction takes place among one art system and at least two audience members. There are two forms of direct public communication from the audience to the art system and multiple instances of such communications can take place. Firstly, the audience can move across the floor to initiate and participate in the interaction, and express themselves in a form of one-to-one communication. Secondly, the audience can touch each other to trigger a different response from the art system in a form of many-to-one communication.

There are two forms of direct public one-to-many communication from the art system to the audience. Firstly, the art system projects a Voronoi diagram based on the positions of the audience as a response to their actions and provides shared control and a means for expression. Secondly, when audience members touch each other, the art system reconfigures the Voronoi diagram as a response and to create connections between the audience. Multiple instances of such communication can take place.

There are two forms of public many-to-many communications among the audience and multiple instances of such communications can take place. Firstly, the audience can move towards or away from each other to shape the Voronoi diagram in a form of mediated communication via the art system. Secondly, the audience can converse with each other through speech and body language in a form of direct communication, allowing them to exchange information and coordinate actions.

Spatial Sounds (100 dB at 100 km/h) (2000) by Marnix de Nijs and Edwin van der Heide

Here the interaction takes place among one art system and at least one audience member. As there are different interaction modes, we describe each separately. In mode 1, there are two forms of direct public one-to-many communication from the art system to the audience. Firstly, the art system rotates to inform the audience the current interaction mode and gather inputs for interaction. Secondly, the art system plays audio to enhance the perception of its movement. Once it detects an audience member, it changes the audio to inform them of the detection and initiates the interaction in a form of direct public one-to-one communication. Multiple instances of such communication can take place. When the audience approaches the art system, they are registered by the art system which allows them to participate in the interaction and trigger a switch of interaction mode. This creates a form of direct public one-to-one communication to the art system, while multiple instances of such communication can take place.

If the art system keeps detecting audience, it switches to mode 2. In this mode, it first makes a full rotation and scans the space to gather inputs for the interaction and inform the audience the current interaction mode in a form of direct public one-to-many communication to the audience. Then it selects one audience member for two forms of direct public one-to-one communication. Firstly, the art system follows the selected audience member to initiate interaction and isolate them. Secondly, it generates audio based on its rotation speed to enhance the perception of its movement. The chosen audience member can move to direct the art system and participate in the interaction in a form of direct public one-to-one communication to the art system. During this interaction, other audience members can approach the art system while their presence is registered to trigger a switch of interaction mode, effectively in a form of direct public one-to-one communication to the art system. Multiple instances of such communication can take place. Note that although the art system attempts to follow one specific audience member, it may lose track of them if they move away or if another person appears between them and the art system. If no person can be detected for some time, the art system reverts to mode 1. If the art system keeps on detecting audience members, it switches to mode 3.

In mode 3, there are two forms of direct public one-to-many communication from the art system to the audience. Firstly, the art system rotates at a fixed speed to indicate the current interaction mode. Secondly, it generates audio based on its rotation speed to enhance the perception of its movement. Once an audience member is detected, there are two forms of direct public one-to-one communication from the art system to the audience and multiple instances of such communication can take place. Firstly, the art system changes direction to initiate the interaction and provide a game for the audience. Secondly, it plays audio upon detecting an audience member to inform and enhance the perception of its action. Meanwhile, the audience can move to direct the art system, participate in the game and trigger a switch of interaction mode. This creates a form of direct public one-to-one communication to the art system and a form of mediated public one-to-one communication between the audience members via the art system as they can 'pass' it around to express themselves. Multiple instances of both communications can take place. If the audience moves away, the art system reverts to mode 2. If the audience remains in front of the art system, it switches to mode 4.

In mode 4, there are three forms of direct public one-to-many communication from the art system to the audience. Firstly, the art system speeds up to indicate its behaviour change and urge the audience to move away. Secondly, once the audience moves away, the art system slows down as a response and to indicate its behaviour change. Lastly, the art system generates audio based on its rotational speed to enhance the perception of its movement. The audience can move away as a response and to comply with the art system. This creates a form of direct public one-to-one communication to the art system and multiple instances of such communication can take place. After some time, the art system reverts to mode 3.

In all modes, the audience can converse with each other through speech and body language to exchange information and coordinate actions. This creates a form of direct public many-to-many communication and multiple instances of such communication can take place.

Lights Contacts (2009) by Scenocosme

Here the interaction takes place among one art system and at least two audience members. There are four forms of public many-to-one communication from the audience to the art system that depend on the actions they take. Firstly, one or more audience members can touch the sensor ball in a form of direct communication to activate the art system and serve as a prerequisite for the interaction. Subsequently, other audience members can touch the 'activating' audience, or the other way around, to trigger the response of the art system, initiate and participate in the interaction and express themselves. This creates both a form of mediated communication from the subsequent audience to the art system via the 'activating' audience and a form of mediated communication from the 'activating' audience to the art system via the subsequent audience. Lastly, once the art system is activated, the activities of both groups of audience are registered by art system.

This creates a form of direct public many-to-one communication from all audience to the art system.

There are two forms of direct public one-to-many communication from the art system to the audience. Firstly, the art system generates audio based on the audience's contact with each other as a response and to provide feedback about their touch intensity and a means for expression. Secondly, the art system alters the colours and behaviours of the lights in line with the audio to enhance the audience's perception of it.

There are two forms of public many-to-many communication between the audience. Firstly, as the lights and audio directly reflect their contact with each other, they can communicate with each other via the light and audio by altering their touch in a form of mediated communication via the art system. Secondly, the audience members can converse with each other through speech and body language to exchange information and coordinate actions in a form of direct communication. Multiple instances of such communication can take place.

Body Movies (2001) by Rafael Lozano-Hemmer

Here the interaction takes place among one art system and one or more audience members. There is a form of direct public many-to-one communication from the audience to the art system and multiple instances of such communication can take place. The audience cast their shadows on the projection and the camera captures and tracks the shadow contours, allowing them to initiate and participate in the interaction.

There are four forms of direct public one-to-many communication from the art system to the audience. Firstly, the art system displays the shadows of the audience to provide feedback about their movements and a means for expression. Multiple instances of such communication can take place. Secondly, the art system displays the tracking interface to provide feedback and inform the audience of their performance and the status of interaction. Thirdly, when a shadow overlaps with a portrait, the art system activates a hotspot and plays an audio clip in response to provide feedback about the achievement. Multiple instances of such communication can take place. Lastly, once all hotspots are activated, the art system blacks out the projection and updates the portraits as a response and initiates a new session of interaction.

There are two forms of public communication between the audience and multiple instances of such communications can take place. Firstly, the audience can communicate with each other through their shadows to express themselves and perform in the interaction. This creates a form of mediated many-to-many communication via the art system. Secondly, the audience members can also converse with each other through speech and body language to exchange information and coordinate actions in a form of direct many-to-many communication.

4.2 Comparing the Selected Artworks

The relational model provides a systematic framework for describing the various forms of co-located interaction in the selected artworks. It breaks down interactions by examining the actions performed by interacting elements and the various forms of communication resulting from these actions. This approach not only helps us describe interactions but also allows for consistent comparisons between them. In total, we identify five common themes from the similarities and differences between the selected artworks and discuss them individually below.

Types of Reactions

In all selected artworks, some actions of the elements serve as responses to other elements' actions. We can say that the elements can act and react to each other. This is the case primarily between the art system and the audience. Upon examination, we observe several different types of reactions.

To start, reactions performed by the art system can provide feedback about the effects of audience actions on the art system and the interaction. For instance, in World Skin, the art system changes the display of the landscape according to the actions of the audience. In Brainball, Body Movies, and Spatial Sounds (100dB at 100km/h), the art system uses visual or auditory displays to inform the audience of the status of interaction or its recognition of them. Besides that, reactions of the art system can provide continuous sensory cues on the actions of the audience. This is most obvious in Zoom Pavilion, Boundary Functions, and Lights Contacts, in which the art systems translate information such as the audience's mutual distances and touch intensity into visual and/or audio cues. Lastly, the reactions of an art system can in turn trigger or initiate the interaction. For instance, in Zoom Pavilion and Spatial Sounds (100dB at 100km/h) the art system reacts to the presence of the audience with visual and/or audio and physical responses to initiate interaction.

Conversely, reactions of the audience are mainly to comply with or perform actions demanded by the art system. For instance, in Randomly Generated Social Interactions and modes 2 and 4 of Spatial Sounds (100dB at 100km/h), the audience follows instructions or prompts from the art system to engage in specific actions. It is important to note that reactions are not isolated events but rather part of a continuous process of exchange between actions and reactions. As a result, a reaction to one element can trigger further reactions or influence subsequent actions of that element.

Influences Between Actions

Besides action-reaction relationships, the actions performed by elements can also influence each other in other ways. Firstly, an action can act as a prerequisite for subsequent actions, necessitating the element to execute the action prior to or concurrently with other actions. As in Brainball, a participant must press a button to start the game. In Lights Contacts, the audience must first touch and maintain contact with the sensor ball to activate the art system. Furthermore, an action can disrupt the execution of other actions. In Brainball, the display of real-time brainwave data and direct communication between the audience can hinder the performance of individual players in the game. Lastly, an action can facilitate or enhance other actions, often in parallel to each other. For instance, in World Skins an audience member navigates the landscape to facilitate others taking photos, and the art system plays audio to enhance their experience. The use of audio to enhance the perception of other actions is also seen in Zoom Pavilion and Spatial Sounds (100 dB at 100 km/h), while in Lights Contacts it is light that enhances the perception of the audio.

Roles of Direct Communication between the Audience

As a characteristic of co-located interaction, the audience can communicate with each other directly. This communication can take the form of verbal exchanges and non-verbal means, such as gestures, facial expressions, and touch. Direct communication among audience members enables them to exchange information and coordinate their actions. By sharing their experiences and knowledge of the interaction's mechanics, the audience can propose new ideas, insights, and comments on each other's performance. Here direct communication can facilitate the audience to take action. This dynamic is present in all artworks but Brainball, where direct communication can actually disrupt the players' composure, potentially affecting their ability to succeed in the game.

Moreover, direct communication among the audience can also become part of the interaction. Notably, in Randomly Generated Social Interactions, the art system instructs the audience to communicate directly with one another, making this form of communication a central aspect of the interaction. Similarly, in Boundary Functions and Lights Contacts, physical touch of audience members triggers responses from the art system, making direct communication a key mechanism driving the interaction.

Creating Connections among the Audience

A recurring theme in all selected artworks is the art system's active role in establishing or enabling connections among the audience during the interaction. We observe several distinct approaches to achieve this.

Firstly, connections can be created in an arbitrary manner. For instance, in Randomly Generated Social Interactions, the art system instructs the audience to engage and converse with one another. Similarly, in Zoom Pavilion, the art system randomly pairs the faces or bodies of audience members, facilitating chance encounters.

Secondly, in some instances the art system creates interdependencies among the actions of the audience: the actions of one audience member may affect or depend on the actions of others. There are different ways in which art systems utilize this approach. The art system can draw upon the differences in people's behaviours. In Brainball, the movement of the ball is determined by the relative states of relaxation of two audience members, rather than being controlled by a single individual. Alternatively, an art system may react to actions that require multiple persons to perform together. For instance, the audience must touch each other to trigger responses from the art system in Lights Contacts and Boundary Functions. Yet another manifestation is when the art system enables shared control over its responses among the audience, as in Zoom Pavilion, where the art system draws a line between two audience members that varies as they move closer and further apart. Similarly, in Boundary Functions the Voronoi diagram depends on the positions of all audience members.

Another strategy employed by art systems to foster connection is by encouraging collaborations among the audience. As demonstrated in World Skin and Lights Contacts, the audience members play different roles and have different functions, requiring them to work together in the interaction.

A final approach observed is seemingly counterintuitive: the art system creates connections between audience members by isolating an individual from the rest. In such instances, the art system deliberately focuses on a single person, placing them under a spotlight. By doing so, it amplifies the selected individual's actions, triggering them to express and connect with others. This effect is evident in Zoom Pavilion, where the art system randomly selects and magnifies the face image of an audience member. Similarly, in mode 2 of Spatial Sounds (100dB at 100km/h), the art system randomly chooses an audience member to follow. This deliberate 'isolation' also evokes an emotional connection among the audience, as the possibility of being chosen looms over all its members.

Forms of Mediated Communication

In all works, there is at least one form of mediated communication, where one element communicates with another element through a third element. Most prevalent is the audience's mutual communication via the art system. In works such as World Skin, Body Movies and Zoom Pavilion, the art system provides a stage and means for expression to the audience. Alternatively, it can translate information about audience actions into additional sensory cues to be used by the audience to communicate with each other—as in Zoom Pavilion, Boundary Functions, and Lights Contacts, the distances between the audience members are translated into (audio-)visual cues.

Alternatively, audience members can communicate to the art system via another audience member. As shown in Lights Contacts, the art system only responds when two groups of audience members touch each other.

A third intriguing form of mediated communication is observed in Randomly Generated Social Interactions, where the art system generates content and instructs audience members to act it out, thereby conveying the information to others. In this scenario, the art system communicates to the receiving audience members via the acting audience member who becomes in effect a 'tool' for expression.

5 Discussion and Conclusion

As shown in the previous section, the relational model enables us to provide detailed descriptions of a diverse range of co-located interactions. We carefully selected eight artworks that represent different forms of co-located interaction based on different combinations of *audience constellation, participation symmetry, activity type,* and *initiator.* Using the relational model we described their co-located interactions following the actions of the audience and art system in terms of various forms of communication. This allows us to specify the role each element plays in the interaction and how they influence and relate to each other. The main concepts used in the model—the identification of elements, their actions and forms of communications to each other—were shown to be universal and applicable to all artworks. More importantly, the model's approach provides a frame for analysis that can be adapted to examine and compare different forms of co-located interaction.

A key aspect of the relational model is identifying the roles of the actions performed by the elements. In doing so, we observe different ways in which the audience and the art system can act and react to each other, revealing patterns in these relationships. Additionally, it highlights the influences between actions of the elements that are necessary for the interaction and establishing various mutual relationships. We can envision creating new forms of interaction by composing different types of action-reactions and influences between actions for different elements. An intriguing avenue for future investigation is a comprehensive exploration of existing relationship typologies and their prevalence across different elements in interactive artworks.

The relational model not only enables us to dissect and examine interaction in terms of various forms of communication but also provides a template for conceiving and creating new relationships among the elements. It shows that an element can both influence other elements directly and mediate the communications between other elements. One related insight of co-located interaction that was revealed by our analyses pertains to its inherently social nature. By definition, a co-located interaction requires multiple co-located audience members to participate simultaneously and the interplay between them becomes central to the overall interactive experience. To achieve this, an art system can employ diverse strategies to foster connections between audience members, draw upon the direct communication between co-located audience members, and devise novel forms of mediated communication.

The relational model does not make prior assumptions about behaviours of an art system and the audience, and attempts to describe them in equal terms. This perspective opens up possibilities to conceive new ways of participation. For instance, the audience can be 'used' as a tool for expression and communication. Nevertheless, we also notice the differences between the behaviours of art systems and audiences. For the art system, its actions are usually concrete and definable as they are often programmed and scripted. While for audiences, their actions are more open and diverse, and variable on an individual basis. As a result, we often use general terms to describe the actions of the audience. However, we also see opportunities here to create art systems that are more 'human' and can surprise or even make mistakes; such as in Spatial Sounds (100dB at 100km/h), where the art system may misidentify the audience.

Meanwhile, when speaking of the audience, we only consider those who are actively participating in the interaction and not those who are merely observing. In some artworks, there is no clear distinction between participating and observing audience members, as in Zoom Pavilion, Spatial Sounds (100 dB at 100 km/h), and Body Movies. While in other works, there is a clear distinction between participating and observing audience. This is the case for both fluid and fixed *audience constellation*, as shown in Brainball, Randomly Generated Social Interactions, World Skin, and Boundary Functions. In these works, the presence of spectators can potentially influence the participating audience. It may be valuable to include this influence in the model too.

We have selected artworks that were mainly developed in the Western cultural context, created between 1997 and 2016 and with documentation available. Our classification dimensions focused on parameters defining audience participation. In the future, it would be interesting to apply the relational model to more recent artworks, use different classification methods, such as aspects defining audience experience and include different cultural contexts.

To conclude, we demonstrated the effectiveness of the relational model in describing and comparing different forms of co-located interaction in interactive art. The model provides a systematic approach to examine co-located interaction and uncover the various relationships and influences between interacting elements. Moreover, it allows us to compare different artworks and reveal patterns of co-located interaction. Through the analysis, we have identified insights about co-located interaction and opportunities for creating new forms of communication and interaction. While our focus has been on co-located interaction, the same process can be applied to study other types of interaction and audience participation. The relational model shows a concrete step forward from the dialogue model to further understand interaction in interactive art.

References

1. Xu, D., Lamers, M., van der Heide, E.: Towards a Relational Model of Co-located Interaction in Interactive Art. In: Proceedings of the 28th International Symposium on Electronic Art (ISEA), Paris, France (2023)
2. Schraffenberger, H., van der Heide, E.: Audience-Artwork Interaction. Inter. J. Arts Technol. 8(2), 91–114 (2015). https://doi.org/10.1504/IJART.2015.069550
3. Guljajeva, V.: From Interaction to Post-participation: The Disappearing Role of the Active Participant. PhD thesis. Estonian Academy of Arts, Tallin, Estonia (2018)
4. Edmonds, E., Turner, G., Candy, L.: Approaches to interactive art systems. In: Proceedings of the 2nd International Conference on Computer Graphics and Interactive Techniques in Australasia and South East Asia, pp. 113–117. ACM, Singapore (2004). Doi: https://doi.org/10.1145/988834.988854
5. Mubarak, O.: Designing and Modeling Collective Co-Located Interactions for Art Installations. PhD thesis. CNAM, Paris, France (2018)
6. Bell, S.: Participatory Art and Computers: Identifying, Analysing and Composing the Characteristics of Works of Participatory Art that Use Computer Technology. PhD thesis. Loughborough University of Technology, Loughborough, England (1991)
7. Ars Electronica Archives. https://archive.aec.at, (Accessed 1 July 2023)
8. Archives of Digital Art. https://digitalartarchive.at, (Accessed 1 July 2023)
9. Hjelm, S.: Research+Design: the Making of Brainball. Interactions 10(1), 26–34 (2003)
10. Germanidis, A.: Randomly Generated Social Interactions. https://vimeo.com/156882347, (Accessed 26 July 2023)
11. Benayoun, M.: World Skin. https://benayoun.com/moben/1997/02/12/world-skin-a-photo-safari-in-the-land-of-war, (Accessed 26 July 2023)
12. Lozano-Hemmer, R.: Zoom Pavilion. https://www.lozano-hemmer.com/texts/manuals/zoom_pavillion.pdf, (Accessed 26 July 2023)
13. Snibbe, S.: Boundary Functions. https://www.snibbe.com/art/boundaryfunctions, (Accessed 1 July 2023)
14. van der Heide, E.: Spatial Sounds (100 dB at 100 km/h) in the Context of Human Robot Personal Relationships. In: Lamers, M.H., Verbeek, F. J. (eds.), Human-Robot Personal Relationships (HRPR 2010). LNICST, vol. 59, pp. 27–33. Springer, Heidelberg (2011). https://doi.org/10.1007/978-3-642-19385-9_4
15. Scenocosme: Lights Contacts. https://www.scenocosme.com/contacts_installation_en.htm, (Accessed 26 July 2023)
16. Lozano-Hemmer, R.: Body Movies. https://www.lozano-hemmer.com/body_movies.php, (Accessed 26 July 2023)

Touching the Untouchable: Playing the Virtual Glass Harmonica

Astrid Pedersen$^{(\boxtimes)}$ ⓘ, Morten Jørgensen ⓘ, and Stefania Serafin ⓘ

Aalborg University, Copenhagen, Denmark
{aekp19,mojarg19}@student.aau.dk, sts@create.aau.dk
https://melcph.create.aau.dk/

Abstract. Cultural heritage museums around the world are embracing new approaches to redefine their mission and roles by combining preserved knowledge with immersive technologies. This paper focuses on a virtual reality (VR) installation project within The Danish Music Museum, which showcases a glass harmonica. The project aims to provide visitors with an immersive and interactive experience, resurrecting a forgotten instrument and presenting its history, sound and interaction. Through a qualitative evaluation at the museum, it was found that the installation establishes a connection between the virtual instrument and the physical glass harmonica on display, offering an engaging and enjoyable experience. However, challenges such as disturbances caused by bystanders affecting the functionality of hand tracking were observed. Overall, the project proposes a VR experience, bringing a historical instrument back to life, offering visitors at The Danish Music Museum an immersive encounter with cultural heritage.

Keywords: Glass Harmonica · Museum · Virtual Reality · Physical Models · Cultural Heritage

1 Introduction

Cultural heritage museums around the world are redefining their mission and roles by combining consolidated and preserved knowledge with novel forms of experience [15,27]. The emergence of immersive technologies has provided new approaches to overcome the physical environmental limitations of cultural heritage exhibitions and proposes new opportunities to virtualize and augment the visitor experience. The use of multimodal and interactive capabilities of VR transforms visitors into proactive explorers of cultural heritage by providing an educational, entertaining, escapist, and aesthetic experience within a fully immersive virtual environment (VE) [15,26,27].

While traditional exhibitions often limit visitors to passive observation, VR allows close inspection and interaction with artefacts, inviting them to actively participate [24,33]. Museum professionals' perceptions of the use of VR

A. L. Brooks (Ed.): ArtsIT 2023, LNICST 565, pp. 218–233, 2024.
https://doi.org/10.1007/978-3-031-55312-7_16

technology found that this emerging technology had created a fundamental change, necessitating a rethinking of traditional concepts [24]. Trunfio and colleagues produced similar findings, not only suggesting an overall increase in visitor satisfaction and experience, but also posits the emergence of novel visitor profiles, drawing on John Falk's museum visitor experience model [8,27].

This project takes part in the larger initiative of The Danish Music Museum *Music History - Taken out of the Box*. Drawing inspiration from the idea of freeing instruments from their display cases, this paper introduces a VR installation that allows visitors to experience the glass harmonica in a fully immersive environment, as depicted in Fig. 1. The installation is situated within the exhibition hall at The Danish Music Museum in Copenhagen, deployed standalone on the Meta Quest 2 using hand-tracking. The real exhibited glass harmonica hails from the 19th century, and though it remains in good condition, it is regrettably non-operational in compliance with preservation regulations. As musical instruments are interactive in nature and difficult to fully appreciate through visual inspection alone, this presents challenges to the museum. This project seeks to explore how the use of an interactive virtual replica of the glass harmonica can enrich the visitor experience and educate them about the instrument. The objective of the installation is to familiarise visitors with the historical background and interaction of the glass harmonica through an interactive VR experience. This paper provides a comprehensive overview of the museum installation project, covering all aspects of its development and execution. It examines the related work in the field to provide context and background for the project. It describes the design and implementation of the glass harmonica, and, finally, discusses the evaluation of the project.

Fig. 1. Glass harmonica displayed in a virtual environment (VE) inspired by the 18th century.

Glass instruments have a rich history that can be traced back to the early days of glassmaking. Among the earliest types of glass instruments were those that produced distinct tones through the striking of differently sized glass bowls. An early account of musical glasses appears in 1741 with the invention of the glass harp, an instrument invented by Richard Puckeridge; consisting of stemmed glasses filled with water, altering the frequencies of the sounds produced. It was not until 1762 that Benjamin Franklin introduced the glass harmonica in its modern form [14]. Franklin's design, heavily influenced by Puckeridge, uses glass bowls spun by a horizontal axle, with one side dipping into a trough of water. Sound was produced from exciting the bowls by touching the rim of the rotating glass with a wet finger [22]. The glass harmonica quickly gained popularity among artists such as Mozart and Beethoven, but association with negative events led to a decline in popularity during the 18th century [14]. In current times, what once was an instrument seemingly destined for permanence was by 1830 a forgotten instrument, a museum piece.

2   Background

As younger generations are more exposed to rich media and interactive content, this is increasingly expected in recreational and educational facilities [19,28]. The benefits of incorporating digital technologies into cultural heritage do not limit themselves to experiential matters. In [5], Chong et al. describe how various studies and efforts incorporate digital technologies as a preservation medium. Their findings reveal positive results for engaging and further motivating greater awareness of society's cultural content. Thus, the author highlights virtual heritage as essential to reconstruct the past and safeguard the current heritage digitally. In [24], the authors acknowledge the need to acquire new knowledge and strategies to implement and evaluate these technologies in museum settings. Their paper offers a holistic assessment of the current use of VR through interviews with museum professionals, who found that the most common use of VR is to allow visitors to experience "impossible" spaces or time travel; providing historical context for the exhibits. This is reflected in the multitude of works describing VR experiences depicting historical sites for museums [4,9,10,29].

Although research suggests advantages of virtual cultural heritage, the impact of VR applications on the spatial and social experience of museums remains a concern for museum professionals. The disruption caused by the loss of visual connection between visitor actions and their physical environment raises questions about the benefits of incorporating virtual elements into museum spaces [18,24]. As the digital space in VR remains concealed from those outside the headset, the process of people watching suddenly lacks meaning. Symptomatic of this are reflexive concerns about how the visitor might appear to others if they inhabit this digital space themselves [18,24]. This has sparked efforts to explore methods for facilitating co-located asymmetric interaction between head-mounted display (HMD) users and bystanders (non-HMD users) [21].

2.1 Virtual Reality Musical Instruments

Digital Musical Instruments (DMIs) are defined as systems enabling gestural control of sound production through controllers mapping gestures to sound parameters [30]. As the focus on VR systems has grown, the field of musical instruments for VR has expanded, and various authors have surveyed different aspects of this broad field. In [23], Serafin et al. surveyed VR musical instruments, differentiating between virtual musical instruments (VMIs) primarily focusing on sonic emulation and virtual reality musical instruments (VRMIs) which incorporate visual elements through head-mounted displays or immersive systems. They also presented nine design principles aimed at enhancing functionality and creating engaging and natural VR experiences. The importance of minimising latency is further emphasised in a study presented in [17], which found that low latency is desirable for VRMIs. Furthermore, in [3] 3D interaction techniques for musical control in VEs are explored, and [1] provides an overview of recent musical work in VR. Certainly, the field of VRMIs has much to offer in terms of how to design well for multimodal VR, especially considering design for public spaces.

According to [5], VR has been recognised as a preservation medium. Consequently, there is an ongoing effort to create VRMIs for historical instruments, as it enables visitors to interact while simulating instrument acoustics [2,16,20]. When looking at VRMIs in museum contexts, these often suffer from a lack of evaluation within actual museum settings [12,25,31,32]. As discussed in [5,24], VR's immersive capabilities effectively provide historical context to exhibited artefacts via narratives or representational virtual environments. Historical appreciation of forgotten instruments goes beyond narration, as it is closely intertwined with the interactive experience of playing them. Therefore, faithful replication of acoustic qualities, gestures, and movements can enhance understanding and appreciation.

3 Design and Implementation

The objective of the installation is to familiarise visitors with the historical background and interaction of the glass harmonica through an interactive VR experience. Based on the notion that the glass harmonica should transgress its display case and become playable in VR, the connection between the virtual and the real instrument plays an important role. The VR experience should augment the visitor experience by adding a layer of narrativity and historical context through the VE, in addition to faithfully replicating the aesthetics, acoustics and real interaction of playing the instrument. The application was targeted a Meta Quest 2 used as a standalone device. Because of the reduction in hardware maintenance and allowing natural interaction, this project utilises the hand-tracking capabilities offered by the Meta Quest 2, enabled through Oculus Integration for Unity. This section will delve into the specific design decisions that were made and the reasoning behind them, along with the implementation of these.

The design of the installation was collaboratively undertaken through co-design activities over the span of several months, involving two graduate students and a professor from Aalborg University, Copenhagen, as well as three museum employees. The research team took on various aspects of the development process, focusing on technical factors such as application development and usability evaluation. Concurrently, the museum personnel, with their expertise in public engagement and pedagogy, provided valuable information on what would be effective within a public and educational setting. A 3D sketch of the installation is shown in Fig. 2. The VR installation was strategically positioned to face the display case with the exhibited glass harmonica, ensuring that when visitors removed the HMD, they were immediately confronted with the real instrument. The installation was designed to aesthetically integrate with the surrounding exhibition space. A screen was mounted on the wall, displaying a looped mute screen-capture of the VR experience to accommodate bystanders. It was originally intended to provide a live-stream of the HMD view, aiming to grab attention and provide entertaining for queuing visitors, but further technological development will be needed for this.

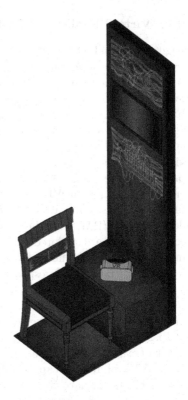

Fig. 2. Isometric 3D visualisation of the installation.

3.1   Interface

To educate the visitor about how to play the original counterpart, the interaction has to be replicated close to its nature. First, the instrument affords control of the sound by touching the rim of the rotating glass bowls with the index finger, as shown in Fig. 3, implemented using the *PokeInteractor* from Oculus Interaction SDK for Unity. The interface offers polyphonic playing, as the user can excite two bowls simultaneously. The speed of rotation is set to an approximated fixed speed, opposite to user-controlled in an original instrument. The design of the virtual replica is illustrated in Fig. 1, with the assigned notes displayed in Fig. 4.

Visual Feedback

The VR experience was built using Unity version 2021.3.11f1 with the universal rendering pipeline, and the virtual glass harmonica was modelled using Blender version 3.3.1, replicating the instrument exhibited at The Danish Music Museum.

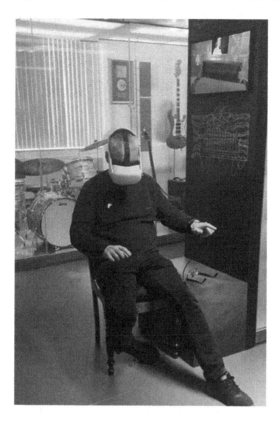

Fig. 3. Participant interacting with the final installation at the museum (the charger is not constantly attached)

The VE was designed to resemble the study room of its inventor, Benjamin Franklin, from the 18th century, according to the time of invention, as depicted in Fig. 1. The application was implemented as a standalone VR application to run on the Meta Quest platform, at a device refresh rate of 72 Hz. As suggested in [13], to compensate for the absence of haptic cues, visual feedback was implemented by modifying the colour of the bowl with which the user interacts, as shown in the VR application screenshot in Fig. 5. Furthermore, throughout the experience the user can initiate animated elements which were made with Blender and Cartoon Animator 5.

Auditory Feedback
The audio is delivered through the built-in headphones of the Meta Quest 2 headset. This decision was made to minimise the need for additional hardware within the museum setting. The immersive soundscape includes atmospheric sounds such as the wind outside and the crackling of the fireplace. The auditory feedback for the instrument was synthesised using a physical model of a glass

Fig. 4. Illustration of the interface with notes and corresponding frequency displayed in Hz.

harmonica, acquired from the FAUST Physical Modelling Toolkit, which was compiled and exported as a Unity audio plugin. The physical model produces the sound of playing glass cylinders of different sizes using a technique called banded waveguides as described in [7]. Using the *FaustVariableController* script, the parameters can be modified either in the Unity inspector or by scripting using *getters* and *setters*. The physical model exposes the following parameters:

- Basic Parameters (frequency, gain: 0.8)
- Physical Parameters (base gain: 1.0, bow pressure, excitation selector: 0, integration constant)
- Nonlinear Filter Parameters (modulation frequency, nonlinearity)
- Reverberation (reverb gain: 0.9, room size: 2)
- Spat (pan angle: 0.5, spatial width: 0.5)

The frequency of each bowl can be seen in Fig. 4. Unfortunately, it was not possible to measure the frequency from the original instrument, the frequency used were therefore estimated based on samples from other glass harmonicas. For this project, the user-controlled parameter was *bow pressure*, the parameter in the physical model related to applied force, when contact was initiated with the bowl. Addressing the challenge of creating meaningful mappings of movement to sound, as mentioned in [11], while the user kept finger contact with the bowl, the bow pressure would be mapped to a constant of 0.85. This value was estimated listening to recordings of other glass harmonicas, as the attack time, determined by the physical model, was perceived to be too long in the lower end bow pressure values. By setting a higher constant bow pressure, the sound would intrude more rapidly, resulting in a closer resemblance to the real instrument. Furthermore, adjusting the parameters of *room size* and *reverb gain* parameters to their maximum value was found to attenuate the sound more authentically, when the user lifts the finger off the glass bowl. The remaining parameters were similarly estimated by instrument recordings.

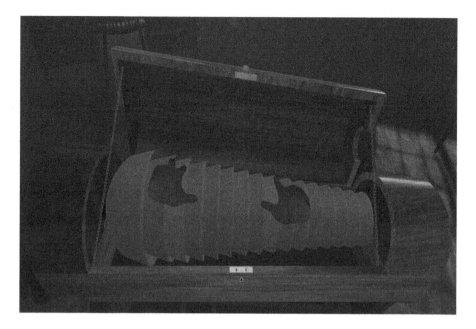

Fig. 5. Visual feedback when player touches the glass bowls.

3.2   Narrative Elements

The narrative elements of the experience extend beyond a historically relevant VE, but also include a storyline narrated by an animated portrait of Benjamin Franklin, as shown in the VR application screenshots in Fig. 6. The storytelling component provides information about the origin, popularity, and causes of the eventual decline of the instrument, accompanied by audio clips featuring Mozart's "Adagio for Glass Harmonica" and other relevant sounds that align with the narrative content (applause from audience). Additionally, animated figures are incorporated to further enhance the storytelling experience, such as ghosts emerging from the bowls, as the instrument was viewed as haunted by many. The narrator not only provides context for the user, but also offers instructions, prompts them to engage with the story, compliments their playing, and serves as a guide throughout the experience. Within the VE, there are two labelled buttons located near the instrument (see Fig. 6), which allow the user to initiate the storytelling or a visual tutorial that demonstrates the interaction using animated hands similar to Fig. 5. The tutorial element enables visitors to have a successful experience without requiring the assistance of museum personnel by providing clear guidance for independent navigation and interaction with the instrument.

Fig. 6. Left: button activating the narrative and tutorial. Right: talking portrait of Benjamin Franklin

4 Methodology

This section provides an overview of the approach and strategies employed during the installation development. Collaborative design, known as co-design, was chosen as the primary method, involving active collaboration with museum personnel. The following steps outline the strategies employed throughout the process:

1. **Initial consultations:** Meetings were held with museum personnel to understand their requirements, objectives, target audience, and physical space considerations for the installation.
2. **Ideation and prototyping:** The design of the virtual space was conceptualised through brainstorming sessions and drawing inspiration from source material suggested by museum personnel. Their valuable input contributed to the refinement of ideas.
3. **Informal testing:** The prototypes were subjected to informal testing at Aalborg University, Copenhagen, to collect feedback and identify potential usability flaws. Adjustments were made based on feedback and discussions were held with museum collaborators for significant design changes.
4. **Implementation and evaluation:** The final design was implemented and installed at the museum. The implementation mainly took place at Aalborg University, Copenhagen, while the surrounding components, such as the wall-mounted screen, were installed by personnel from the museum. The installation was evaluated through observations and interviews with museum guests. The museum personnel would record, summarize, and document the responses in an Excel spreadsheet.

5   Evaluation

The VR installation was established within the exhibition hall as a permanent component of their exhibit. The evaluation was carried out by three museum professionals who briefly interviewed visitors during museum opening hours, covering three days. There were 22 participants between the ages of 6 and 78 (Mdn 31.0, SD 23.9). Unfortunately, the gender of the participants was not recorded. They had visited museums between 0 and 8 times a month and 84% had previous knowledge of The Danish Music Museum before visiting that day. The participants were selected based on their voluntary interaction with the installation. During the interactions, museum staff observed participants, noting noteworthy behaviors. Subsequently, participants were invited to engage in brief interviews, during which their responses were recorded, summarized, and documented in an Excel spreadsheet. The following questions were defined and asked by museum personnel:

1. What words can describe your experience?
2. To what extent were you motivated throughout the experience?
3. What did you think was good/bad?
4. What do you think was difficult/easy/too easy?
5. Did you experience the involvement of multiple senses during the experience?
6. Did you experience the installation alone or with others?
7. Did you perceive a connection between the object in the display case and the experience?
8. Would you expect there to be other similar installations in the exhibition?
9. We have the hypothesis that one can get closer to objects by conveying them in new ways. What do you think about this?
10. If you had to give us one piece of advice the next time we build an installation at the museum, what would it be?
11. Is there anything else you would like to tell us?

5.1   Observations

Participants who interacted with the VR installation were observed to spend up to 10 min (Mdn 4.0, SD 1.9). Though the time spent was recorded by the museum staff and can therefore vary in reliability. Initially, most participants showed hesitancy. However, as time progressed, they became more relaxed and began to enjoy the experience; expressed by laughter and smiles. Most of the time, multiple individuals (up to five), often families, would participate in the experience together, actively helping each other with the HMD and interactions. Bystanders would also gather around, observing the screen and engaging in conversations with the person immersed in VR, and would frequently capture photographs of each other. While immersed, participants would exclaim words such as: "That is fun" and "Really cool" and encourage each other to try. Additionally, it was observed that younger participants would verbally respond to the narrator in the VE. However, it was observed that children would frequently playfully touch the fingers of the immersed participants, this would startle the people wearing the HMD and disrupt hand tracking, resulting in a poorer experience.

5.2 Interviews

Looking at the responses related Question 2, 68.2% explicitly stated that they where motivated or *highly* motivated throughout the whole experience. This is reflected in the response of most participants to Question 1, describing the experience as *fun*, *impressive* and *different*. When asked about difficulty (Question 4), most participants found the interaction to be easy. This was supported by observations, as even the youngest participants had a successful time playing and enjoying themselves. Concerning Question 3, about what was good/bad, some participants expressed they did not succeed in creating sound, leading to frustration and feeling "impatient". Some participants were likewise frustrated due to technical errors, particularly one of the buttons not responding correctly during the initial day of testing. This was fixed after the first day.

The social aspect of the experience was evident as nearly all participants answered that they experienced the installation with others (Question 6). As mentioned, some visitors would touch the hands of the immersed person, disrupting the tracking. One participant explicitly mentioned this in the interview, describing it as their "hands flying around", while another participant found it very uncomfortable. The challenges with hand tracking were exacerbated by the installation's location in a heavily trafficked walkway, leading to instances of tracking loss. Participants experienced this as their virtual hands got "stuck" or "poorly calibrated", expressing that they would want improved hand tracking (Question 10).

When asked Question 7 about the perceived connection between the installation and the display case, the majority (73.6%) of the participants confirmed a clear connection between the real glass harmonica and the interactive installation upon exiting the virtual experience. This is supported by observation, as participants would approach the real instrument and demonstrate how it should be played, using the correct gestures they had learnt in the VE. When asked Question 9, all participants similarly expressed that the installation brought them closer to the instrument; expressed as "I absolutely think so. Touching things creates understanding and connection", and "Yes, one gets a sense of what it would do", and "Many of the instruments hanging here are dead, so they become accessible again here (in VR)".

6   Discussion

This section will present a comprehensive analysis of several components of the glass harmonica VR installation project. These components include technical issues, overall impression, and the connection established between the virtual representation and the displayed glass harmonica.

First, it should be noted that staff familiarity and expertise with the technology are important factors when implementing and evaluating VR for museums, as emphasised in [24]. In the case of The Danish Music Museum, the staff were already used to working with VR, and were responsible for setting up and closing down the application. Furthermore, as the limited sample size hinders the ability to establish trends, the findings merely provide suggestions, as drawing conclusions would necessitate a larger number of participants. The evaluation questions were formulated by the museum team, and it is evident that questions 7 and 9, in particular, are phrased in a manner that may elicit biased responses.

Although the interactive experience was novel for participants, some exhibited familiarity with VR technology. Consequently, certain expectations arose, such as the desire to stand up and move around within the virtual environment. A technical solution was therefore implemented, having the VR view be occluded with a sign to sit back down if the HMD y-position exceeded a specified offset from the floor. Furthermore, physical movement in the HMD xz-plane was mitigated by utilizing the pass-through guardian on the device. In addition, it is important to note that the orientation of the VE is established the moment participants mount the HMD, necessitating them to sit down and face forward to ensure the accurate positioning of the virtual instrument. Exploring technical solutions to enhance system predictability and reliability would be valuable in addressing the absence of a reset-orientation button on the Touch controllers, thus reducing the potential for user errors. Currently the solution implemented after testing was to install a sign telling people to sit down before mounting the HMD.

Although hand tracking seemed to generally function well, the heavily trafficked location and movement in front of the cameras on the HMD led to occasional tracking loss, rendering interaction impossible. Experiencing these errors had a detrimental effect on the overall experience of the VRMI, in correspondence with [23]. To combat this issue, markings on the floor could indicate the virtual space to discourage bystander interference, though completely negating the issue is difficult with hand tracking.

According to the concepts discussed in [18], the social experience is important in museum settings. The findings underline the importance of implementing the live stream from the HMD to facilitate cooperative exploration and potentially mitigate feelings of impatience. Furthermore, this observation indicates the relevance of the asymmetric interaction between HMD wearers and bystanders in museum settings.

The VR installation received positive feedback and prompted suggestions for including additional instruments. During the interviews, the participants expressed significant importance in the storytelling aspect of the experience, highlighting its role in creating an enjoyable experience and effectively guiding them throughout. The connection between the virtual and the exhibited instrument was evident among many participants, as they engaged in gestures that mimicked playing while describing the instrument to their peers, all while recounting the narrative from the virtual environment. This observation supports the decision to prioritise faithful replication of the interaction in the context of musical heritage and the preservation power of VR. However, it should be noted that some younger participants, in particular, tended to resort to dynamic movements resembling striking piano keys if they were unable to produce sound within a short period of time. Therefore, the slow nature of the interaction may not be suitable for visitors who prefer more dynamic and active experiences; such as replicating traditional percussion instrument, where the player strikes the instrument.

The choice of VR as the technical solution suggested by the museum staff can be supported by looking at the responses for Question 5, as around half of the participants (47.0%) explicitly stated that they believed that they experienced a sense of touch, despite the absence of haptic stimuli. They acknowledged this perceptual illusion, as they were aware of the lack of haptic feedback, but the other sensory modalities contributed to the perception of touch. The phenomenon of pseudo-haptic illusion, which is evoked by the use of cross-modal illusions, has been subject to comprehensive investigation in prior research [6]. Although it did not constitute the central area of emphasis for this project, it aligns with the museum staffs conception that VR could be a technical solution for the lack of touch in museum setting. While VR offers a captivating and immersive experience, it's essential to acknowledge its limitations. The unreliability of the current hand-tracking, absence of haptic feedback and the potential for isolation from other visitors are concerns. Alternative solutions, such as touch screens, may provide better accuracy and ease of use, with the added advantage of haptic feedback. While VR undoubtedly elicits a 'wow' effect, it's important to recognize that this effect can be short-lived. A critical view on the technology employed in this project is not only valuable but also necessary for future enhancements and to ensure an inclusive and engaging experience for all museum visitors.

7   Conclusion

In conclusion, this paper has discussed a project aimed at virtually resurrecting and presenting the long-overlooked Glass Harmonica instrument from the collection of The Danish Music Museum. A VR installation that provides visitors with an understanding of the sound, interaction, and history of the instrument, through co-design. This paper provides a comprehensive view of the development and execution of the project. The evaluation of the virtual glass harmonica was carried out at the museum, which provided valuable insights into the user experience of the VR installation. However, it is important to acknowledge the limitations stemming from the small sample size and the occasional incomplete responses from participants. These factors underscore the need for future research to expand upon our findings.

The results indicates that the VR installation was engaging and enjoyable for the majority of visitors, as the participants smiled and laughed, as well as positive statements recorded in the interviews. Nevertheless, the experience wasn't without challenges, as disruptions caused by bystanders led to a loss of hand-tracking functionality for the immersed participant. From observations, we acknowledge the significance of taking into account the social dimension of museum visits. Consequently, future iterations of this project will incorporate the live-streaming feature, aiming to enhance the connection between VR users and onlookers.

Nonetheless, the project successfully established a connection between the virtual instrument and the glass harmonica in the display case, for the majority of the participants. By employing faithful interaction replication, acoustics, and storytelling, this project proposes a VR experience, bringing a historical instrument back to life, and letting visitors at The Danish Music Museum touch the untouchable.

Acknowledgements. We extend our gratitude to Signe Beckmann for her valuable contribution to the design of the visual installation. We also express our appreciation to the museum personnel for their support and assistance throughout the project. The authors thank The Augustinus Foundation for funding the project presented in this paper.

References

1. Atherton, J., Wang, G.: Doing vs. being: a philosophy of design for artful vr. J. New Music Res. **49**(1), 35–59 (2020)
2. Bellia, A.: The virtual reconstruction of an ancient musical instrument: the aulos of selinus. In: 2015 Digital Heritage. vol. 1, pp. 55–58. IEEE (2015)
3. Berthaut, F.: 3d interaction techniques for musical expression. J. New Music Res. **49**(1), 60–72 (2020)
4. Büyüksalih, G., Kan, T., Özkan, G.E., Meriç, M., Isın, L., Kersten, T.P.: Preserving the knowledge of the past through virtual visits: From 3d laser scanning to virtual reality visualisation at the istanbul çatalca inceğiz caves. PFG-J. Photogramm., Remote Sens. Geoinform. Sci. **88**(2), 133–146 (2020)
5. Chong, H.T., Lim, C.K., Rafi, A., Tan, K.L., Mokhtar, M.: Comprehensive systematic review on virtual reality for cultural heritage practices: coherent taxonomy and motivations. Multimedia Systems, pp. 1–16 (2022)
6. Collins, K., Kapralos, B.: Pseudo-haptics: leveraging cross-modal perception in virtual environments. Senses Society **14**(3), 313–329 (2019)
7. Essl, G., Serafin, S., Cook, P.R., Smith, J.O.: Musical applications of banded waveguides. Comput. Music. J. **28**(1), 51–63 (2004)
8. Falk, J.H.: Identity and the museum visitor experience. Routledge (2016)
9. Gaitatzes, A., Christopoulos, D., Roussou, M.: Reviving the past: cultural heritage meets virtual reality. In: Proceedings of the 2001 Conference on Virtual Reality, Archeology, and Cultural Heritage, pp. 103–110 (2001)

10. Häkkilä, J., et al.: Visiting a virtual graveyard: designing virtual reality cultural heritage experiences. In: Proceedings of the 18th International Conference on Mobile and Ubiquitous Multimedia, pp. 1–4 (2019)

11. Holland, S., Mudd, T., Wilkie-McKenna, K., McPherson, A., Wanderley, M.M.: Understanding music interaction, and why it matters. New directions in music and human-computer interaction, pp. 1–20 (2019)

12. Jia, Q.: Application of Chinese traditional musical instruments. Sci. Social Res. **3**(1) (2021)

13. Jordà, S.: Interactive music systems for everyone: exploring visual feedback as a way for creating more intuitive, efficient and learnable instruments. In: En Proceedings of the Stockholm Music Acoustics Conference, August, pp. 6–9 (2003)

14. King, A.H.: The musical glasses and glass harmonica. In: Proceedings of the Royal Musical Association. vol. 72, pp. 97–122. Cambridge University Press (1945)

15. Lee, H., Jung, T.H., tom Dieck, M.C., Chung, N.: Experiencing immersive virtual reality in museums. Inform. Manage. **57**(5), 103229 (2020)

16. Lukasik, E.: Reconstructing musical instruments in mixed reality. In: New Trends in Audio and Video/Signal Processing Algorithms, Architectures, Arrangements, and Applications SPA 2008, pp. 201–206. IEEE (2008)

17. Mäki-Patola, T., Laitinen, J., Kanerva, A., Takala, T.: Experiments with virtual reality instruments. In: Proceedings of the 2005 Conference on New Interfaces for Musical Expression, pp. 11–16 (2005)

18. Parker, E., Saker, M.: Art museums and the incorporation of virtual reality: examining the impact of vr on spatial and social norms. Convergence **26**(5–6), 1159–1173 (2020)

19. Recupero, A., Talamo, A., Triberti, S., Modesti, C.: Bridging museum mission to visitors' experience: activity, meanings, interactions, technology. Front. Psychol. **10**, 2092 (2019)

20. Roda, A., DE POLI, G., Canazza, S., Sun, Z., Whiting, E.: 3d virtual reconstruction and sound simulation of old musical instruments. Archeologia e Calcolatori (1) (2021)

21. Rogers, K., Karaosmanoglu, S., Wolf, D., Steinicke, F., Nacke, L.E.: A best-fit framework and systematic review of asymmetric gameplay in multiplayer virtual reality games. Front. Virtual Reality **2**, 694660 (2021)

22. Rossing, T.D.: Acoustics of the glass harmonica. J. Acoust. Society America **95**(2), 1106–1111 (1994)

23. Serafin, S., Erkut, C., Kojs, J., Nilsson, N.C., Nordahl, R.: Virtual reality musical instruments: state of the art, design principles, and future directions. Comput. Music. J. **40**(3), 22–40 (2016)

24. Shehade, M., Stylianou-Lambert, T.: Virtual reality in museums: exploring the experiences of museum professionals. Appl. Sci. **10**(11), 4031 (2020)

25. Syukur, A., Andono, P.N., Hastuti, K., Syarif, A.M.: Immersive and challenging experiences through a virtual reality musical instruments game: an approach to gamelan preservation. J. Metaverse **3**(1), 34–42 (2023)

26. Theodoropoulos, A., Antoniou, A.: Vr games in cultural heritage: a systematic review of the emerging fields of virtual reality and culture games. Appl. Sci. **12**(17), 8476 (2022)

27. Trunfio, M., Lucia, M.D., Campana, S., Magnelli, A.: Innovating the cultural heritage museum service model through virtual reality and augmented reality: the effects on the overall visitor experience and satisfaction. J. Herit. Tour. **17**(1), 1–19 (2022)

28. Vermeeren, A.P.O.S., et al.: Future museum experience design: crowds, ecosystems and novel technologies. In: Vermeeren, A., Calvi, L., Sabiescu, A. (eds.) Museum Experience Design. SSCC, pp. 1–16. Springer, Cham (2018). https://doi.org/10.1007/978-3-319-58550-5_1

29. Villena Taranilla, R., Cózar-Gutiérrez, R., González-Calero, J.A., López Cirugeda, I.: Strolling through a city of the roman empire: an analysis of the potential of virtual reality to teach history in primary education. Interact. Learn. Environ. **30**(4), 608–618 (2022)

30. Wanderley, M.M., Depalle, P.: Gestural control of sound synthesis. Proc. IEEE **92**(4), 632–644 (2004)

31. Willemsen, S., Paisa, R., Serafin, S.: Resurrecting the tromba marina: A bowed virtual reality instrument using haptic feedback and accurate physical modelling. In: 17th Sound and Music Computing Conference, pp. 300–307. Axea sas/SMC Network (2020)

32. Wu, L., Su, W., Ye, S., Yu, R.: Digital museum for traditional culture showcase and interactive experience based on virtual reality. In: 2021 IEEE International Conference on Advances in Electrical Engineering and Computer Applications (AEECA), pp. 218–223. IEEE (2021)

33. Zouboula, N., Fokides, E., Tsolakidis, C., Vratsalis, C.: Virtual reality and museum: an educational application for museum education. Int. J. Emerg. Technol. Learn. (iJET) **3**(2008) (2008)

Technology as a Means of Musical and Artistic Expression: A Comparative Study of Nono's Prometheus and Pink Floyd's Concert in Venice in the 1980s

Elena Partesotti[✉]

University of Campinas (UNICAMP), NICS, Campinas, Brazil
eparteso@unicamp.br

Abstract. This study delves into the integration of technology into musical art, focusing on Luigi Nono's Prometheus and Pink Floyd's live performance in Venice. Despite their differences, both events represent pivotal moments in musical evolution, necessitating detailed analysis. This article explores the intersection of embodied cognition, music, and audience experience in Nono's opera Prometeo and Pink Floyd's concert, revealing the fascinating integration of technology into music. Nono's Prometheus, performed in 1984, and Pink Floyd's 1989 concert in Venice showcase distinct yet intertwined paths in the realm of live electronic music. This analysis considers the innovative approaches of both events, highlighting their impact on musical expression and the audience's perception. By exploring these iconic performances, this study sheds light on the intricate relationship between technology, music, and the human experience, illustrating the profound implications of their fusion for the future of musical creativity and audience engagement.

Keywords: Luigi Nono · live electronics · Venice · Pink Floyd · Renzo Piano

1 Introduction

This study compares some of the musicological and technical aspects of two live musical events, one of classical and the other of popular music, that took place in Venice, Italy, in the 1980's. These two very different cultural events revealed some interesting similarities concerning the way in which they were devised and implemented. Furthermore, Nono's *Prometeo* serves as a prime example of embodied cognition, merging mind and body through innovative sound spatialisation, and similar embodied experience is found in Pink Floyd's live concert. The aim of the present research was to focus on the use that was made of technology for the two concerts, and the innovative thinking behind said use.

Conducting this type of research has brought to light an important feature of every musical and cultural phenomenon, i.e. the existence of a link between the strategies for

A. L. Brooks (Ed.): ArtsIT 2023, LNICST 565, pp. 234–250, 2024.
https://doi.org/10.1007/978-3-031-55312-7_17

using technology and the most often studied aspects of this phenomenon, aspects relating to the set of technologies acquired over the course of time and to the traditional approach to the concert, all episodes that play a part in the history of music. These situations can also give rise to a phenomenon of custom that profoundly influences the history and traditions of a culture, and of musical culture in our case.

The aim of this study, thus, was to focus on the technological aspects of this phenomenon, which is often disregarded due to ignorance or superficiality. I have also sought to combine marginal and ephemeral notions into a single document that could be consulted and revised by other scholars. From the material obtained for the analysis, there also emerged an aspect of the modernity of such research, and a consequent awareness of the experimental nature of this work. I consequently make no claim to have exhausted the field of investigation or even the material available for analysis, but merely to have sketched out a preliminary model for future, more thorough studies. The study method chosen for this purpose was based on seeking and filing as much information as possible, drawn from academic research, specialist and non-specialist journals, interviews, and website. Moreover, in order to keep the technical perspective as objective and up-to-date as possible, an effort was made to track down the largest possible number of opinions and different kinds of sources, wherever the situations permitted. As a consequence, although a huge quantity of material was collected, some information had to be rejected if the sources could not be considered entirely reliable and objective.

The events considered belong to two clearly distinct musical categories: the so-called classical music and popular music. Luigi Nono with his *Prometheus* ("*Prometeo, tragedia dell'ascolto*") and the Pink Floyd with a concert that was part of their *Another Lapse tour* produced two extremely different musical events, given the type of music and the type of audience involved, but what they had in common was a highly sophisticated use of technology and an impact on the city of Venice, where both these events took place.

Countless studies and discussions have focused on the distinction between classical music and popular music, but - given the character of this investigation, which aims mainly to further analyse the technological aspects of musical expression - my attention was concentrated mainly on the use made for these two sound events of the same electronic technology, albeit without neglecting an issue that emerged from an analysis of the material considered, i.e. the importance of the different musical concepts that lay behind their use of this technology. Luigi Nono's work breaks away from the convention that requires a frontal listening experience, with the audience placed in a separate area from the musicians. He represented his Prometheus inside a church, placing the audience at the center of this world of sound coming from a specifically-designed wooden construction occupied by the musicians. Though it may not seem so at first glance, this was done to facilitate an open, multidirectional listening experience. The Pink Floyd's performance, on the other hand, relied essentially on the conventional idea of a stage placed in the open air and paradoxically everything contributed to creating the usual frontal presentation of sound and lights. The artists took two opposite approaches: Luigi Nono wanted to combat the traditional, frontal representation, while the Pink Floyd aimed to exploit it to the full. Another aspect that emerged from our research is the contrast between Nono's principle of economy and a shared usage of the technological media, as opposed to the principle of accumulation and individual usage of the technological media adopted by the Pink Floyd.

Nono was committed to using only a few technological functions but to exploit all their possible combinations; the Pink Floyd pursued a different objective, fully exploiting all the possibilities deriving from the juxtaposition of these technologies. As for the spatialisation of the sound, Luigi Nono and the Pink Floyd both made use of high-end technologies, obliging each of their listeners to choose their own individual sound space, by means of the Halaphon (Nono), or by reinforcing the mass frontal listening modality with principles drawn from quadraphonics (Pink Floyd).

Both the concerts were criticised at length in the newspapers and were a source of debate and sometimes of political discussions: Nono was criticised for having created a costly work that was difficult to understand, the Pink Floyd were accused of attracting a crowd of 200,000 spectators, who were guilty of damaging and polluting the city.

2   New Sonorities

Around the 1950s, there arose a compelling need to merge musical expression with the burgeoning scientific-technological realm. This synergy gave birth to live electronic music, a genre characterised by improvisation and constant evolution. Not confined to classical music, this genre found its footing both in classical and popular music domains, each enriching it with distinctive features. These live electronic performances, driven by impromptu technical choices, depended on the dynamic interaction of space, creating unique and ever-changing auditory experience.

Among musicians and composers of this period was born the idea of performing real musical workshops, where the combination of electronics and music could create splendid stylistic innovations. Only in the year 1955, the Studio di Fonologia della RAI established in Italy, more precisely in Milan. The Studio aimed to address and exploit electronic resources as well as the processing of recorded sounds. Many musicians operated together on his project; Luigi Nono - who embarks on his path towards live electronics- was one of them.

Live electronic music gradually paved the way to a growth in the number of live concerts, because this is a genre closely linked to the use of improvisation, or the so-called work in progress [1]; it is thus different from traditional instrumental music [2]. By live electronics I mean a set of technologies and electronic instruments that enable sounds to be processed and controlled starting from live acoustic sources. This tool (live electronics) has been used in two ways, i.e. both to expand the possibilities of instrumental music and to serve as an electronic instrument per se. Live electronics are in themselves a form of work in progress, or real-time elaboration and creation, which is why it they demand the presence of a musician and a composer; and both these figures interact with what they hear.

2.1   Live Electronics: Bridging Classical and Popular Music Through Sonic Innovation and Spatial Exploration

Live electronics is not an instrument that belongs to classical music; it is usually found in the live concerts of popular music groups. In classical music, we can see that there is a composer in the role of conductor of the orchestra and/or sound director, while in popular music the musician-composer is also the director of his own sounds at one and the same time.

Despite the existence of a precise program for its execution, a live performance using live electronics will be different every time because of the impromptu technical choices that are made in each case. One of the indispensable elements enabling this variability is the use of space. The attitude of a composer required to choose the space in which to represent a work may go in two directions: he or she may opt for a given place because it has certain special characteristics, or decide to change some of the parameters of the concert and the instruments involved to adapt them to a place. Either way, it is of fundamental importance to consider the space in order to fully exploit the sound effects.

Already by the end of 1970, live electronics was no longer considered a novelty and in the musical world it was consequently occupying a space in both classical and popular music, but with different characteristics.

In the context of classical music, Luigi Nono performed his Prometheus already in 1984, while in the world of popular music, the Pink Floyd brought a live show to Italy in 1989. Both held their concerts live in Venice during the 1980s, using live electronics for different purposes. These artists also have another aspect in common that emphasises their affinity, and that is the need to experiment, to seek new sonorities and a new music that does not necessarily have to appeal to the audience, but that can make new inroads into the world of music.

Their performances both immediately aroused a great deal of debate in the artistic world of the time, and some criticism from various quarters. In the end, however, these artists came to occupy a place in history for the grandiosity of their projects.

3 The Research Behind Prometheus

Nono's Prometheus transcended conventional boundaries, embodying an innovative and ongoing critical analysis of musical language. This masterpiece, unshackled from traditional theatrical constraints, presented a revolutionary fusion of sound and color, captivating audiences with its visceral impact. At the heart of Prometheus was a visionary partnership between Nono and the Institut de Recherche et Coordination Acoustique/Musique (IRCAM), enabling the exploration of live electronic music. Through intricate band-pass filters and inventive spatialisation techniques like the Halaphon, Prometheus pushed the boundaries of live electronic manipulation, immersing listeners in a sonic universe. Nono gave his first representation of Prometheus in 1984 at the church of San Lorenzo in Venice. The character of Prometheus represents a constant search that leads to the character finding, fixing and transgressing the limits; he is the incarnation of restlessness and curiosity about what is unknown and "something else" [1]. Nono combined this with his own personal need to conduct an ongoing, innovative critical analysis of the language, in order to be able to "Cross the threshold and wander beyond the doors opened up by modern technology, to draw on new perceptions"[1] p.367 [3]. But it also represents the "tragedy of listening" because it means "constantly seeking all these secrets of sound

---

[1] "Varcare e vagabondare oltre le porte aperte anche dalla tecnologia d'oggi, per attingere nuove percezioni." [3].

and space" p. 335 [4], a search that Nono conducted meticulously, studying all the spaces of the churches in Venice - a city for multidirectional listening - arriving at the church of San Lorenzo that was to host the premier of a historical event. Nono saw Venice as an acoustic environment suitable for transmitting sounds, a city whose waters and canals carried the sound waves.

For Luigi Nono, space was the essential element of the compositional process: in composing a work, every artist should imagine it set in a given space in order to bring it to life; the space thus becomes an instrument in itself, a voice, "[..] an integral part of the musical event" p. 494 [5]. It is because of this innovative use that he makes of space that it is impossible to establish a score suitable for performing anywhere, and his work is continuously transformed in relation to the architectural space where it is to be performed. At San Lorenzo, Luigi Nono sought to identify the various opportunities that the spaces in the church gave him to change and manipulate his score in a process typical of live electronics, a topic that he had studied in depth in Freiburg.

The musical production was arranged into four orchestral groups: soloists, choir, instrumentalists and live electronics, which managed 24 loudspeakers [5]. These components were placed in Renzo Piano's Ark, a wooden structure organised vertically and horizontally, where the movement of the sound spreads both internally and externally. As Prof. Vidolin explained "[..] live electronics [...] takes these natural sounds [..] and manipulates them, treating them and transmitting them in a continuous movement, or on different planes of sound, outside the space for listening to them [..]. Sometimes there were even synthetic sounds under the floor of the auditorium, sometimes in the middle of the hall" p. 379 [6].

In Prometheus, the use of light and color also takes on a function related to the sound [7]. As the composer himself stated "the stage box cancels the static nature of the visual element [...] precisely for a different use of the auditory element, [...] with new acoustic diffusion techniques" p. 210–15 [8].

Nono uses color as a tool closely connected to the music, but without allowing it to override the sound or misrepresent to the ear what it is hearing. Quite the reverse, the color has to further investigate the force of the listening experience, so it must not distract the audience. That is why in Prometheus he uses a cold, inconsistent light, that is non- descriptive. The movements of the light must rely on very slow changes of shade, taking as long as a quarter of an hour to go from black to grey and proceeding independently, not following the music. The person chosen as lighting director was Emilio Vedova, who - being a painter - was a great expert on colours. The initial plan for Luigi Nono's "Prometheus" included a color projection inspired by the theories of Goethe, Runge, Itten, and Kandinsky. However, this idea was abandoned during rehearsals, leaving Emilio Vedova with the sole responsibility of managing the lighting. At this point, Nono was resolute in his vision: "Prometheus" was not to be an opera, but a sound-based tragedy interacting with two distinct spaces, devoid of any staged or visual elements [9].

Nono's color scheme, which is well-documented in sketches and interviews, involved thirteen colours, each with twelve different shades. Unfortunately, Nono's annotated edition of Goethe's "Farbenlehre" has been lost.

The concept was to navigate through various color shades and use them, along with the music, to 'explore the space'. This idea is mentioned in a letter from Nono to Piano dated December 6, 1983 [9].

4 Luigi Nono's Innovation

Prometheus, devoid of traditional theatrical embellishments. Rooted in ancient mythological themes, Prometheus redefined the essence of listening, challenging established norms with its avant-garde brilliance. Nono's intricate use of live electronic music involved detailed phases of electronic treatment facilitated by band-pass filters and cutting-edge spatialisation techniques. However, beneath its technical complexity, Prometheus remained one of his most important work, transcending the confines of traditional music and ushering audiences into an immersive auditory universe. Nono produced Prometheus with an unconventional theatricality that already existed in his previous works ('Intolleranza', 'Al gran sole carico d'amore') and he came to reject the visual dimension apart from the use of color as an introspective guide for the audience.There is no libretto for his work: the text is actually a synthesis of different texts in Italian, German and Ancient Greek on the legend of Prometheus. The text draws from "Aeschylus, Pindar, and Hesiod". Prometheus is the bearer of fire, seen by some as the last of the old gods battling against the young gods, symbolising the contrast between the conservative and the innovative way of thinking. Nono's work therefore aims to stimulate innovation, an innovation that includes the use of the new electronic technologies in a construction designed exclusively for the audience's introspective listening experience. Prometheus is not acted theatrically, with actions taking place on stage before an audience, nor is it based on any particular story; these are the characteristics that make Prometheus the most important of Nono's works. The eyes make way for the human ear to take in a polyphonic work that deals with the mythical past, present and future; a work capable of creating a new sound spatiality around the audience with fragmentary sounds that are sometimes disconnected from the original text.

In an interview Luigi Nono explained how the projects relating to the visual plane of Prometheus were born: Cacciari saw the opera as an archipelago of numerous islands; and Renzo Piano designed these islands suspended in space [10]. To show how to navigate between them, there would therefore be a path of coloured light projected onto the walls and onto the audience. Initially, the interest in the sound/color combination visually represented within the opera attracted attention and produced great results. Despite the originality of the first Prometheus project, Nono soon abandoned these ideas, however, driven by the need to make more space for listening - a need that Luigi Nono called an anti-visualistic syndrome pp. 70–71 [10].

4.1 A Massive Sound Box

In the context of Prometheus, the collaboration with Italian philosopher and writer Massimo Cacciari holds significance. Massimo Cacciari, renowned for his philosophical insights and interpretations of culture, provided a fundamental intellectual contribution to Luigi Nono's work. His philosophical vision permeated the conceptualisation of Prometheus, enriching it with profound meanings and philosophical implications. Through his collaboration with Nono, Cacciari contributed to shaping the theoretical and conceptual framework of the opera, offering a unique interpretation of mythological themes and the exploration of sound and space. The partnership between Nono and Cacciari created an interdisciplinary dialogue that further enriched the work, transforming Prometheus into an artistic creation imbued with philosophical depth and human sensitivity. As Cacciari claims: "We needed a place for this particular listening experience. An instrument that would make it possible, that [..] would detach the listening from the "seeing" normally dominant in concert halls and opera houses. [..] A place where the listening is not distracted and can reflect on its own origins" p. 21 [11].

This is how the philosopher defined the importance of constructing an instrument designed to contain the whole event. The architect Renzo Piano was employed to build what many nicknamed the "Ark", but that he himself defined as "a massive sound box" p. 59 [13]. The construction stemmed from the need to simulate a sort of archipelago, where there were five (musical) islands. When you are on an island forming part of an archipelago and you gaze all around you, you will never be able to embrace the whole system because the human visual field is limited. Instead, you will be able to perceive and imagine sounds that are behind you [12, 13]. That is how the Ark was born, like an archipelago with the audience at its center, surrounded by musical islands that they can never see all at once, though they can perceive their presence thanks to the music.

Piano was inspired by the idea of an uncompleted wooden boat, in which he had built a "sound box" more than 3 m above the floor. The construction consisted of four scaffolding walls extending parallel and perpendicular to the walls of the church, on the sides of which there was a system with three levels of overlapping walkways, on which the musicians and soloists move about. Being placed on different heights, the three levels were connected by stairways. In various parts of the construction screens were also installed, by means of which the director Claudio Abbado and the deputy director Roberto Cecconi simultaneously directed the instrumentalists. Prometheus gave priority to the spatial nature of listening, which is why the 400 members of the audience were accommodated in the middle of the Ark, on chairs that could turn both clockwise and anticlockwise to some degree, and tilt forwards and backwards: the audience was thus given an instrument for seeking out the sound, in a search that is the characterising element of the opera, and that makes the audience at one with its protagonist, Prometheus.

4.2 Live Electronics in Prometheus

"[..] with live electronics I think the listener has to have a much more active role. The live electronics system broadcasts composite acoustic signals in the hall, but these vagabond sounds vary in quality, they are transformed and combined, and they must also be connected together by the listeners, instead of simply passing through them" p. 83 [14]. This consideration of Nono's explains why he used live electronics. In Prometheus, the listeners are not just given a composition, they themselves are cast in the composer's role to some degree, immersed in the space of the sounds. The opera involves the use of the 4i system: this is a computer-based for sound synthesising system that stemmed from co-operation between the CSC (Centro di Sonologia Computazionale) in Padova, the LIMB (Laboratorio per l'Informatica Musicale della Biennale) in Venice and the IRCAM (Istituto di Ricerca e Coordinazione Acustico/Musicale) in Paris.

The system comprises a 4i numerical processor, a PDP-11/34 digital processor that handles the controls process, and a console of manual inputs suitable for live performances (Fig. 1).

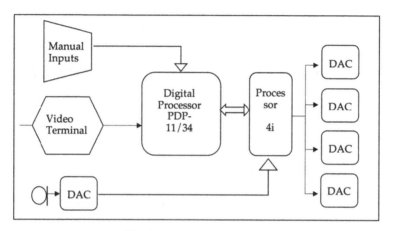

Fig. 1. 4i system and its parts.

The 4i numerical processor includes: 64 oscillators, 32 envelope generators, 96 oscillators, 32 timers, 150 adders, a 64K memory for the sampled waveforms, which can contain up to 16 different functions of any type, and 4 output channels each with 90 dB dynamics. All these units can be interconnected with one another in an infinite variety of ways to obtain different sound processing and synthesising outcomes. So we can apply additive technology (obtained by connecting several oscillators in parallel), amplitude modulation (by multiplying two or more waveforms), frequency modulation (by connecting 2 or more oscillators in series), non-linear distortion (by distorting a periodic wave, which is usually a sinusoid, with a polynomial frequency), subtractive technology (obtained by connecting several filters, i.e. linear combinations of delayed sums and multiplications, in series and/or in parallel), and reverberation (obtained by delaying lines, exploiting the memory of the functions).

In an interview with [15], Nono said that the 4i was used to produce synthetic sounds where live electronics was unable to do so, and these sounds were combined in real time with the sounds that live electronics then converted to obtain a mobile sound [15], in which the two signals mingled and combined with one another. Using the functions of the 4i also made it possible to reach levels of pianissimo as low as 0.50 dB, below the threshold perceived by the human ear, and the same applied to the high notes.

Live electronics involves electronically treating "natural" sounds and consists in listening to an original instrumental sound through loudspeakers; this sound can then be mixed with the electronically modified version of that instrumental sound, so it is not the musician who manipulates the sound. All the equipment used to treat sound electronically (microphones, loudspeakers, and live electronics per se) are considered as instruments and they consequently occupy a visible place in the concert hall; the sound technicians and musicians also co-operate as peers.

The electronic treatment of sound consists of three phases:

(1) Sound conversion

Ring modulator - Harmonizer (audiocomputer) - Vocoder (voice coding system).

In the harmonizer (Fig. 2) the natural signal can be shifted one octave higher and two octaves lower; it also has a "reverse" function. Nono used such a transposition to create micro-intervals in two female soprano voices that would have been impossible to achieve live. The input signal was converted from an analogical into a digital signal so that the data could be processed numerically before it was converted back from digital to analogical to make the sound audible.

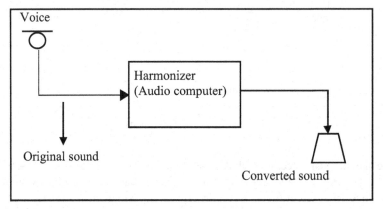

Fig. 2. Representation of the Harmonizer in Prometheus that varies the pitch of the sound without changing its duration and transposes the high soprano, the male choir and the bass flute.

(2) Sound selection

Only a certain field of sound frequencies is selected from a given sound by means of special third-, second- and fifth-order filter banks. It is worth noting that it is important at this point to select which sound to "enter" and improve. We can obtain a filter bank by aligning several pass-band filters and, depending on the filters' range of action, we can speak of second-, fifth- or third-order filters.

(3) Gate

The opening of the gate is governed by and directly proportional to the intensity of the microphone pulse, and the loudspeaker can be used to hear the live or converted sound of the instrument (Fig. 3).

The microphone becomes an instrument by means of which musicians can interact with one another and, at the same time, they can interact with their own sound.

A Halaphon creates a spatialisation and a continuous dynamicity of the sound in space by means of the loudspeakers (Fig. 4).

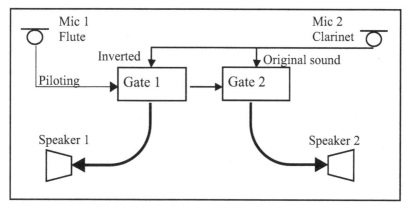

Fig. 3. Representation of the Halaphon employed in the Prometheus

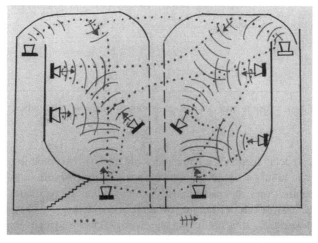

Fig. 4. Reproduction of Nono's draft-memo on the spatial movement of the sound of Prometheus, at the church of San Lorenzo[2].

The electronic instrument is therefore capable of interacting harmoniously with the acoustic instrument.

A great deal of importance is also attributed to the sound engineer (or sound technician), who - on a par with the director of the orchestra - must adjust and correctly articulate the sound sources within the space.

The section dedicated to *Hölderlin* is an example of the fascinating use of live electronics in Prometheus.

The instruments used are two sopranos, a double bass clarinet, a bass flute, and two reciting voices. In this piece, the voices of the two sopranos enter two different delay

---

[2] http://www.jia-tokai.org/archive/sibu/architect/2003/0310/ongaku.htm

circuits (*delay* with accumulation[3]) of 4" and 8" respectively, and subsequently undergo two different spatialisation processes calculated through the Halaphon (Fig. 5).

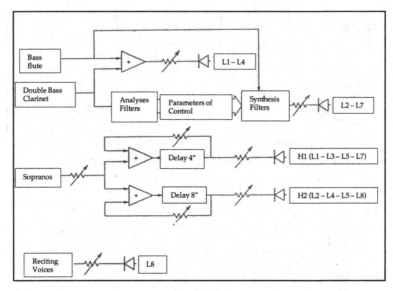

Fig. 5. Filtering bench for the analysis of signals transformed into pilot pulses of proportional size.

Halaphon and delay feedback are manually adjusted by two potentiometers: the first allows the user to adjust a sound through the speakers at two different speeds, while the feedback controls the "repetition density" of the signal; by regulating it, therefore, the increase/decrease of the overlapping of items is decided. The two spatialise voices follow two opposite paths thus producing an obsessive chorus that envelops the viewer. The sound of the double bass clarinet enters a first filter bench, where it is analysed. The filters of the second bench (the synthesis bench) are then subsequently opened through the pilot impulse coming from the analysis bench; the opening of them corresponds to the intensity of the pilot impulse that generates it.

This type of manipulation applied to the two instruments can be defined as cross treatment [6]. In it, the so-called pilot sound is analysed in real-time and one or more parameters are extracted from it to control the sound processed, coming from the analysis bench.

For the rest, the filter banks for analysis and synthesis are the same[4]; the linear connection that usually binds them is, however, reversed in this passage. The filters that operate in the low registers are associated with the high ones and vice versa. Consequently, the sound impulses coming from the contrabass clarinet will drive the sounds

---

[3] Repetition of the sound for a certain number of times.

[4] Two banks made up of 48 bandpass filters, the bandpass filter is a system that lets all frequencies between two values pass: the lower cutoff frequency (at low frequencies) and the upper cutoff frequency (at high frequencies).

of the bass flute, opening the filters that highlight the high harmonics of the flute sound. Finally, the two reciting voices are amplified.

5 Light Show and Psychedelics: Visual Perception as a Mnemonic Support

In the realm of music performance, Pink Floyd dominated, captivating audiences with their mesmerising light shows and psychedelic visual elements. This amalgamation of auditory and visual stimuli crafted a mnemonic landscape, channeling music into the minds of spectators. The synchronisation of lights with music, although not universally synonymous with their concerts, became an integral component of their live performances. Their innovative fusion of lights and sounds underscores a deliberate effort to enhance the listening experience, expertly blending technology and artistic expression. Indeed, right from their very first show, the Pink Floyd demonstrated an interest in and attention to their visual impact on their audience, as we can see from the covers of their albums and their video clips, where the image becomes a mnemonic support for a spoken message.

For the Pink Floyd, every concert was an opportunity to use powerful new scenic effects to stun their audience. The group made use of a vast array of lights and motorised lasers, and a great variety of colors to achieve an effect of continuous movement. In fact, very often the lights moved in step with the rhythm of the music of their songs, so that the light show provided a mnemonic support for the music: the audience would automatically link the musical message to the lighting effects. The Pink Floyd will go down in history for the grandiosity of their light shows (a term that became synonymous with their concerts), by means of which they also established certain habits with their fans, such as the surprise element that typically came at the end of each concert. The concert in Venice ended with an explosion of fireworks connected a little earlier to the traditional fireworks to celebrate the Festa del Redentore.

From the perspective of their usage of light, Luigi Nono and the Pink Floyd would seem to have nothing in common. In actual fact, despite their different approaches, both the composer and the pop group were attempting to reinforce this sound scenario, adopting two different styles: the former aimed to strengthen the sound effect through spatialisation; the latter combined the stunning effect of light in their musical representation. Both relied heavily on live electronics to achieve these goals.

5.1 Technical Organisation, Quadrophony and Live Electronics

Pink Floyd choose for the public a traditional-frontal experience of music as regards the division of musicians and the audience: the sounds come from a stage placed in front of the spectator. Despite this choice, the group intends to help listen to the public through satellite amplification stations.

The sound engineer Jerry Wing of Britannia Row describes[5] the system they provided for the concert as follow: a Maryland Sound Incorporated or a sub box made up of 4x15-inch cones with frequency response from 30 to 80 Hz and in a high frequency box

---

[5] Jerry Wing, personal e-mail communication with author.

composed of 4x12-inch cones and a flared horn in turn powered by a TAD 4001 (a compressor), which extends the frequency up to 18kHz. Wing also writes that these speakers were powered by a rack of SAEP500 and CREST 7001 amplifiers all around 1000 watts per channel; each rack also has 4 amplifiers. The speakers were suspended thanks to straps: the first group of speakers (for the high frequencies) was hung with a strap, the next one attached to the first with other straps and so on for the other 40 heads (speakers for the high frequencies) and 40 bass subs for general PA. The engineer adds that for the quadraphonics stations the speakers used were Turbosound TMS3. Each combination of speakers, for treble and sub, was powered by around 3000-W amplifiers which gave around 120,000 watts to the main system per gig.

The latest information on the concert concerns the sound engineer Buford Jones who uses 2 Yamaha PM 4000s and a specially created Quad mixer to control the surround system. They probably aimed to apply the principle of quadraphonics to their live shows, and this is another aspect that relates them to the not very successful attempt to use the Ark that Nono commissioned from Renzo Piano, which involved more than mere quadraphonics, it aimed for a poly-centricity. Although the supply of these quadraphonics installations was confirmed by Britannia Row of London, it is hard to say how, and how much they were actually used at the Pink Floyd concert. From the images recorded live in Venice, it is unfortunately impossible to distinguish more than one quadraphonic

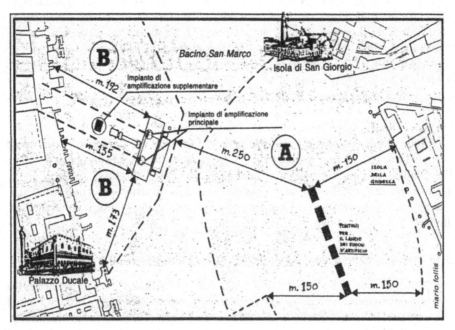

Fig. 6. Technical map of the concert of Venice, published in the daily newspaper "Il Mattino di Padova" [15].

station, though it is certain that a "supplementary amplification system"[6] was used (Fig. 6), located about 100 yards away from the front of the stage used for the concert in Venice. Based on information given by the sound engineer Wing (Britannia Row), we can assume that the Pink Floyd's quadraphonic system was conceived for the purpose of surrounding and immersing the audience in the music and in the group's visual spectacle.

6   Embodied Experience in Prometeo and Pink Floyd Concerts

Luigi Nono's Prometheus stands as an embodiment of musical cognition, a pivotal concept in cognitive psychology. Nono's work deeply intertwines with embodied cognition, where the mind is not separate from the body but relies on continuous interaction between the body, the environment, and experience. Nono didn't merely compose music; he sculpted a sensory experience engaging hearing, sight, and even physical movement.

For the audience, Prometheus offers total immersion in a sonic world that transcends mere listening. The innovative use of sound spatialisation creates a three-dimensional sonic environment enveloping listeners. Sounds appear to emerge from various directions, channeling the listener into a unique auditory journey. This sensory involvement transforms listening to Prometheus into a corporeal experience where the audience is not merely a passive spectator but an active participant. The innovative use of sound spatialisation, coupled with reclining seats facilitating movement, intensifies concentration, transforming simple listening into a complete immersion. Through this, the audience lives what Prometheus lives, forging a unique connection between the opera, the audience, and the surrounding space.

Yet, the experience doesn't end with the audience. Musicians, immersed in the intricate system of the Ark, become integral parts of this incarnate experience. Their interactions with instruments, the surrounding space, and the audience are fundamental to the performance. Musicians' gestures, their physical presence within the Ark's environment, and responses to sonic nuances intertwine their experience with that of the listeners, creating a unique synergy between performers and audience.

In the context of Pink Floyd, embodied experience takes on a different yet equally engaging form. Their remarkable blend of psychedelic lights, enveloping sounds, and energetic performances transformed their concerts into multi-sensory experiences. Audience members didn't just hear the music; they felt it deep within, as vibrations from powerful speakers reverberated through their bodies. Moreover, the undeniable star of the show was the sea, enabling the audience to experience music while on a boat, amplifying the sensory impact.

In both contexts, embodied experience becomes a bridge between art and human perception. Music becomes more than a sound wave hitting the ear; it evolves into an entity that embraces the entire body, involving the entire human experience in a symphonic dance of sounds, lights, and movements. This connection between Nono's work, embodied cognition, and sensory immersion in Pink Floyd concerts reveals the power of art in shaping not only our listening but also our perception of the world around us.

---

[6] Alberto Vitucci, "I Pink Floyd suonano", Il Mattino (Padova), July 14, 1989, accessed December 12, 2020.

7   On Embodied Cognition and Live Electronics

Certainly, live electronics represents a fundamental evolutionary step in contemporary music. When we view these works through the lens of embodied cognition, we enter a territory where music is a total experience, engaging every aspect of our being.

Looking at the technological aspect, live electronics offers artists the ability to manipulate and shape sound in real-time. This isn't merely an exercise in musical creation but an active interaction between the artist and the instrument, often transforming the stage into a live sonic laboratory. The artist's physical presence, hand movements, use of electronic controls – all of these are part of artistic expression. These physical gestures become an integral part of the performance, an extension of musical intention that goes beyond traditional instrumental execution.

Simultaneously, the audience is drawn into a unique sensory journey. This perspective transforms music into an equally inclusive experience, accessible to all, eliminating distinctions between the audience and the artist. The sounds can surround listeners from every angle, creating a three-dimensional immersion. Music is no longer confined to the stage; it spreads throughout the entire space, interacting with the architecture of the environment and involving the audience in a synergy of sounds, movements, and tactile sensations.

This perspective of embodied cognition should not be merely an academic analysis but a guide for the listener. The significance of examining various music and technology-related events through the lens of embodied cognition has already been underscored by [16, 17]. It invites listening to music with the entire body, not just with the ears. Thus, works related to live electronics are not merely compositions to contemplate but multisensory experiences that engage the body, mind, and soul, forging deep connections between the artist, the audience, and the surrounding environment. In this context, music isn't just heard; it's felt, seen, touched, and lived at profound levels, revealing the power of art in integrating technology into our human experience.

8   Conclusions

In this article, I have examined two events that took place in Venice. The results of this research brought to light substantial differences between the two concerts considered. Although both events used live electronics, the first difference concerns the approach to their use. Nono employed a technology that was quantitatively limited to just a few elements, but exploited them in every possible combination. Conversely, a glance at the table relating to the Pink Floyd shows that their technological instrumentation was much more abundant and detailed, giving the impression of a cumulative use of the available material.

Other substantial differences concern the use of space and the interaction between artist and audience. Nono adopted a closed, circumscribed space inside a church; he commissioned an architectural construction with specific functions for the purpose of listening to sounds, on which he entirely focused the audience's attention; his aim was to achieve an immersive spatial involvement and he tended to eliminate any presence of images. The Pink Floyd, on the other hand, almost exploited the whole city of Venice, but for a paradoxically traditional fruition, where the frontal dimension was emphasised by the previously-described light show and by an unusual use of quadraphonics.

Nono's construction was intended for a limited audience (there were approximately 400 seats available), where they were expected to remain seated while they awaited the sound event. On the other hand, the chairs allow the audience to turn some degree, in order to follow the sounds that surrounded them. The audience was new to the type of event, and it had been given no opportunity to know what to expect beforehand.

The Pink Floyd took a very different approach to its audience: the show was in the open air and it could be attended by anybody free of charge. The audience was consequently free to move around and interact with the sound. Unlike the case of Nono's concert, the pop group's performance was based on its interest in promoting a song album that was already on the market, and that their audience had therefore already heard.

It is important to emphasise that the reactions of the audience attending Prometheus were very different from those of the audience at the Pink Floyd's live performance. In the former case, the audience's comments were discordant, and often critical; in the latter case, the crowd of fans was enthusiastic (as was self-evident from the numbers who attended).

Another fundamental difference lies in the dynamic features of the two performances. Nono explored the whole dynamic spectrum, touching on sonorities that were sometimes hardly audible, sometimes extremely loud, but that generally demanded total silence and great concentration. Within a closed space, Nono created a sound panorama of such wealth that it reflected that of the city of Venice pulsing outside the concert hall doors, with which it was implicitly in conversation. The Pink Floyd concert, on the other hand, tended to use only the higher dynamic levels, to the point of risking penalties for exceeding the volume threshold imposed by Italian law; it sought to emphasise an artificial sound to wipe out the sound panorama of the city.

Both performances focused on a different spatial involvement from the usual, exclusively frontal fruition, but only Prometheus completely succeeded in this intent: the concept behind it lies in taking the spatial effect of the sound into account during the compositional process, considering it as the main aspect.

In conclusion, the two events considered here seem to share the ambition of occupying a position on a level with the high-end technologies of the time (the 1980s), as if they were riding the wave of the same "wave of the times". A more detailed study of the use made of such technologies nonetheless enabled to identify different (and sometimes contrasting) conceptions of the role of the audience, the sound space, the musical work of art, and its relationship with the place where it is represented.

In conclusion, exploring the integration of technology into music, embodied in Nono's Prometheus and Pink Floyd's electrifying live show, reveals a fascinating mastery of innovation and creativity. Despite distinct yet intertwined paths, both events resonate with a common aspiration: immersing the audience in a sonic odyssey guided by live electronic music. Nono's Prometheus stands as a testament to meticulous composition, where every note resonates with profound meaning, while Pink Floyd's visual and auditory ingenuity redefine the concert experience. These distinct yet interwoven paths underscore the limitless potential of technology in shaping the future of musical expression and embody the concept of embodied cognition, where the mind's interaction with the environment plays a pivotal role in artistic creation. In this innovation, the legacies of Prometheus and Pink Floyd remind us of the boundless possibilities when art and technology harmonise.

While in this paper I introduces the concept of embodied cognition, in the final section, a more detailed exploration of how these artists harnessed embodied cognition to enhance the audience's immersive experience would undoubtedly enrich the discussion. Further exploration of these perspectives would provide valuable insights into the profound ways in which music, technology, and cognition converge in live performances.

Acknowledgements. This research was supported by São Paulo Research Foundation – Brazil (FAPESP) – Grants 2016/22619–0.

References

1. Favaro, R., Pestalozza, L.: Storia della musica. Milano: Nuova Carisch s.r.l (1999)
2. Chadabe, J.: Preserving performances of electronic music. J. New Music Res. **30**(4), 303–305 (2001). https://doi.org/10.1076/jnmr.30.4.303.7485
3. Tamburini A.: Dopo Prometeo. Intervista di Alessandro Tamburin (1984). In: Nono, L., (ed.) Scritti e colloqui, vol. II, p. 364. Ricordi-LIM, Milano-Lucca (2001)
4. Sinigaglia, A.: Nono: con i suoni di Prometeo reinvento l'uomo. In: Ida De Benedictis, A., Rizzardi, V., (eds.) Le Sfere Scritti e Colloqui. Lucca, vol. 2, p. 335 (2001)
5. Nono, L.: Prometeo. Tragedia dell'Ascolto. Score. Milano: Ricordi (1984)
6. Vidolin, A.: Interpretazione musicale e signal processing. Padova, Centro di Sonologia Computazionale (2009)
7. Nono, L.: Musica e Teatro, Scritti e Colloqui. In: Ida De Benedictis, A., Rizzardi, V., (eds.) Le Sfere Scritti e Colloqui. Lucca, vol. 1, pp. 210–215 (2001)
8. Nono, L.: Prometeo, Scritti sulle opere. In: Ida De Benedictis A., Rizzardi, V. (eds.) Le Sfere Scritti e Colloqui, Lucca, vol. 2, p. 494 (2001)
9. Nielinger-Vakil, C.: Prometeo. Tragedia dell'ascolto (1975–85). In: Nono, L., (ed.) A Composer in Context, Music since 1900, pp. 191–316. Cambridge University Press, Cambridge (2016). doi:https://doi.org/10.1017/CBO9780511842672.008
10. Restagno, E., Nono, L.: EDT, Torino (1987)
11. Cacciari, M.: Verso Prometeo, tragedia dell'ascolto. In: Cacciari, M., (eds.) Verso Prometeo. Milano: Ricordi, p. 21 (1984)
12. Cacciari, M.: Verso Prometeo, tragedia dell'ascolto. In: Degrada, F., (eds.) Luigi Nono e il suono elettronico. Milano (2000)
13. Piano, R. 1984. "Prometeo: uno spazio per l'ascolto". In M. Cacciari, eds. Verso Prometeo. Milano: Ricordi, p.59
14. Piccardi, C.: Il messaggio totale di Luigi Nono. In: Rivista quadrimestrale Musica/ Realtà, vol. 27, p. 83 (1988)
15. Cecchinato, M.: Il suono mobile. La mobilità interna ed esterna de suoni. In: Borio, G., Morelli, G., Rizzardi, V., (eds.) La nuova ricerca sull'Opera di Luigi Nono. Venezia: Fondazione Giorgio Cini, pp. 135–136 (1988)
16. Partesotti, E., Peñalba, A., Manzolli, J.: Digital instruments and their uses in Music therapy. Nord. J. Music. Ther. **27**(5), 399–418 (2018)
17. Partesotti, E.: Extended Digital Music Instruments to Empower Well-being Through Creativity (Chapter 13). In Creating Digitally. Shifting Boundaries: Arts and Technologies - contemporary applications and concepts, pubblicato da Springer (2023)

Soundscape to Music: Experiences in an Additive Manufacturing Open Lab

Laureana Stelmastchuk Benassi Fontolan$^{(\boxtimes)}$ (iD)

CTI Renato Archer, Campinas, São Paulo, Brazil
lfontolan@cti.gov.br

Abstract. The CTI Renato Archer's Open Laboratories have their own ambience, the lo-fi soundscape mostly hidden in the routine of researchers is an important part of the act of experiencing the work space and trying to reach a higher level of awareness about the surroundings through sound is part of a more compelling path to understand the space and how the interactions between individuals and machines develop. By recording the soundscape during routine work fabricating objects using Additive Manufacturing (AM) and conducing experiments, this work aims to bring to the foreground the sound layers emanating from the machines and materials used at the 3D Print Open Lab (LAprint) and the Micro and Nanoimaging Open Lab (LAimage), blending it with musical instruments to generate compositions. The musical instruments used to add to the soundscape tracks were a Clarinet, a Baglama, a PVC flute and Mouth Harps. For all instruments, 3D printed objects were fabricated and used as accessories, such as clarinet reeds and guitar picks. All recordings were made with a TASCAM DR-07X and treated with Audacity 1.3 software. The detachment from the vision perspective of the space by isolating sounds provides a more accurate analysis of the machine components activated in each routine. In addition to the art produced, the recordings can be useful for teaching, equipment troubleshooting and it has a particular important role for presenting technologies to people with blindness and visual impairment.

Keywords: Additive Manufacturing · Field Recording · Audio Art

1 Introduction

Music and Additive Manufacturing (AM) developed a tight relation, as it became a popular fabricating method more examples of applications in music can be found for readily available technologies such as Fused Filament Fabrication (FFF) [1,3,12,15]. More restricted AM technologies, such as Electron Beam Melting (EBM), were also used to fabricate musical wind instruments showing very particular characteristics and a new biomorphic concept of titanium alloy flute was proposed [8]. AM fabricates parts by adding layers of material consecutively using a digital 3D model as the base for design [14] and is considered a digital fabrication method also known as Rapid Prototyping and 3D Printing.

Supported by the funding agency CNPq.

Besides the potential for fabricating musical instruments and accessories, an AM laboratory has it's own ambience. Sounds emerge from vacuum pumps, compressed air lines, fans, motors, materials tinkling and parts being sanded. All the machinery symphony hides behind human interactions, only emerging to attention when a distinguished noise is not expected or the lack of noise disrupts the work routine, meaning that the sound or the lack of the sound in this context is only noticed when it's a problem.

The Renato Archer Information Technology Center (CTI Renato Archer) is a Brazilian research center located in Campinas, state of São Paulo. It supports academic and industrial research with AM technnologies such as Selective Laser Sintering (SLS), EBM, FFF and Polyjet through the 3D printing Open Lab (LAprint). CTI Renato Archer has another Open Labs available, such as the Micro and Nanoimaging Open Lab (LAimage) which was also involved by having it's ambiance recorded.

In a lo-fi soundscape possibility of a research center with heavy machinery, full of workers and close to the road connecting key cities in São Paulo state, the audio obtained can be analysed and separated in three components: figure, background and field [2,13] to consciously define the portions of the audio file which are going to be suppressed or enhanced depending on the intended effect. The figure is the center of attention to the listener, the background comprises other elements in the soundscape and the field is the place. Three aspects were further analysed: onsetness, loudness and pitchness. Those aspects were used previously as descriptors for soundscape design in the context of algorithmic composition and simulated evolution [5,10].

The interdisciplinarity of the author arises from the inability to give less importance to a particular activity, insulating certain subjects by defining them as hobbies, sentencing it to never have a leading role in life. The pharmacist Msc in biotechnology always curious about electronics and open technologies eager to absorb and share knowledge, now shares her adventurous experiments in field recording and sound art. This work aims to bring from the background the sound layers emanating from the machines and materials used at LAprint and LAimage, emphasizing and transforming it into compositions. To transport the soundscape where the researcher is immersed every day, the author takes inspiration in Throbbing Gristle, Jean-Michel Jarre, Hermeto Pascoal, Thomas Rohrer field recordings and many others.

2 Methods

All sounds were recorded using the portable digital recorder TASCAM DR-07X with dual unidirectional condenser microphones for A/B or XY stereo recording techniques and a Sennheiser HD 65 TV headphone. The recorder was mounted on a Manfrotto adjustable Tripod when necessary and a windscreen was used when recording outdoors. The following musical instruments were played and sampled: a Clarinet, a Baglama, a PVC flute and Mouth Harps. For all instruments, 3D printed objects were fabricated and used for specific functions, such as clarinet reeds and guitar picks. One of the Mouth Harps employed was fully 3D printed [7].

The 3D Printer used to fabricate parts is a Prusa Mendel i3 FFF equipment, part of the author's personal collection, assembled with the E3D REVO^{TM} SIX hotend employed for filament extrusion of PLA, PETG and ABS.

Machine sounds were generated by the AM equipment available at LAprint and the imaging equipment available at LAimage, both open laboratories the author uses to develop her current lines of research focusing on process optimization.

The WAV samples were sorted and organized using Sony Sound Organizer 2 and treated with Audacity 1.3 for the final compositions. For analyzing files and composing, the sound system employed a TEAC AG-D8900 AV Digital Home Stereo Surround Sound Theater Receiver and two Wharfedale Modus Five speakers using Amphenol audio connectors.

3 Results

The following sections will discuss results regarding sampling strategies, equipments and materials used for recording and interesting insights about sound handling and composing.

3.1 Sampling

Some of the LAprint's sounds were generated by the AM equipments Sintertation® HiQ^{TM} (3D Systems), Arcam EBM Q10plus (GE Additive), Fortus 400mc (Stratasys) and Object Connex 350 (Stratasys) and for LAimage sounds were recorded from the the the X-Ray Microcomputed-tomography equipment, Bruker Skyscan 1272 and the Scanning Electron Microscope SEM-FEG Mira 3 XMU - Tescan. Figure 1 shows a portion of the LAprint, depicting EBM and SLS AM equipment recorded for the present work. The strategy for TASCAM recordings used the sampling A-B configuration to capture wider audio space, while X-Y focuses on the sweet spot for a tighter audio field.

The audio recorded directly from the machines was not generated for the exclusive purpose of composing it in the future, the recordings were captured while the author was performing experiments in the AM context, using the recording act as a self observation of the author's research activity, perhaps to have the soundscape as fond remembrance of the past. Dividing the elements in the recordings as Figure, Field and Background, the Figure is always a specific louder noise starting or ending a routine, the Field are the subtle noises generated from the main equipment recorded and others in each laboratory (the Background).

The usage of musical instruments and 3D printed objects have a more intentional role in composing, using it deliberately to fill the soundscape recorded making the compositions whole in the author's point of view. All of the recordings were manipulated and remixed freely, although the soundscape recordings ware more carefully manipulated to remain recognizable.

Fig. 1. Panoramic picture taken inside LAprint depicting EBM and SLS AM equipment. Available at [4]

3.2   Reduce, Reuse and Recycle

To optimize the LAprint use of material, the 3D modeling and process planning strategies of AM were carefully designed to reduce the waste of material, mainly in the form of support structures used to fabricate parts. For FFF the support structures are generated to hold material in place in regions where overhangs are below 45°C. For EBM the supports have an additional role, to heat energy dissipation reducing curling or warping defects. The titanium alloy (Ti6Al4V) structures used in the sampling for high pitch sounds were support structures usually discarded after cleaning parts fabricated by EBM. Those parts were used as is for generating sounds or assembled together. Figure 2 shows two types of support structures generated for EBM AM that were used to generate sounds.

Although the recycling of plastics used for FFF AM is common, using printed parts for recycling and inserting it in circular economy strategies [6,9] for metal Powder Bed Fusion (PBF) technologies such as EBM the most discussed topic on recycling is the continuous use of the powder for a number of process depending on the applications, but the fabricated parts can't be processed back to powder and reused. Thus re-purposing EBM supports reduces the waste generated by LAprint.

3.3   Compositions

Navigating through many audio files was facilitated using Sony Sound Organizer 2 software, as the files were sorted, they were used as input to Audacity 1.3, software in which each track was manipulated to be shortened, extended or had filters and effects applied to it as necessary. Each track was analysed and processed taking into account the field, background and place characteristics and onsetness, loudness and pitchness aspects, highlighting or hiding parts of the tracks.

Fig. 2. EBM generated supports. (A) Cone type supports in white under the colored cylindrical models; (B) block type supports isolated from the main models, excerted from cylindrical surfaces and green from rectangular surfaces and; (C) the fabricated supports after having it removed from the main parts. (Color figure online)

4 Conclusions

The compositions created reflect the expression of art in which the machine manipulator can interpret it's functions and deals with the digital fabrication context in an oblivious reality, usually ignored or not important enough to the daily routine, interpreting this parallel and powerful connection to the space where the machines are situated and one can choreographically interact with them. The symphony of sounds emanating from each laboratory is unique and pervades the minds of the professionals who work in such place, capturing the influence of the machines' noises brings a new awareness of the space, with the sound components that were always present, resembling the definition of acoustemology [11].

The detachment from the vision perspective of the space by isolating sounds provides a more accurate analysis of the machine parts and which components are activated in each routine. In addition to the art produced, the recordings can be useful for teaching, equipment troubleshooting and it has a particular important role for presenting technologies to people with blindness and visual impairment, offering a way to interpret machine functions through sound. Future work involves the capturing of the soundscape recording as a tool for inclusion focusing on diversity and equity.

Acknowledgement. For all the shared excitement around music and technology, the author thanks ALPH for inspiring and supporting the subtle person/machine interactions making it possible to amplify and express it in the form of this work.

References

1. Barinque, J.K.J., et al.: Development of an ergonomically - designed violin chinrest using additive manufacturing. Adv. Sustain. Sci. Eng. Technol. **4**(2), 0220211 (Nov 2022). https://doi.org/10.26877/asset.v4i2.13011
2. Batista, I., De Paula Barretto, F.: Developing an synthetic binaural interactive soundscape based on user 3d space displacement using opencv and pure data. In: HCI International 2018 - Posters' Extended Abstracts, vol. 851, pp. 231–236. Springer International Publishing (2018)
3. Cavdir, D., Wang, G.: Designing felt experiences with movement-based, wearable musical instruments: from inclusive practices toward participatory design. Wearable Technol. **3**, e19 (2022). https://doi.org/10.1017/wtc.2022.15
4. CTI Renato Archer: LAprint (Aug 2023). https://www1.cti.gov.br/colab/language/en/laprint.html
5. Fornari, J., Maia, A., Manzolli, J.: Soundscape design through evolutionary engines. J. Braz. Comput. Soc. **14**(3), 51–64 (2008). https://doi.org/10.1007/BF03192564
6. Hettiarachchi, B.D., Brandenburg, M., Seuring, S.: Connecting additive manufacturing to circular economy implementation strategies: links, contingencies and causal loops. Int. J. Prod. Econ. **246**, 108414 (2022). https://doi.org/10.1016/j.ijpe.2022.108414
7. jprodgers: Thingiverse - printable Dan Moi project (2011). https://www.thingiverse.com/thing:9377
8. Kolomiets, A., Grobman, Y., Popov, V.V., Strokin, E., Senchikhin, G., Tarazi, E.: The titanium 3D-printed flute: new prospects of additive manufacturing for musical wind instruments design. J. New Music Res. **50**(1), 1–17 (2021). https://doi.org/10.1080/09298215.2020.1824240
9. Mohammed, M., Wilson, D., Gomez-Kervin, E., Petsiuk, A., Dick, R., Pearce, J.M.: Sustainability and feasibility assessment of distributed E-waste recycling using additive manufacturing in a Bi-continental context. Addit. Manuf. **50**, 102548 (2022). https://doi.org/10.1016/j.addma.2021.102548
10. Moroni, A., Manzolli, J., Zuben, F.V., Gudwin, R.: Vox Populi: an interactive evolutionary system for algorithmic music composition. Leonardo Music J. **10**, 49–54 (2000). https://doi.org/10.1162/096112100570602
11. Novak, D., Sakakeeny, M. (eds.): Keywords in Sound. Duke University Press (Apr 2015). https://doi.org/10.1215/9780822375494
12. Rodríguez, J.C., Del Rey, R., Peydro, M.A., Alba, J., Gámez, J.L.: Design, manufacturing and acoustic assessment of polymer mouthpieces for trombones. Polymers **15**(7), 1667 (2023). https://doi.org/10.3390/polym15071667
13. Schafer, R.M.: The tuning of the world: toward a theory of soundscape design. University of Pennsylvania Press (1977)
14. Volpato, N.: Manufatura Aditiva: tecnologias e aplicações da impressão 3D, 1st edn. Edgard Blucher Ltda, São Paulo, Brasil (2017)
15. Zvoníček, T., Vašina, M., Pata, V., Smolka, P.: Three-dimensional printing process for musical instruments: sound reflection properties of polymeric materials for enhanced acoustical performance. Polymers **15**(9), 2025 (2023). https://doi.org/10.3390/polym15092025

Presenting the Testimonial in Multimedia Documentaries

Kenneth Feinstein[(✉)]

Sunway University, No. 5, Jalan Universiti, 47500 Bandar Sunway, Selangor Darul Ehsan,
Malaysia
kenf@sunway.edu.my

Abstract. This paper looks at two multi-media documentary projects one in Portugal and the other a developing work in Malaysia. It will look at relationship of the documentary to the bearing witness of the subject. Form the will look at the ethical underpinnings of this. From there we will look at how this ethical relationship between the subject bearing witness and the audience changes into a new form within the installation. We will see how the new form of installation documentary is based on an idea of presence and how that leads to the work bringing forward a relationship of the viewer to the witness as a one ethical responsibility to the other. From there we will look at two examples of installation documentaries one completed in Lisbon in 2022 and the other in Sarawak, Malaysia is an ongoing project to see how these considerations are used to design new works.

In looking at these two projects we will see concepts of direct testimonial found in documentary come into direct conflict with the interviewees sense of governmental repercussions. We will ask if it is possible to create a documentary work that has a greater sense of presence while having to face the need of anonymity by interviewees. From this we will draw some conclusions and look forward to how this experience can help in future works.

Keywords: documentary · installation · media art · interface · testimony · performativity · otherness

1 Introduction

This paper is looking into how we can use installation design as a way of creating a new form of documentary. We will look at what are the distinguishing factors regarding a documentary work, how they have been defined in the cinematic tradition and how installation design allows for a different form of documentary to exist. From this we will look at a specific example of such a work and then a speculative design for a future project. Within the analysis of these two works we see how interface design can be used not only for a more experiential interaction, but how it can be fundamental in the conception and design of the work itself. We will address issues regarding how to ethically present the voice of the subject in differing political and free speech environments and see how interface design must take center stage to address these issues.

© ICST Institute for Computer Sciences, Social Informatics and Telecommunications Engineering 2024
Published by Springer Nature Switzerland AG 2024. All Rights Reserved
A. L. Brooks (Ed.): ArtsIT 2023, LNICST 565, pp. 257–268, 2024.
https://doi.org/10.1007/978-3-031-55312-7_19

2 The Documentary as a Form

As a form of non-fiction narrative, the documentary is unique. History is based on the linear text. It defines our understanding of events in a cause and effect relationship [1]. The narrative voice is found in the historian author. While documents and testimonials of participants may find a place in the work, they are always supplemental to the authorial narrative. With a written text the voice of the participant or subject of history is always at a remove. At best it is a transcript of them talking, but their visage and voice must be by definition missing. Even if the historian is claiming to be a witness to events, they are still subject to the literary form. It is only with the technical image that a sense of presence can be brought to bear on historical events. Beginning with Roger Fenton's images of the Crimean War in 1856, we can see what the aftermath of a battles looks like and begin to feel as if we are experiencing the events as they are. As journalistic photography and cinema were invented, the historical gave way to the documentary. The role of the witness changed. We as viewers were given access to the direct testimony of the witness as well as being able to witness events as they unfolded for ourselves. Where history is a linear narrative derived from the distilling of documentation, the documentary centers on the presentation of the documentation directly to us. It is designed to bring the real to us as a way of presenting us the truth. This truth is found in the unfolding of events as they occur. This is why they tend to present us events in some form of linear time. Most documentaries use linear time as the main structure of the narrative from. We see this more strongly in direct cinema but find it in other forms such as the expository or voice of God film.

2.1 Various Mmodes of Documentary

In *Representing Reality*, Bill Nichols defines the various types of documentary as four different modes namely, "expository, observational, interactive, and reflexive" [2]. Later this was amended by Stella Bruzzi so that interactive became participatory and a performative mode was added [3]. Each mode has a different approach to how we are presented with reality in the film, but all are based on some sort of presentation of documentation. In the expository mode which is driven by the "objective" voice of the (male) narrator, we still will have images that verify the narration. The images in John Huston's *Report from the Aleutians* reinforce the narration. Eyewitness testimonials are also used to reenforce the overall narrative by they are still supplemental to the narration [4]. More recently we find this mode used in the multipart documentaries of Ken Burns. In a series like *The Civil War*, Burns uses voice actors to read testimonial documentation, letters and diaries, from eyewitnesses to events that predate cinema. [5]. Yet these testimonials are supplemental to the narrative voice. This expository mode is the most problematic in terms of presenting the testimonial as the core of the documentary. In form it is the closest to written history as the film is developed around the written text.

As the technology allowed for handheld cameras which enabled synchronous sound and images to be filmed, the observational mode became predominate. This mode is possibly best known by films such as *The War Room* [6] or *Salesman* [7] where the filmmakers follow their subjects recording them in their daily lives. In this form the filmmaker tries to eliminate their presence from the film leaving the viewer to experience events as if they were there. The camera eye becomes the surrogate for the viewer's eye. Where the expository depends on the narrator and interviewees to be the witness form which we arrive at the truth, the observational mode places us in the action making us the witness. This form eschews an externally imposed narration as much as possible and just allows events to unfold. It gives us access to events as they happen, but this doesn't mean that the underlying motivations for various people's actions are available to us. We do not get to know how the actor's perceive events and how this modifies their actions. This form is based on the idea that seeing is believing and that truth is always found in the outside actions before the camera. The observational film tends to center on a specific (time limited) event. This gives the film its narrative structure. It allows the film's narrative to resemble a fiction narrative by reinventing the documentary as a three-act play. For example, in *The War Room* the event is the 1992 presidential election. We have a structure that builds from the first primary in New Hampshire through to the climax of election night. Here we have a form that on first sight appears to give us a raw view of reality as it unfolds. What happened and said is unscripted, yet there is a great deal of editorial control by the filmmakers both in the placement of camera and especially in structuring a narrative. The filmmaker is not physically present in the film, but it is their views that shape and define what we take as real and true. Testimony is presented to us, the camera is unflinching in what it sees, yet that doesn't mean that it doesn't have a point of view. It is the director and editor who have the last say in how we experience this reality. Because we do not always perceive events that we live through in the same way, an issue for observational documentarians is the possibility of the subjects feeling betrayed by how they are presented in final work. Since the film can only show us the physicality of events as they unfold the perspectives of the participants are not communicated, the whys of their actions, and subjects can feel that without that understanding their actions are not fairly presented. It is this last issue that helped lead to the last mode of documentary that needs to be discussed here, the performative mode.

A major problem with the objective mode of documentary film is that subjects can feel betrayed by how they are portrayed in edited film. They feel that their actions are not given context and may appear very different from how they saw the situation. Even in a performative film like *The Fog of War*, the subject can feel manipulated by the editing. Those are choices and risks these filmmakers have made. Both styles also tend to thematize the subject, making them part of an argument made by the filmmaker.

What distinguishes the performative mode is that the presence of the documentarian is acknowledged. This can be in the form of them being visible actors in the work, such as Michael Moore in *Roger & Me* [8] or more of an off-camera presence like Errol Morris in *The Fog of War* [9]. Within this structure we see that the testimony presented is a performative act. Where the observational film pretends that the presence of the camera doesn't affect the events depicted, in the performative mode how the camera defines what we see and hear is central to the form. The voice of the subject is given to us as the main form of documentation. The filmmaker's job is to tease out the thoughts of the subject giving us a chance to understand why they act the way that they do. *The Fog of War* is a good example of this mode of documentary. The film is based on extended on camera interviews with Robert McNamara. While other images are used as illustration of the voiceover, it is McNamara's testimony that is the core of the film. McNamara gets to present his testimonial; it is his voice that we mainly hear. Morris, who remains off camera, questions what McNamara says. He will try to get more introspective reactions from his subject, but at the end of the day we are left with the voice of the subject of the film as our primary voice. We as viewers get to determine if he is an unreliable narrator, but it is still his voice that we are presented with. This direct interview style that Morris uses is framed by the director, he chooses what to intercut and how to edit the conversation. As a style it is trying to overcome what was perceived as the weakness of the observational mode, the inability to get into the thoughts of the subjects. Even with the more hybrid style that Michael Moore uses where he becomes the subject of the observational mode and the interlocker for the performative mode, it is in the direct interviewing of each sections subjects that Moore tries to bring out the reasoning of the interviewee. The voicing of the subject, their knowing performance before the camera allows us into a different idea of what the truth is compared to the observational mode of documentary. In Morris' work the subject is in direct conversation with the documentarian, they are very aware of their role as both witness and performer as they are placed in an unnatural situation. They are seated in front of a camera, and this allows them and us to understand the performative aspect of their role. In this style of filmmaking, we are confronted with the face of the subject. We see them as they reason out what they understood to be the forces behind their actions. We enter into a direct relationship with them through this concentration on their face and voice. Reacting to them as real people first and deciding what we think of their statements second. As such we enter into an ethical relationship with them as other. This relationship demands that we bear the burden of being witness to their testimony.

2.2 The Ethics of Presenting the Voice of the Other

In the different modes of documentary, we have looked at there are a few central tenants across all forms. First is that we are learning something about reality. What we see before us is real or in the case or re-enactments based on the real. Second is the idea of testimony. What is unique about the recorded sound and image is that it allows us to experience events as they unfold, and it allows us to see and hear the testimony of those involved in the events. This means that we enter into a direct relationship with the witness as a *being-with* them. One where we are taking up a responsibility to them by bearing witness to their testimony. Emmanuel Lévinas writes about how the other precedes us and defines us, and that this means that we have an ethical responsibility to the other. To be able to be responsible to the other we first have to experience them as real. By this I mean we must see them as another in all its reality before we can turn their testimony into meaningful language. Lévinas says that we first must experience the other as being nonthematizable and it only after that twhen we can turn what we have seen into signifiers of meaning [10].

If we look at the documentary *Shoah* [11], in one section we follow Simon Srebrnik a survivor of the first phase of the Chełmno Extermination camp. Claude Lanzmann, the director, brings him back to the village of Chełmno where he meets members of the local community. They all gather around in front of a Catholic church where a service is just getting out. They all speak fondly of Srebrnik and seem to be happy to see him. When suddenly a woman starts telling a story in which she claims that a rabbi speaking to his fellow Jews who have been rounded up by the Nazi's claims that their fate is justified as they as a people are responsible for killing Jesus. This is all shot in one take, we see Srebrnik amongst them as this tale of pure antisemitism is said before him. What we see is his face, his expression and his inability to respond to her racism in that situation. In this moment we are witnessing him in his otherness. We are seeing him and his reaction as testimony. Our reaction comes from our being in relation to him in his humanity first. Whatever understanding of how this scene relates to history and racism comes after the initial reaction. Our first reaction is one of feeling the pain that we see as the camera zooms in on his face. In that moment we know what he feels and that language will not be sufficient to express it. This is what Lévinas means about our facing the other and our responsibility to it.

In the different modes of documentary, the directors are trying to find ways of presenting evidence so that we can make judgements on what we are seeing. In all these forms there is the belief that the real can be presented and that we must witness it. This core tenant is the enactment of our responsibility to the other. This is core to the idea of testimony. Within each particular mode there are different approaches to this idea, but it is the underlying factor in all documentaries. In making a documentary film any of these modes as well as the reflective and participatory modes are free to be used, but all these forms were created for the cinematic experience. They assume that we are seeing the film on a screen that we are seated before. They are built around the expectation of a linear viewing of the work either in a cinema or on a television. With this expectation certain ways of constructing a narrative are assumed. Arguments are built in a linear fashion and are strictly defined by the editing of the work. This is what we expect from film. Even as it is presenting us with the testimony of the subjects or directly showing us events as they unfold, the film is structured by its editing to become an understandable narrative. As we move away from the cinematic form and begin to make documentaries for experiential spaces or to be interactive how we structure the work changes. How we tell stories change and so does the relationship of the audience to the work. The importance of the testimony, our being-with the other as an experience becomes more front and centre. The space becomes a part of the experience of the work, we move away from a framed image on a wall into the creation of an entire work as a unified whole. With this the design of the space as an interface becomes important. We will now look at two works, one completed and the other still being designed as examples of how experiential design is used as an ethical tool to safely present the voice of the other. We will see how design becomes part of the creation process in order to put the viewer into relation with the other.

3   The Lisbon Project

In the summer of 2022, I was awarded an artist-in-residency from Carpentarias de São Lázaro an non-profit art centre in the Martim Muniz/Mouraria area of Lisbon Portugal. The brief of this residency was to create a work that related to the neighbourhood where the art centre was located. This area has a large immigrant community. It is made up of people from all over the world. Where most cities have specific ethnic enclaves, this area is home to a very mixed group of people. As you walk through the streets a Senegalese hair salon will be next door to a Nepalese restaurant and a Bangladeshi travel agent.

The nature of the work and its subject was determined after observing the area and talking to both members of the art centre and people in the community. From observing the physical area, I saw that there were many Roman Catholic religious tiles embedded into the facades of buildings. Seeing these objects as signifiers of an older Lisbon, one that was no longer relevant to the current situation, it raised the question of how does one assimilate into the culture this long-established culture? The first response was to make montages where the Catholic iconography was replaced with images and reliefs from South and Southeast Asia. From there the following research questions were established. How assimilated do immigrants and second-generation people feel in Lisbon? What is their relationship to their new culture and how much of the old culture do they feel they can retain? What are the daily examples of negative or positive reactions they get moving through the culture? Language was another import issue explored. All of this was aimed at seeing if migrants, rich or poor, felt that Lisbon was home for them. In order to accomplish the project, it had to be designed according to several principles. It was important that the work centre on people living and working in the area. That it be focused on their testimonials, centring on their life experiences. In order to do this a series of interviews with a variety of people were organised. The art centre and I reached out to a variety of people and organisations, including community groups, international student organisations and friends of friends. From this we able to find fourteen people to interview. A series of questions were developed. These questions were determined from conversations with immigrants from former colonies and other countries. They also reflected personal experiences as foreigners living in a country that is not always comfortable with immigration. These questions were further refined through suggestions of the interviewees and after observing certain topic that recurred in conversation.

A series of interviews were arranged at the studio space in the art centre or in a few cases online. Our first interviewee spoke only Portuguese, and this changed the dynamic of the interview as I was not able to conduct it. It was still based on the set of questions established, but as the documentarian my input into the interview was limited. Beyond this I did not budget for a translator to transcribe the text for subtitles. One of the members of the art centre was kind enough to do this for me, but it did mean that we limited all our future interviews to English speakers. This constraint was less of a problem as most people were comfortable speaking in English and a few were more comfortable in English than Portuguese.

The questioning was done to try to find as wide a variety of experiences as possible. It was not planned with a specific outcome in mind. I tried to ask questions that could elicit responses of inclusion or exclusion, but I let the conversations take their own course. Subjects felt free to recount their experiences negative and positive. They did not have a problem with appearing on camera. Interviews were about from fifteen to forty-five minutes in length. They were then cut up into clips ranging from half a minute to three minutes long. Approximately ten clips were used per interviewee.

The work was designed to be a multi-screen installation. The desire was to move away from the traditional documentary and make the encounter with the project and its testimonials express the idea of being in the presence of the other. In order to emphasise this, it was decided that the images needed to be presented as a series of clips randomly displayed on the three screens. The randomisation of the clips was to remove the editorial control of the documentarian and allow the voices of the subjects interact with each other in a free fashion. Meaning was created by how various clips interact with each other, how they reinforced or contradicted each other. The interview clips were projected on a larger middle screen while clips of the area and the montages were shown on the side screen. Again, these images were randomly displayed. The screens were at varying heights and distances from the viewer (see Fig. 1). Images from each of the screens were creating and recreating context through this randomisation process. The visual hierarchy of traditional cinema was removed as much as possible. Viewers could walk around the space or just sit on benches provided. The sound was primarily from the interviews, but the sound of the other clips also played at a lower volume to help create a sense of ambience.

Fig. 1. View of installation at Carpentarias de São Lázaro

A major problem with the objective mode of documentary film is that subjects can feel betrayed by how they are portrayed in edited film. They feel that their actions are not given context and may appear very different from how they saw the situation. Even in a performative film like *The Fog of War*, the subject can feel manipulated by the editing. Those are choices and risks those filmmakers have made, but those forms also tend to thematize the subject, making them part of an argument made by the filmmaker. As stated earlier that was part of what I was trying to avoid in this project. The randomization of the clips allowed for the removal of a heavy-handed editorial voice and allowed the images and sound to stand on its own.

The final work was titled, *My Head is My only House Unless It Rains* and was show in August of 2022. The title came from a song by Captain Beefheart. In observing the audience and having conversations with viewers, I determined that the design of the work did communicate a feeling of presence with the subjects. The audience tended to stay with the work for about twenty to thirty minutes at a time. The reactions were positive in that they expressed having a feeling of intimacy with the work. Interestingly, subjects who saw the work felt that they were fairly represented by it. As a work that was designed to place the testimonial at its core the reactions of the audience and subjects indicated that it was successful.

Coming from this the next question was if this approach both in focus and design could be developed in future work. What elements in the work are adaptable and what parts of the format need to be changed as new works are developed. The work can be seen as the prototype for a method of installation-based documentary as well as a way of presenting works focused on bearing witness and the presence of the witness to an audience. As a methodological prototype we can look at the work in its form and structure. In the latter view we are seeing how installation-based non-fiction work can continue the traditions of the documentary and apply them to a new form. The work was an attempt to fuse the two issues together in a single work. As much as this piece was able to address these issues in a coherent work, the next step is to develop new projects along these theoretical and design standards. The subject matter of the new works does not have to be the same as the Lisbon project, nor do they have to be different. What does have to be taken into consideration is the relationship of the installation design to the subject matter, the audience's experience and how does it allow for the testimony of the subjects to be central to the work. Moving forward a new as of yet unnamed project has been proposed to develop these issues in relation to the situation of indigenous people migrating into urban areas in the state of Sarawak, Malaysia.

4   Sarawak Project

In 2023 the work *My Head is My only House Unless It Rains* was show at the University of Technology, Sarawak's (UTS) SBE Gallery. Here the physical form of the installation had to be adapted to the physical space and the technical considerations of the gallery. The side images were presented on large television monitors and the central images were on a single projection a pull-down screen. This allowed me to see certain degree of flexibility in installation and how much it changes the experience. A smaller space allowed for a more intimate relationship between the viewer and the work. With the images being closer to eye level. There was an excitement to seeing such a different way of presenting a documentary work. As part of this interest the research office of UTS asked me if I would be willing to collaborate with members of their faculty on a version of the work that spoke to issues found in Sarawak. This version of the work is intended to adapt the same general subject, that of assimilation, to the East Malaysian context.

Sarawak, located on the northern part of the island of Borneo is a state in Malaysia, but its ethnic make-up is distinct from the rest of the country, referred to as peninsula Malaysia. Where the peninsula has a Malay ethnic majority, East Malaysia, the states of Sarawak and Sabah, are made up of a variety of indigenous peoples such as the Dayak and Iban. They have different customs and religious traditions. Peninsula Malaysia is religiously Muslim and culturally and historically tied to Indonesia, the Middle East and the Subcontinent. As a state Sarawak is separate from yet dominated by policies determined in Peninsula Malaysia. There is a tendency to try to bring a Malay ideology to life in East Malaysia. As indigenous people move from the countryside into urban areas, they are challenged with a need to assimilate. They find themselves living in situation that is different from traditional village structures, for example moving from more communal long houses into smaller single family-oriented apartments. They find that they have to speak Malay or English instead of their own languages. In these aspects their experience of integrating into urban Malaysian culture is one of assimilation. Even though they are citizens of the country their integration into the Greater Malaysian culture bears very similar hallmarks found with the immigrant.

We agreed that the project would be a collaboration between myself and Annie Jaid Parker, the program head for the BA in Interior Architecture at UTS. First, an overall design for the project had to be conceptualized. In Portugal the main issue regarding the movement of people is around immigrants coming into the country. Issues of diversity of populations and integration revolve around this. In Sarawak there is an issue of assimilating indigenous people into the larger Malay culture. Here the questions that arise from the movement of people into urban spaces are focused on how native people can or should be allowed to maintain their traditional culture while still being part of a developing Malaysia. Much like in Portugal there is a need to adapt to a new language as well as new social structures, but unlike the migrant in Portugal these people are still living in their native land.

In the Lisbon project almost, everyone was fluent in English well enough to be interviewed in it. Many of the South and Southeast Asian subjects were more comfortable in English than in Portuguese. In Malaysia although English is ubiquitous it is still not spoken by many indigenous people, or they are uncomfortable speaking in it. Because of this the planned interviews will have to be conducted in their languages. The research team includes native speakers of Dayak, Iban and other languages. In the final work we still want to hear the voices of the subjects in their native tongues as it communicates much of their testimony. In order for there to be a common language for the project all interviews will be translated into English and presented with English subtitles. English was chosen because it is a more neutral language choice. Given the politics of integration in Malaysia, Malay would be seen as a language representing the abandoning of the native culture. English also allows for the project to be exhibited internationally.

The images in the Lisbon work included random images of the neighborhood and montages shot by the artist/documentarian. These were created to be shown alongside the images of the interviewees. Because the number of language groups is larger, we are expecting to have more screens than the previous installation, the number of images required are greater than needed in the Lisbon version. In order to have more contextual images we will be asking the subjects to provide videos of their lives and situations. Of course, these images will have to be edited so as to provide anonymity to those involved. Images can be completely randomized, or they can be grouped by ethnic groups so that Iban images are only seen of the Iban screen and the same for the Dayak and other groups. This is a good example of how the choices of interface design help enhance the narrative structure of the work while still putting the testimony of the subjects in the fore.

As in the Lisbon project, the images and testimonies will be displayed randomly on the various screens. This will allow the viewer to experience the work without feeling that there is an editorial control over the work. It will allow the viewer to feel a more direct relationship with the images and sound. The testimonials and the images will relate to each other in a tangential way. This means that meaning is found in the unfolding montage. As both the audio testimonials and the images are randomised each viewer will have a unique experience with the work. Installation will be designed to make the testimonials the main focus of the work while still allowing the images to have a strong effect on the narrative.

For the Lisbon project, the narrative was simpler in that I didn't divide the subjects into subgroups as this didn't seem necessary in the Portuguese context. People didn't feel that their speech could have negative effects on their situations. Therefore, it was easier to have the central screen show the subjects. The situation in Malaysia is different the government has been known to use the sedition act as a way to stop or silence critics, international or local. In 2020 AL Jazeera English broadcast an episode of the series *101 East* called *Locked Up in Malaysia's Lockdown*. The filmmakers were questioned by the police and threatened with the prosecution under the sedition act [12]. The filmmakers were eventually sent out of the country and cannot return. Again in 2021 the Freedom Film Festival showed and then posted an animated documentary online about a police beating of underage suspects. The festival directors and the filmmaker were brought in for questioning and again threatened with the sedition act [13]. As the subjects are all citizens of Malaysia there is no anticipated threat to them foreseen. Therefore, we are assuming that that subjects will be willing to appear on screen. If any foreign workers are interviewed, we will have to review how to design the work with off camera interview subjects.

In the Sarawak project a similar structure to the Lisbon work is planned, but since language issues are different in relation to the indigenous populations, we are planning to have different screens reserved for different language groups including a screen for Malay and English. Multiple screens with multiple speakers will change the focus of the work opening it up to a multi-tiered narrative. This will allow the work the various voices to appear more in conversation with each other as their faces will be on different screens at the same time. Depending on the space of the exhibited work is designed these faces could be next to each other or facing each other. It will change how the audience will experience the work as their focus will shift as the various screens change and new faces or images appear. The space can be designed so that the viewers have to physically move through it to encounter the different testimonies and images. This will bring forward conversational nature to the work. The testimonies and images will be randomized so as to emphasize the ideas of presence and our relationship with Otherness found within it. But for a more complex design like this will incorporate Isaac Julian's idea of the multiscreen montage [14] as a way to understand how the shifting attention of multiscreen installations can relate to a narrative structure. In his work the spatialization of a narrative is accomplished through this shifting of attention from screen to screen in order to create a single narrative. By contrast the work proposed here is focused on the relationship between the person bearing witness and the person who this witnessing is addressed to. While it will still have an opened ended structure based on the principles described above in order to make a work that is understandable it is important to understand the lessons learned from Julian's approach.

An important aspect of this new design issue revolves around the presentation of the audio and as well as the subtitling. How is the sound played, does the work allow for multiple tracks to play? Do we allow the voices to play over each other or does only one sound file dominate? If there are multiple subtitles visible which one has its sound played? At the moment there are three solutions proposed. One is that each screen has its own ultra-directional speaker aimed at the area in front of the screen. This will allow multiple sound files to play, and the viewer will only hear the one that they are in range of. This will require the viewers to move through the space and interact with it physically. The drawback with this is that there will be no overall sound for the work. As people walk into the gallery it will appear to be a silent work until the viewer moves through the space. The other option is to program the sound files so that only one plays at a time. This would mean that only one subtitle file would appear in the installation at any given time, but as in the directional speaker option, all screens would still play random images. Thirdly the directional speakers could be used for the testimonies, but a general speaker could play the ambient sounds of the videos at a low volume to create a sense of ambience. How the sound will be presented and programmed will determined in consultation with our sound designer after experimenting with the various options.

5   Conclusion

It was with the filmic documentary that direct experience and testimony was brought into non-fiction storytelling. With this form, viewers were put more directly in relation with events and the voice of the witness. Yet it still was tied to traditional narrative forms. We see this not only in the expository, but also in the objective and participatory modes of documentary. As new forms of documentary developed, we saw the presence of the witness take a stronger role in and some cases the director even becoming the witness. The relationship between subject and viewer found in the bearing of witness is an important part of the filmic documentary, but it remains supplemental to the voice of the director. The idea of presence is implied, but it is not a major aspect of the documentary film. As documentary moves into the installation space, the possibility of new ways of creating non-fiction works open up with the new medium. Within this medium the sense of presence found in the work comes more to the fore and with this bearing witness of a subject to the audience becomes more central.

In both works above, the attempt has been to create this feeling of presence between the subject giving testimony and the viewers of the work. They are designed to create a space where a relationship between the viewer and the other is brought to the fore and that our ethical relationship to the other is manifest. It is hoped that this type of work will bring understanding to the viewer in a way not possible through traditional documentaries.

References

1. Flusser, V.: Towards a philosophy of photography. Reaktion, London (2000)
2. Nichols, B.: Representing reality: issues and concepts in documentary. Indiana University Press, Bloomington (1991)
3. Bruzzi, S.: New documentary, 2nd edn. Routledge, London, New York (2006)
4. Huston, J.: Dastar Corp., Marathon Music & Video (Firm), Copyright Collection (Library of Congress). Eyewitness to war. Color documentaries of WWII disc 3. Distributed by Entertainment Distributing, Eugene, OR (2003)
5. Burns, K., Burn, R., Ward, G.C., McCullough, D.G., Plimpton, G., Robards, J., et al.: The Civil War.PBS Home Video; Distributed by Paramount Home Entertainment, Hollywood, Calif. (2011)
6. Hegedus, C., Pennebaker, D.A.: The war room. Criterion collection 602.The Criterion Collection, United States (2012)
7. Maysles, D., Maysles, A., Zwerin, C., Brennan, P., McDevitt, P., Baker, J., et al.: Salesman. Criterion collection 122. RSDL dual-layer ed.: Home Vision Entertainment, Chicago, Ill. (2001)
8. Moore, M., Beaver, C., Prusak, J., Rafferty, K., Schermer, B., Stanzler, W., et al.: Roger & me. Warner Home Video, Burbank, CA (2003)
9. Morris, E., Williams, M., Ahlberg, J.B., McNamara, R.S., Glass P., @Radical.media (Firm), et al.: The fog of war eleven lessons from the life of Robert S. McNamara. Columbia TriStar Home Entertainment, Culver City, Calif. (2004)
10. Lévinas, E.: Entre nous : on thinking-of-the-other. European perspectives. Columbia University Press, New York (1998)
11. Lanzmann, C., Chapius, D., Glasberg, J., Lubchansky, W., Postec, Z., Ruiz, A., et al. Shoah. Hollywood, CA: New Yorker Video 2003
12. Latiff, R.S., Joseph: Al Jazeera staff grilled in Malaysia over report on migrant arrests. https://www.reuters.com/article/us-malaysia-media-idUSKBN24B0BC (2020). Accessed 25 Dec 2022 (2022)
13. Thomas, J.: Cops raid Freedom Film Network's office, cartoonist's home. https://www.freemalaysiatoday.com/category/nation/2021/07/02/cops-raid-freedom-film-networks-office-cartoonists-home/ (2021). Accessed 25 Dec 2022
14. Julien, I.: Theatrical Fields: Special Brunch and Screening with Isaac Julien and Mark Nash, NTU CCA Singapore Digital Archive, Singapore (2014)

Human at Centre

Animated Pedagogical Agents Performing Affective Gestures Extracted from the GEMEP Dataset: Can People Recognize Their Emotions?

Magzhan Mukanova[1], Nicoletta Adamo[1(✉)], Christos Mousas[1], Minsoo Choi[1], Klay Hauser[1], Richard Mayer[2], and Fangzheng Zhao[2]

[1] Purdue University, West Lafayette, IN 47907, USA
nadamovi@purdue.edu
[2] University of California Santa Barbara, Santa Barbara, CA 93106, USA

Abstract. The study reported in the paper focused on applying a set of affective body gestures extracted from the Geneva Multimodal Emotion Portrayals (GEMEP) dataset to two pedagogical animated agents in an online lecture context and studying the effects of those gestures on subjects' perception of the agents' emotions. 131 participants completed an online survey where they watched different animations featuring a female and a male animated agent expressing six emotions (anger, joy, sadness, disgust, fear, and surprise) while delivering a lecture segment. After watching the animations, subjects were asked to identify the agents' emotions. Findings showed that only one expression of the angry emotion by the female agent was recognized with an acceptable level of accuracy (recognition rate >75%), while the remaining emotions showed low recognition rates ranging from 1.5% to 64%. A mapping of the results on Russel's Circumplex model of emotion showed that participants' identification of levels of emotion arousal and valence was slightly more accurate than recognition of emotion quality but still low (recognition rates <75% for 5 out of 6 emotions). Results suggest that hand and arm gestures alone are not sufficient for conveying the agent's emotion type and the levels of emotion valence and arousal.

Keywords: Affective Body Gestures · Animated Pedagogical Agents · GEMEP database · Emotion Recognition

1 Introduction

The Internet is replete with free online instructional videos featuring an instructor next to a progression of slides, or the instructor is portrayed as a talking head next to a progression of slides, or even in which the instructor provides voice over for a progression of slides. Although the script and slide content of instructional videos may be informationally appropriate, their effectiveness can be diminished because the instructor delivers them in a way that lacks emotional and social sensitivity for the learner. Preliminary research shows that the effectiveness of instructional video depends both on cognitive factors

A. L. Brooks (Ed.): ArtsIT 2023, LNICST 565, pp. 271–280, 2024.
https://doi.org/10.1007/978-3-031-55312-7_20

(such as the content and organization of the presentation) and affective and social factors, such as the degree to which the instructor displays certain emotions via voice, facial articulations and body movements, and the degree to which the learners feel rapport with the instructor (Fiorella, in press; Mayer, 2020; Mayer, Stull, & Fiorella, 2020).

Our research seeks to create an AI-based animation system that takes as input video lectures with human instructors and converts them into animated presentations delivered by affective agents. The pedagogical agents show different emotional styles and social intelligence intended to maintain a connection with the learner.

The research outlined in the paper represents progress toward achieving this objective, as it focused on how agents' emotions can be conveyed through body cues. More specifically, our study investigated whether people could recognize the quality, valence, and arousal of lecturing animated agents' emotions from a set of body gestures (no voice and facial articulations), which were extracted from the GEMEP video dataset (Bänziger & Scherer, 2010), converted to animation data, and applied to the pedagogical agents.

The paper comprises 5 sections: Section 1 is an introduction to the research; Section 2 includes an analysis of prior work on animated pedagogical agents, expression/perception of emotion in animated agents and emotion theories. Section 3 describes the study and Sect. 4 reports the study findings. Discussion and conclusion are presented in Sect. 5.

2   Review of Prior Work

2.1   Animated Pedagogical Agents

Prior research suggests that pedagogical agents (APAs) could lead to increased students' learning (Johnson & Lester, 2016; Martha & Santoso, 2019; Schroeder et al., 2013). Research also indicates that APAs have the potential to improve students' perceptions of online courses (Annetta & Holmes, 2006), while employing multimodal agents can result in more effective learning compared to agents utilizing a single channel (Cui et al., 2017).

Poggiali's study (2018) revealed that students favored "animated videos with agents for easier learning experiences, largely due to enhanced attention retention," highlighting the impact of the agent's personality in increasing engagement levels. Similarly, Mayer and DaPra (2012) observed that students exhibited improved learning outcomes when instructed by a fully embodied agent employing human-like gestures and an appealing voice, compared to an agent lacking these human-like communication elements. Furthermore, Lawson et al. (2021) established that students could identify and react to the emotional expressions conveyed by agents in a STEM-focused video lecture. Other experiments show that the agent's voice and look (Fiorella & Mayer, in press; Mayer, 2021), visual presence (Rosemberg-Kima et al., 2008), non-verbal communication (Baylor & Kim, 2009), and communication style (Wang et al., 2008) can impact learning and motivation. Pedagogical agents equipped with emotional capacities can facilitate simulated social interactions between learners and computers. A comprehensive meta-analysis, which examined results from various studies evaluating the efficacy of affective agents in educational technology-based settings, revealed a substantial and moderate influence of incorporating emotions into APAs on students' ability to stay motivated and to retain and apply the acquired knowledge to other domains (Guo & Goh, 2015).

2.2 Expression/Perception of Emotion in Animated Agents

People use different channels, such as facial expressions, body postures, dynamic gestures, and voice to express emotions. Nonverbal cues play a significant role in communicating emotions and have important functions such as conveying a person's personality, feelings, and attitude (Meyer et al., 2021; Cheng et al., 2019). All these emotional channels are crucial in building an animated agent's personality and creating emotional bonding with the students. Research shows that body gestures can be responsible for up to 93% of the conversation (Larsson, 2014). Among body movements, the perception of emotions from hands only or just one hand was more precise than heads, torso, or arms (Blythe et al., 2023). A study by Ross & Flack (2020) revealed that images without hands impeded recognition of angry and fearful emotions.

Although some studies indicate that basic emotions can be conveyed through body movements and static poses alone, it is not easy to detect emotions solely from these cues. Certain emotions like anger, sadness, and happiness are easily recognizable through body gestures (Karg et al., 2013). However, conveying emotions such as surprise, disgust, and fear from arm movements alone poses significant challenges (Sawada et al., 2003). Atkinson et al. (2004) found that sadness and disgust were frequently misinterpreted for each other. Additionally, Ennis et al. (2013) observed that relying solely on body gestures, without facial expressions, led to inaccuracies in distinguishing emotions with a high level of arousal. Ennis concluded that while identifying happy and angry gestures was challenging, differentiating between sadness and fear was more feasible. Karg's argument emphasizes that conveying emotions with positive valence and low arousal, like contentment, was particularly difficult using gestures alone (Karg et al., 2013). Although gestures might aid in distinguishing between emotions based on valence, movement characteristics appeared to be more effective in conveying the emotion arousal level compared to static poses.

2.3 Emotion Theories/Classifications

There are two dominant emotion theories: the basic emotion theory (Ekman, 2003; Ekman & Friesen, 1969) and the dimensional theory (Russell, 2003). These theories diverge significantly, mainly concerning whether emotions are discrete entities or rather defined by two continuous dimensions—(1) valence, spanning from positive to negative, and (2) arousal, ranging from activation to deactivation. According to basic emotion theory (Ekman, 2003), humans exhibit a small number of emotions (e.g., fear, anger, joy, sadness, surprise, disgust), each displaying a consistent pattern of associated behavioral components.

In contrast, Dimensional Emotion Theory, as proposed by Russell (1980; 2003), asserts that all emotions can be positioned within a circular framework known as the circumplex of emotions, governed by two distinct dimensions: hedonic/valence (pleasure-displeasure) and arousal/engagement (rest-activated) (Russell, 1980; Posner et al., 2005;

Barrett and Russell, 2015). Within this circumplex, the horizontal axis denotes hedonic valence, while the vertical axis denotes arousal levels. Consequently, each emotion's placement within the quadrant reflects varying degrees of hedonic (valence) and arousal attributes (Posner et al., 2005). The study detailed in the paper considered both Ekman's and Russell's classifications of emotion.

3   Description of the Study

The goal of the study was to examine the extent to which the participants were able to identify six emotions displayed by two APAs performing a set of affective body gestures extracted from the GEMEP video dataset. The APAs performed the gestures while delivering a science lecture segment; the agent's voice was muted, and the face was blurred to prevent the subjects from seeing the agent's facial articulations. After watching each animated clip, the subjects were prompted to input the emotion displayed by the agent. The study used a within-subject design. The independent variables were the affective gesture type and the agent's gender; the dependent variable was the participant's emotion recognition.

3.1   Stimuli

The GEMEP database of affective expressions consists of 145 videos of French speaking professional actors portraying 12 different emotions. Each emotion is conveyed by 10 expressions, with each expression performed by a different actor. Two of the 12 emotions, e.g., surprise and disgust, are portrayed using 5 expressions only. Our study focused on six basic emotions, namely anger, fear, sadness, joy, surprise, and disgust. Only four of the most expressive recordings per emotion, two for each actor's gender, were selected from the GEMEP dataset for use in our experiment. The selection of the recordings to be used in the study was done by a group of five experts in animation and acting.

Video-based motion capture software (Deepmotion) was used to extract the actors' body gestures from the videos and convert them to animation data (Fig. 1, left). Autodesk Maya was used to clean up the motion captured data, and Unity cross-platform game engine to apply the affective gestures to the animated lecturing agents. As the videos in the GEMEP database were recorded in a professionally lit environment with a black background and clearly visible actors in the foreground, the animations of arm and hand movements were extracted with acceptable quality. Due to the fast movements of some expressions, the software could not detect the speed and all micro movements fully. To solve this issue, the videos' speed was decreased to 50% to better detect the fine and fast motions. The extracted gestures from the slowed-down videos were more accurate, however some manual editing of the animation data, especially hand and finger motions, was necessary. After the manual corrections, the affective gesture animations were applied to two of the lecturing animated agents produced by Adamo et al. (2022). The duration of each animation clip varied from 2 to 3 seconds, and to have the same length of animations for all stimuli, only 5 seconds of lecturing animation were used. As the study focused on expression of emotions through body gestures alone, the face of the animated agents was blurred, and the voice was muted. To allow for better perception of the playing gestures, two views (front and side views with slight tilt) were included in the stimuli animations. Figure 1 (right) shows two frames extracted from two of the stimuli.

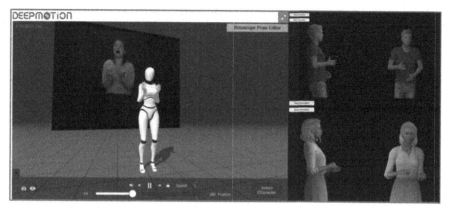

Fig. 1. Left: screenshot from video-based motion capture software showing the video of the actor in the background and the extracted gesture applied to a 3D mannequin in real-time. Right: frames extracted from two of the animation stimuli.

3.2 Subjects

205 subjects participated in the online survey, but only 131 survey responses were complete and therefore considered for further analysis. The age of the participants ranged between 16–53 years old, but most of the subjects were 18–35 years old (92%). 52% of the subjects identified themselves as female, 44% as male, 3% as non-binary and one person declined to declare their gender. Most participants indicated they were Asian, (54%) and White (37%). People with nationality from the United States of America (38%), Kazakhstan (34%), India (8%), and China (8%) made up the majority of the participants.

3.3 Procedure and Evaluation Instrument

The subjects were recruited via email. Those who were willing to take part in the study were given access an online questionnaire. The survey included 26 animation video stimuli which were presented in random order (4 animations x 6 emotions + 2 animations without affective gestures). After watching each video, the subjects were prompted to detect the emotion displayed by the agent. The subjects could select the emotion from a drop-down list with 7 items: joy, sadness, anger, disgust, fear, surprise, other. If the participants selected "other", they could input the perceived emotion. The authors used the feeling wheel to categorize the emotions entered in the text box. The last screen of the survey included a set of demographic questions.

4 Results

Figure 2 shows the emotion recognition rate for each emotion expression. To consider an emotion as accurately recognized, the authors set a recognition rate of 75% or higher based on prior studies on emotion recognition. Only one expression passed the 75% bar. It was an angry emotion expressed by the female agent, which was correctly recognized by 93.1298% of the participants; the remaining animations did not pass the set bar. The second emotion with the highest recognition rate was still anger expressed by the male agent (64.12% recognition rate). Overall, the most recognized emotions were anger (37.4046%–93.1298%), fear (the highest was 58.0153%), and sadness (58.7786%). The emotions recognized with the lowest level of accuracy were surprise and sadness (recognition rate <23% overall).

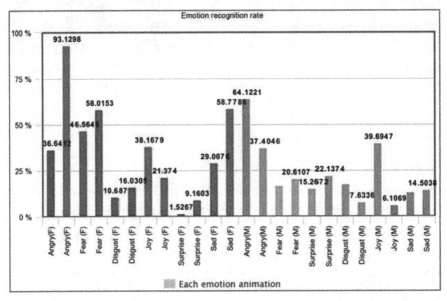

Fig. 2. Bar chart showing the emotion recognition rates. Each emotion had 4 expressions (two expressions performed by the male agent and two performed by the female agent)

Participants recognized the emotions displayed by the female agents more accurately than the emotions displayed by the male agent. However, results of pair sample t-tests comparing each emotion recognition rates for the female versus the male agent showed that the differences were not statistically significant at a significance level set to 0.05 (all p-values >0.05).

A mapping of the results on Russel's Circumplex model of emotion, showed that participants' identification of levels of emotion arousal (high versus low) and valence (negative versus positive) was more accurate than recognition of emotion quality (see Fig. 3). The identification of the levels of arousal and valence was considered correct if the emotion selected by the participant belonged in the same quadrant as the emotion portrayed by the agent. In other words, if the agent was expressing an angry emotion, which has high arousal and negative valence (top left quadrant) and the participant, for instance, input 'stressed' as the perceived agent's emotion, the levels of valence and arousal were considered correct as the 'stressed' emotion belongs in the top left quadrant as well. For two expressions of anger more than 75% of the participants identified the correct levels of valence and arousal: Overall, fear and joy were the second highest, followed by sadness, disgust and surprise respectively.

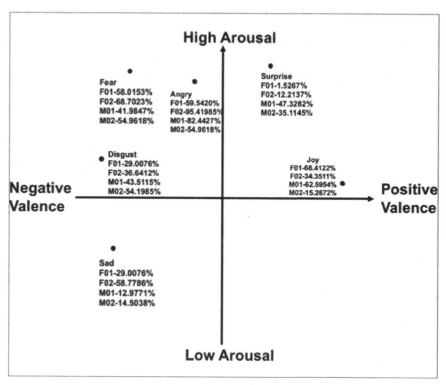

Fig. 3. Percentage of participants who recognized the correct levels of arousal and valence for each of 24 expressions of emotions.

5   Discussion and Conclusion

A significant discussion has emerged regarding whether body poses and movement effectively communicate emotions or merely indicate the intensity of those emotions (Ekman & Friesen, 1969; Lhommet & Marsella, 2015). The findings from the study suggest that, except for one emotion, hand and arm gestures alone are not sufficient for conveying the agent's emotion type or for expressing the correct levels of emotion valence and arousal. Our results align in part with prior work that showed that some emotions, such as anger, can be detected more easily from body cues (Karg et al., 2013) while other emotions, such as surprise, disgust, and fear are the most challenging emotions to convey from arm motions alone (Sawada et al., 2003). In our experiment, anger had the highest recognition rate, whereas disgust and sadness were the most difficult emotions to detect. Further, the results of our study support prior research which indicates that people tend to perceive the emotions conveyed by female agents more accurately than those displayed by male agents.

The study had some limitations that might have affected the results. First, our experiment considered a limited set of affective gestures selected from the GEMEP dataset, hence the results may be due to the inherent features of those gestures. A different group of affective gestures could, for instance, yield higher emotion recognition rates, so it is not possible to state with certainty that body gestures alone are not sufficient for expressing animated agents' emotions.

Second, the affective hand and arm gestures in the GEMEP video database were performed by French actors while most of the subjects were white Americans and Asians. Thus, cultural differences in perception of body gestures might have affected the results. Future experiments should include a larger set of affective body gestures extracted from different datasets and a larger and more diverse pool of subjects. Third, only two animated pedagogical agents were considered in the study. It is possible that the results are dependent in part on the design features and body proportions of the agents. Future research should examine whether the look and visual style of the agents have an effect on perception of the affective gestures.

The findings from our study are important as they advance knowledge on how body gestures contribute to perception/expression of animated agents' emotions. Our research aims to create animated pedagogical agents that can authentically portray personality traits and emotions through lifelike non-verbal cues and speech. This effort seeks to offer a compelling alternative to empathetic human coaches or learning buddies. The study detailed in this paper represents a stride towards reaching this objective.

Acknowledgments. This research is funded by NSF award# 2201019 – "Collaborative Research: Using Artificial Intelligence to Transform Online Video Lectures into Effective and Inclusive Agent-Based Presentations".

References

1. Adamo, N., Mousas, C., Mayer, R.: NSF Award # 2201019 - Collaborative Research: Using Artificial Intelligence to Transform Online Video Lectures into Effective and Inclusive Agent-Based Presentations (2022). https://www.nsf.gov/awardsearch/showAward?AWD_ID=2201019&HistoricalAwards=false&_ga=2.46658222.863696423.1696858525-163693 4363.1694122541
2. Annetta, L.A., Holmes, S.: Creating presence and community in a synchronous virtual learning environment using avatars. Intern. J. Instruct. Technol. Dist. Learn. **3**, 27–43 (2006)
3. Atkinson, A.P., Dittrich, W.H., Gemmell, A.J., Young, A.W.: Emotion perception from dynamic and static body expressions in point-light and full-light displays. Perception **33**(6), 717–746 (2004). https://doi.org/10.1068/p5096
4. Bänziger, T., Scherer, K.R.: Introducing the Geneva Multimodal Emotion Portrayal (GEMEP) corpus. In: Scherer, K.R., Bänziger, T., Roesch, E.B. (eds.) Blueprint for affective computing: A sourcebook, pp. 271–294. Oxford university Press, Oxford, England (2010)

5. Barrett, L.F., Russell, J.A. (eds.): The Psychological Construction of Emotion. The Guilford Press (2015)
6. Baylor, A.L., Kim, S.: Designing nonverbal communication for pedagogical agents: when less is more. Comput. Hum. Behav.. Hum. Behav. **25**(2), 450–457 (2009)
7. Blythe, E., Garrido, L., Longo, M.R.: Emotion is perceived accurately from isolated body parts, especially hands. Cognition **230**, 105260 (2023). https://doi.org/10.1016/j.cognition.2022.105260. Epub 2022 Sep 1. PMID: 36058103
8. Cheng, J., Zhou, W., Lei, X., Adamo, N., Benes, B.: The Effects of Body Gestures and Gender on Viewer's Perception of Animated Pedagogical Agent's Emotions. In: Kurosu, M. (ed.) Human-Computer Interaction. Multimodal and Natural Interaction: Thematic Area, HCI 2020, Held as Part of the 22nd International Conference, HCII 2020, Copenhagen, Denmark, July 19–24, 2020, Proceedings, Part II, pp. 169–186. Springer International Publishing, Cham (2020). https://doi.org/10.1007/978-3-030-49062-1_11
9. Cui, J., Popescu, V., Adamo-Villani, N., Cook, S.W., Duggan, K.A., Friedman, H.S.: Animation stimuli system for research on instructor gestures in education. IEEE Comput. Graph. Appl.Comput. Graph. Appl. **37**(4), 72–83 (2017). https://doi.org/10.1109/MCG.2017.3271471
10. Ekman, P.:. Emotions revealed. New York/London: Times Books (US)/Weidenfeld & Nicolson (2003)
11. Ekman, P., Friesen, W.: The repertoire of nonverbal behavior: categories, origins, usage, and coding. Semiotica **1**(1), 49–98 (1969)
12. Ennis, C., Hoyet, L., Egges, A., McDonnell, R.: Emotion capture: emotionally expressive characters for games. In: Proceedings of Motion on Games (2013). https://doi.org/10.1145/2522628.2522633
13. Fiorella, L., Mayer, R.E.: Principles based on social cues in multimedia learning: Personalization, voice, embodiment, and image principles. In: Mayer, R.E., Fiorella, L. (eds.) The Cambridge Handbook of Multimedia Learning, 3rd edn, pp. 277–286. Cambridge University Press, New York (in press)
14. Guo, Y.R., Goh, D.H.-L.: Affect in embodied pedagogical agents: meta- analytic review. J. Educ. Comput. Res. **53**(1), 124–149 (2015). https://doi.org/10.1177/0735633115588774
15. Johnson, W.L., Lester, J.C.: Face-to-face interaction with pedagogical agents, twenty years later. Int. J. Artif. Intell. Educ.Artif. Intell. Educ. **26**(1), 25–36 (2016)
16. Karg, M., Samadani, A.A., Gorbet, R., Kühnlenz, K., Hoey, J., Kulić, D.: Body movements for affective expression: a survey of automatic recognition and generation. IEEE Trans. Affect. Comput.Comput. **4**(4), 341–359 (2013)
17. Larsson, P.: Discerning emotion through movement – a study of body language in portraying emotion in animation, pp. 6–7 (2014) MS Thesis. Retrieved from: http://www.divaportal.org/smash/record.jsf?pid=diva2%3A723103&dswid=4060
18. Lawson, A.P., Mayer, R.E., Adamo-Villani, N., Benes, B., Lei, X., Cheng, J.: Do learners recognize and relate to the emotions displayed by virtual instructors? Int. J. Artif. Intell. Educ.Artif. Intell. Educ. **31**, 134–153 (2021)
19. Lhommet, M., Marsella, S.: Expressing emotion through posture and gesture. In: Calvo, R., D'Mello, S., Gratch, J., Kappas, A. (eds.) The Oxford Handbook of Affective Computing, pp. 273–285. Oxford University Press, Oxford (2015)
20. Martha, A.S., Santoso, H.B.: The design and impact of the pedagogical agent: a systematic literature review. J. Educ. Online, **16**(1) (2019)
21. Mayer, R.E.: Multimedia learning, 3rd edn. Cambridge University Press, New York (2021)
22. Mayer, R.E., Fiorella, L., Stull, A.: Five ways to increase the effectiveness of instructional video. Educ. Tech. Res. Dev. **68**, 837–852 (2020)
23. Mayer, R.E., DaPra, C.S.: An Embodiment effect in computer-based learning with animated pedagogical agents. J. Exp. Psychol. Appl. **18**(3), 239–252 (2012)

24. Mayer, R.E.: Searching for the role of emotions in e-learning. Learn. Instr. **70**, 101213 (2020). https://doi.org/10.1007/978-3-030-90436-4_38
25. Poggiali, J.: Student responses to an animated character in information literacy instruction. Library Hi Tech **36**(1), 29–42 (2018). https://doi.org/10.1108/LHT-12-2016-0149
26. Posner, J., Russell, J.A., Peterson, B.S. (2005). The circumplex model of affect: an integrative approach to affective neuroscience, cognitive development, and psychopathology. Dev Psychopathol. 2005 Summer **17**(3), 715–734 PMID: 16262989; PMCID: https://doi.org/10.1017/S0954579405050340
27. Rosenberg-Kima, R.B., Baylor, A.L., Plant, E.A., Doerr, C.E.: Interface agents as social models for female students: the effects of agent visual presence and appearance on female students' attitudes and beliefs. Comput. Hum. Behav.. Hum. Behav. **24**, 2741–2756 (2008)
28. Ross, P., Flack, T.: Removing hand form information specifically impairs emotion recognition for fearful and angry body stimuli. Perception **49**(1), 98–112 (2020). https://doi.org/10.1177/0301006619893229
29. Russell, J.A.: Core affect and the psychological construct of emotion. Psychol. Rev. **110**, 145–172 (2003)
30. Russell, J.A.: A circumplex model of affect. J. Pers. Soc. Psychol. **39**, 1161–1178 (1980)
31. Sawada, M., Suda, K., Ishii, M.: Expression of emotions in dance: relation between arm movement characteristics and emotion. Percept. Mot. Skills **97**, 697–708 (2003)
32. Schroeder, N.L., Adesope, O.O., Gilbert, R.B.: How effective are pedagogical agents for learning? a meta-analytic review. J. Educ. Comput. Res. **49**(1), 1–39 (2013)
33. Wang, N., Johnson, W.L., Mayer, R.E., Rizzo, P., Shaw, E., Collins, H.: The politeness effect: pedagogical agents and learning outcomes. Int. J. Hum. Comput. Stud.Comput. Stud. **66**, 98–112 (2008)

The Potential of Holographic Avatars in the Hybrid Workplace: An Industrial/Organizational Psychology Perspective

Nicholas J. Villani[✉] [iD]

Harvard University, Cambridge, MA 02139, USA
nickjvillani@gmail.com

Abstract. Hybrid work, defined as a work arrangement that allows employees to perform work at different locations, will become the most prevalent work model in the next decade. While hybrid work has many benefits, such as the flexibility for employees to work in ways that are most effective for them, it also has drawbacks. Hybrid work is often characterized by ambiguity in work-related information and ineffective communication and may lead to employees feeling disconnected from the rest of the team. Current computer-mediated communication interfaces used in hybrid and fully remote work situations still pose challenges to employees and organizations, such as screen fatigue and lack of non-verbal communication. In this position paper I describe a hypothetical holographic avatar-based two-way communication system that could be used in a variety of hybrid/remote work settings. I discuss how the proposed system could help overcome some of the limitations of current video conferencing tools, fulfill workers' psychological needs and increase workers' motivation.

Keywords: Holographic avatars · Hybrid work · Communication interfaces

1 Introduction

Recent research shows that the main barrier to hybrid/remote working is the need for direct interaction (Wong et al., 2022). With the time spent working remotely expected to increase in the next decade, many employees will need more immersive forms of digital interaction. Holographic interaction systems could provide a solution to this problem by providing a more authentic and immersive form of digital communication.

"Holographic communication refers to real-time capturing, encoding, transporting and rendering of 3D representations, anchored in space, of remote persons shown as stereoscopic images or as 3D avatars in extended reality (XR) headsets that deliver a visual effect similar to a hologram" (Essaili et al., 2022, p. 2). Compared to current video conferencing tools in which workers communicate with each other through small video windows displayed on a flat screen, holographic avatars can convey the nuances

A. L. Brooks (Ed.): ArtsIT 2023, LNICST 565, pp. 281–289, 2024.
https://doi.org/10.1007/978-3-031-55312-7_21

of fully three-dimensional non-verbal communication and provide a sense of presence that leads to more authentic and believable human interactions.

Once confined to the field of science fiction, holographic avatars have now become possible thanks to increased bandwidth and ubiquitous availability of Augmented Reality (AR) devices and due to recent advances in the fields of 3D modeling and animation, AR, and Mixed Reality (MR), game and holographic technologies, Artificial Intelligence (AI), multimodal signal processing and affective computing.

In this position paper I describe a hypothetical mixed reality two-way holographic avatar-based communication system that could be implemented with existing technologies and discuss how the proposed system could overcome the limitations of current communication interfaces in hybrid work settings, increase work performance and improve workers' wellbeing. In Sect. 2, I examine the benefits and disadvantages of hybrid work and current video conferencing platforms. In Sect. 3, I discuss avatar-based communication interfaces and avatar perception. In Sect. 4, I provide a high-level description of the proposed communication system; In Sect. 5 I discuss its benefits, limitations, and possible barriers to adoption.

2   Hybrid Work: Benefits and Disadvantages

Hybrid work is defined as a work arrangement that allows employees to perform work at different locations. According to a McKinsey survey of 100 executives across different industries and countries the hybrid work model will become more common in the future. Most of the surveyed executives expect that most employees will be on-site between 21 and 80 percent of the time, or one to four days per week (McKinsey, 2021).

According to a poll by Gallup of 8,090 hybrid workers (Gallop Workplace, 2022) some of the major advantages of hybrid work include an increase in perceived work-life balance, increased efficiency, the flexibility to choose where and when someone works, less burnout, and greater productivity. Some disadvantages reported in the poll were reduced access to resources and equipment, less connection to organization culture, decreased team interaction, impaired working relationships, and reduced cross-functional communication and collaboration, which was noted as the greatest challenge to hybrid work.

Wong et al. (2022) have substantiated that hybrid work exhibits inherent shortcomings in terms of ambiguity and inefficient communication. In the context of hybrid work environments, the transmission of work-related information predominantly relies on summaries, reports, and video presentations, as opposed to direct experiential learning, informal interactions with colleagues, and observation. Consequently, employees may find themselves devoid of essential contextual information and clarity, which can significantly impact their decision-making and problem-solving abilities. The inefficiency of communication in hybrid work is a well-documented concern (Chang, 2023). According to the well-established Mehrabian's rule, 55% of communication is conveyed through visual cues, such as body language and facial expressions, while 38% relies on vocal cues, including accent, pitch, tone, laughter, and yelling, leaving only 7% attributed to verbal cues like words and connotations (Mehrabian & Ferris, 1967). Although diverse communication media like emails, chats, phone calls, and videoconferencing facilitate communication in hybrid work settings, they often lack the crucial nonverbal cues necessary for effective communication.

The Gallop report (Gallop Workplace, 2022) also showed that hybrid workers spent most of their time in the office in meetings with coworkers and supervisors and on collaborative work with colleagues, whereas when working from home, workers were primarily focused on independent work tasks and reported far less time spent in meetings/collaborative activities with coworkers, customers, and supervisors. One factor contributing to this divide in activities in the office versus time spent at home is the absence of a socioemotional connection while video conferencing. Many workers would rather have meetings in person, as video conferencing does not provide the same level of personal connection and understanding that in-person meetings do.

Current video conferencing platforms such as Zoom, Microsoft Teams and Cisco Webex, which allow users to connect in real time through audio and video pose challenges to employees and organizations. First, video conference meetings can feel impersonal and even with video enabled, employees are simply communicating with screens, and this may prevent employees from developing authentic connections with their coworkers. Second, staring at a flat screen for hours is physically and psychologically draining. Third, one of the most important elements of in-person meetings is how we read others. Video conferencing can inhibit our minds' ability to process movements and body language. We do not gesture, move, or convey emotions as effectively on screen, nor do we read them as well, either. The lack of body language influences effective delivery of the intended message and interferes with close and trusting relationships and social presence among workers in hybrid work settings (Park, 2020).

If there were a way to increase the effectiveness of video conferencing calls by eliciting a greater socioemotional connection between coworkers, resulting in more authentic and efficient meetings whilst at home, the positive aspects of working in the office (such as increased connection and collaboration) could be harnessed from one's home resulting in greater hybrid work potential.

3   Avatar-Based Communication Interfaces

Avatars are a common feature in numerous computer-mediated communication and interaction interfaces, and their prevalence is on the rise (Wu et al., 2021). Various definitions of the term "avatar" are being used, encompassing a spectrum that extends from 2D static icons to highly detailed human replicas. For the purposes of this paper, the term "avatar" is employed to refer specifically to lifelike 3D animated digital human characters.

"Among the first judgments made of an avatar in a digital [or mixed reality environment] is determining its ….humanity" (Nowak & Fox, 2018, p. 35). Research studies have shown that people are likely to establish stronger connections with avatars that show human features (Sheehan & Sosna, 1991). More specifically, HCI scholars argue that 'agency', 'anthropomorphism', and 'realism' are the three most important characteristics affecting people's perceptions of avatars (Nowak & Fox, 2018). Agency refers to the degree to which an entity is perceived to be human, anthropomorphism refers to an entity having human form and/or behavior and realism refers to having accurate form and/or behavior.

While the terms 'avatar' and 'agent' are often used interchangeably, they typically refer to distinct entities. In general, an avatar is a virtual human character whose actions are guided by a human operator, whereas an agent is a virtual human whose actions are determined by a computer algorithm. Research comparing agents to avatars has demonstrated that representations controlled by humans or perceived as being controlled by humans tend to be more persuasive than those controlled or perceived to be controlled by autonomous agents (Nowak & Fox, 2018). Based on these empirical findings, I posit that a future effective interaction system centered around avatars should incorporate holographic avatars that are partially under the control of their human counterparts, such as remote employees, possibly facilitated through handheld devices (refer to Fig. 1).

Prior research also shows that a higher level of form and behavioral anthropomorphism may lead to increased user engagement, social presence, and interaction effectiveness (Bailenson et al., 2006; Kang & Watt, 2013). Anthropomorphism is also related to social influence; digital representations that show a high degree of anthropomorphism are likely to be more persuasive (Guadagno et al., 2007). Drawing on these research findings, I envision a system of animated avatars that not only have human appearance, but also speak, move, gesture and express emotions in the same ways as humans.

Avatar realism can be defined on different levels. For instance, in regard to appearance, Steed and Schoereder (2015) define avatar realism as faithfulness of its representation, while they define behavioral realism as natural, accurate physical behavior. Some research studies suggest that the photorealism of an avatar may enhance the feeling of co-presence in AR systems, e.g., the belief that the remote person is actually there (Orts-Escolano et al., 2016). Pakanen et al. (2022) argue that in professional settings, such as educational or work environments, the avatar should look like its human counterpart and be clearly recognizable, as people need to be sure that the person is the one that they expect to be in order to trust them.

Hence, a system of holographic photorealistic avatars that look and behave exactly like their human counterparts might be the best solution. However, there are also advantages in using avatars whose appearance is different from their human correspondents. For instance, it is widely understood that individuals with disabilities often experience workplace bias and discrimination. Yet, when they can choose an identity and embodiment in VR/AR environments, the results can be extremely positive (Davis & Chansiri, 2019).

4 Holographic Avatars in the Hybrid Workplace: A Vision for the Future

Recent advances in 3D modeling and animation, AR/MR, game and holographic technologies, AI, multimodal signal processing, and affective computing together with increased bandwidth and ubiquitous availability of AR devices are paving the way for the next generation communication/interaction interfaces in the workplace.

Figure 1 illustrates a two-way holographic avatar-based interaction system that could be used in a variety of work contexts to overcome the limitations of current video conferencing systems. The employee working remotely controls the holographic avatar (on the job site) via speech input through a handheld device. The avatar model, which could be a high-fidelity replica of its human counterpart (e.g., a digital twin), can be created using a variety of methods including manual sculpting in a 3D modeling software, photogrammetry techniques, or 3D scanning. The avatar rig, including the skeleton and animation controls can be manually created in an animation software or generated automatically using auto rig tools. The avatar animation is generated by performing speech analysis: (1) phonemes are identified from the speech and mapped to the corresponding visemes (mouth shapes) to generate the avatar lip synch animation;

(2) data-driven approaches, such as machine learning and deep learning are used to generate conversational gestures and facial articulations on-the-fly from speech input (Asakawa et al., 2022); and (3) affective/expressive gestures are identified based on speech prosody analysis and rule-based methods (Adamo et al., 2021) and retrieved from a database of previously captured gestures of the employee–the employees are videotaped at different times, their affective/expressive gestures are extracted from the videos and transformed into animation data which is stored in a gesture database. The result is a human-controlled full-body animated holographic avatar that is viewed by the onsite employees as a holographic projection (Arcao et al., 2019) or through an AR headset, such as Microsoft Hololens. The avatar speaks, moves and gestures exactly like their human counterpart.

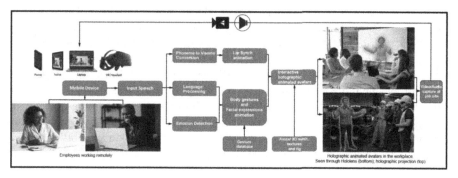

Fig. 1. Diagram illustrating a hypothetical mixed reality two-way holographic avatar-based communication/interaction system in hybrid work settings.

A 360-degree camera capturing what is happening on the jobsite/office sends back a continuous video/audio stream to the remote employee. The remote workers can display the video/audio signal on a laptop, tablet or phone or they can view the other employees who are on the job site or in the office through a VR headset, which allows for a more immersive experience. This two-way communication can be implemented using a game engine, such as Unity.

5   Discussion

The proposed mixed reality holographic avatar-based communication system could be applied to many work contexts. When training employees from remote locations, a holographic avatar may provide a more intimate context for employees to learn, eliciting the feeling that someone was actually there to teach in physical form, thus providing a socioemotional connection that may result in increased learning. When recruiting or interviewing many candidates at once or one on one, a holographic digital twin of the interviewer may create a more personable experience for applicants resulting in less dropout rates. When having employment or termination conversations, an avatar may contribute to feelings of ease and stave off perceived indifference from a company to an employee. Employees often have feelings of insignificance when being let go through a screen, whereas the avatar may help to create a more personable and candid situation in which the employee doesn't harbor feelings of resentment due to perceived indifference on the part of the company.

The proposed mixed reality holographic avatar interaction system will help overcome the problems of ineffective communication and ambiguity discussed in Sect. 2. By communicating using visual (face and body), vocal, and verbal cues, the holographic avatars will allow for more efficient communication. Further, by conveying the feeling that the remote workers "are really there" on the job site, the holographic avatars will afford more direct experiences and interactions and hence, less ambiguity in work-related information.

In accordance with self-determination theory, workers exhibit autonomous motivation when their three fundamental psychological needs are met: autonomy, relatedness, and competence (Deci & Ryan, 2000; Gagné & Deci, 2005). Autonomy is defined as "the experience of acting with a sense of choice, volition, and self-determination" (Stone et al., 2009). When workers have the freedom to decide when, where, and how they work, their psychological need for autonomy is fulfilled within the workplace. The second psychological need, relatedness, is described as "the experience of having fulfilling and supportive social relationships" (Stone et al., 2009). When workers perceive a lack of meaningful connection with their colleagues and a lack of support for their work, their satisfaction of the psychological need for relatedness diminishes, potentially leading to a decline in motivation. Competence refers to "the belief that one possesses the ability to influence important outcomes" (Stone et al., 2009). In a hybrid work environment often characterized by ambiguity, if workers lack clarity due to the absence of direct and authentic interactions with their co-workers, they may develop doubts about their competence to successfully carry out their work. This uncertainty can erode their autonomous motivation towards their tasks (Deci & Ryan, 2000; Gagné & Deci, 2005).

The avatar-based two-way interaction system will enhance workers' (1) *competence*–the remote employee, such as for instance a factory supervisor, can 'teleport' to the job site, monitor the situation and provide expert advice where it is needed by communicating in a more direct and effective way with the onsite workers through a full body three-dimensional avatar that speaks gestures and conveys emotions. (2) *Autonomy*–by controlling their holographic avatar via speech through a standard mobile device, the employees can perform their work effectively and establish a sense of presence from any location. (3) *Relatedness*–through holographic avatar teleportation, which allows for verbal/vocal/visual communication and an enhanced feeling of immersion, presence and immediacy, the remote workers can establish stronger and more authentic social connections with the onsite employees.

Holographic avatar-based communication has also the potential to improve diversity, equity, and inclusion in the workplace. Recent research suggests the importance of choice in avatar representation for creating inclusive work environments for people with disabilities and underrepresented groups in general (Davis & Chansiri, 2019). On the other hand, some researchers caution that avatar-based virtual work environments do not constitute unconditional and neutral spaces and workplace avatars whose appearance can be customized by the employee *"may produce a new locus for bias to flourish"* (Martin, 2012, p. 605). Another possible drawback of avatar technology is the feeling of uncanniness that could be experienced when the avatars try to replicate the appearance and behaviors of their human counterparts (Mori & Mcdorman, 2012).

5.1 Limitations and Future Work

One limitation of the proposed system is that the benefits of the holographic avatars are only available on the jobsite, as the remote worker still sees and interact with the other employees through a flat screen or a VR headset (which allows a more immersive experience than a flat screen). Based on the presented motivation and introduction, it would also be important for the remote worker to experience the advantages of holographic presence. This could be accomplished in a future development of the system by creating a shared holographic workspace in which all employees, represented by their digital twins, can share content, and communicate with each other through speech, facial expressions, and body language.

Another limitation is that the avatars described in the proposed system are not completely human-controlled. In Sect. 3, I argue that avatars controlled by humans are more persuasive than those controlled by autonomous agents. However, the avatars in the proposed system are only partially controlled by humans, as some gestures and facial expressions are generated on the fly from speech input. One might argue that the algorithmically generated gestures and facial articulations might not actually match the ones of the real human and thus lead to a discrepancy between audio and behavior, adding to ambiguity, rather than improving clarity and communication. One way to solve this problem would be to motion capture larger sets of gestures and facial articulations of each employee, store them in a database and retrieve all gestures from the database rather than generate them on the fly. This would reduce the possibility of the avatar's gestures not matching the expressive style of the real human and would decrease the risk of uncanniness and ambiguity.

Additional potential barriers to the implementation of holographic avatars in hybrid work settings include high initial cost of infrastructure, bandwidth limitations, reluctance to adapt to new trends and lack of technical know-how. I anticipate that these barriers will be overcome by future technology advances and by a more technically prepared future workforce.

Despite these potential obstacles, the holographic avatar system presented in this paper represents a step forward toward future highly technological work environments in which the differences between human-computer interactions and human-to-human interactions are becoming less and less pronounced. Further, in the future of work in an increasingly digital world, holographic avatars may contribute to the creation of more inclusive workplaces by providing increased access and creative opportunities for people who are often marginalized as 'misfits'.

References

Adamo, N., et al.: Multimodal affective pedagogical agents for different types of learners. In: Russo, D., Ahram, T., Karwowski, W., Di Bucchianico, G., Taiar, R. (eds.) IHSI 2021. AISC, vol. 1322, pp. 218–224. Springer, Cham (2021). https://doi.org/10.1007/978-3-030-68017-6_33

Arcao, J.A., Carl Cadag, V., Martinez, V., Roxas, E., Serrano, K.J., Tolentino, R.: Holographic projection of 3D realistic avatar that mimics human body motion. In: Proceedings of IEEE 1st International Conference on Innovations in Information and Communication Technology (ICIICT), Chennai, India, pp. 1–6 (2019). https://doi.org/10.1109/ICIICT1.2019.8741454

Asakawa, E., Kaneko, N., Hasegawa, D., Shirakawa, S.: Evaluation of text-to-gesture generation model using convolutional neural network. Neural Networks 15, 365–375 (2022). https://doi.org/10.1016/j.neunet.2022.03.041

Bailenson, J.N., Yee, N., Merget, D., Schroeder, R.: The effect of behavioral realism and form realism of real-time avatar faces on verbal disclosure, nonverbal disclosure, emotion recognition, and copresence in dyadic interaction. Presence: Teleoper. Virt. Environ. 15(4), 359–372 (2006). https://doi.org/10.1162/pres.15.4.359

Blascovich, J., Loomis, J., Beall, A.C., Swinth, K.R., Hoyt, C.L., Bailenson, J.N.: Immersive virtual environment technology as a methodological tool for social psychology. Psychol. Inq. 13(2), 103–124 (2002). https://doi.org/10.1207/S15327965PLI1302_01

Chang, H.: Business communication post corona: strategies for the hybrid workplace. Bus. Commun. Res. Pract. 6(1), 1–3 (2023). https://doi.org/10.22682/bcrp.2023.6.1.1

Davis, D.Z., Chansiri, K.: Digital identities – overcoming visual bias through virtual embodiment. Inf. Commun. Soc. 22(4), 491–505 (2019)

Deci, E.L., Ryan, R.M.: The 'what' and 'why' of goal pursuits: human needs and the self-determination of behavior. Psychol. Inq. 11(4), 227–268 (2000). https://doi.org/10.1207/S15327965PLI1104_01

Essaili, A.E., et al.: Holographic communication in 5G networks. Ericsson Technol. Rev. 2022, 2–11 (2022)

Gagné, M., Deci, E.L.: Self-determination theory and work motivation. J. Organ. Behav. 26(4), 331–362 (2005). https://doi.org/10.1002/job.322

Gallop Workplace. The Advantages and Challenges of Hybrid Work (2022). https://www.gallup.com/workplace/398135/advantages-challenges-hybrid-work.aspx

Guadagno, R.E., Blascovich, J., Bailenson, J.N., Mccall, C.: Virtual humans and persuasion: the effects of agency and behavioral realism. Media Psychol. 10(1), 1–22 (2007). https://doi.org/10.1080/15213260701300865

Kang, S.-H., Watt, J.H.: The impact of avatar realism and anonymity on effective communication via mobile devices. Comput. Hum. Behav. 29(3), 1169–1181 (2013). https://doi.org/10.1016/j.chb.2012.10.010

Martin, N.T.: Diversity in the virtual workplace: performance identity and shifting boundaries of workplace engagement. Lewis Clark Law Rev. 16(2), 606–646 (2012)

McKinsey & Company. What executives are saying about the future of hybrid work. McKinsey Insights May 17, 2021 (2021). performance/our-insights/what-executives-are-saying-about-the-future-of-hybrid-work#/

Mehrabian, A., Ferris, S.R.: Inference of attitudes from nonverbal communication in two channels. J. Consult. Psychol. 31, 248–252 (1967). https://doi.org/10.1037/h0024648

Mori, M., Macdorman, K.F.: The Uncanny Valley: The Original Essay by Masahiro Mori. IEEE Spectrum (2012)

Nowak, K.L., Fox, J.: Avatars and computer-mediated communication: a review of the definitions, uses, and effects of digital representations. Rev. Commun. Res. **6**, 30–53 (2018). https://doi.org/10.12840/issn.2255-4165.2018.06.01.015

Orts-Escolano, S., et al.: Holoportation: virtual 3D teleportation in real-time. In: Proceedings of the 29th Annual Symposium on User Interface Software and Technology, New York, October 2016, pp. 741–754 (2016). https://doi.org/10.1145/2984511.2984517

Pakanen, M., Alavesa, P., van Berkel, N., Koskela, T., Ojala, T.: Nice to see you virtually: thoughtful design and evaluation of virtual avatar of the other user in AR and VR based telexistence systems. Entertain. Comput. **40**, 100457 (2022). https://doi.org/10.1016/j.entcom.2021.100457

Park, C.H.: Diagnosis: assessing virtual group dynamics and providing feedback. In: Virtual Group Coaching to Improve Group Relationship: Process Consultation Reimagined, 1st edn, pp. 71–94. Routledge (2020)

Sheehan, J.J., Sosna, M. (eds.): The Boundaries of Humanity: Humans, Animals, Machines. University of California Press, Berkeley (1991)

Steed, A., Schroeder, R.: Collaboration in immersive and non-immersive virtual environments. In: Lombard, M., Biocca, F., Freeman, J., IJsselsteijn, W., Schaevitz, R.J. (eds.) Immersed in Media, pp. 263–282. Springer, Cham (2015). https://doi.org/10.1007/978-3-319-10190-3_11

Stone, D.N., Deci, E.L., Ryan, R.M.: Beyond talk: creating autonomous motivation through self-determination theory. J. Gen. Manag. **34**(3), 75–91 (2009). https://doi.org/10.1177/030630700903400305

Wong, S.I., Berntzen, M., Warner-Søderholm, G., Giessner, S.R.: The negative impact of individual perceived isolation in distributed teams and its possible remedies. Hum. Resour. Manag. J. **32**(4), 906–927 (2022). https://doi.org/10.1111/1748-8583.12447

Wu, Y., Wang, Y., Jung, S., Hoermann, S., Lindeman, R.W.: Using a fully expressive avatar to collaborate in virtual reality. Front. Virt. Real. **2** (2021). https://doi.org/10.3389/frvir.2021.641296

Construction of Immersive Art Space Using Mirror Display and Its Evaluation by Psychological Experiment

Ryohei Nakatsu[1]([⊠]), Naoko Tosa[1], Yunian Pang[1], Satoshi Niiyama[2],
Yasuyuki Uraoka[3], Akane Kitagawa[3], Koichi Murata[3], Tatsuya Munaka[3],
Yoshiyuki Ueda[1], Masafumi Furuta[3], and Michio Nomura[1]

[1] Kyoto University, Sakyo, Kyoto 606-8317, Japan
nakatsu.ryohei@gmail.com, {tosa.naoko.5c,pang.yunian.2c,
ueda.yoshiyuki.3e,nomura.michio.8u}@kyoto-u.ac.jp
[2] AGC Inc., Yokohama 230-0045, Kanagawa, Japan
satoshi.niiyama@agc.com
[3] Shimadzu Corporation, Seika-cho, Soraku-gun, Kyoto 619-0237, Japan
{uraoka-y,kitagawa.akane.5jf,murata.koichi.9ar,munaka,
m_furuta}@shimadzu.co.jp

Abstract. How art appreciation affects the human mind is an interesting question. Several studies have already been conducted on art's calming and inspiring effects on the human mind. As an extension of this, whether art can enhance people's creativity is a fundamental and interesting question. If art appreciation can enhance people's creativity, a new function of art will be discovered. It is well known that displaying media art, such as video art, in a large space can provide a deeply immersive experience, as with projection mapping. A deep sense of immersion can contribute to enhancing creativity. Therefore, it is desirable to research to evaluate whether creativity is aroused by displaying and viewing media art in a large space. One way to make a small space look vast is to construct a space using mirrors. We have designed and constructed an immersive space surrounded by mirror displays with the functions of both a mirror and a display. Firstly, this paper describes the specific method. In addition, using art created by one of the authors, we conducted a psychological experiment using 40 subjects to compare and evaluate her art and geometric figures displayed in the space. The results show that the immersive space and her art combination have characteristics that motivate people's minds and arouse creativity.

Keywords: Immersive Space · Art Space · Mirror Display · Psychological Evaluation

1 Introduction

Art can enrich people's minds, heal their hearts, and inspire them [1, 2]. Art can be considered the ultimate VR with the power to immerse people. Much emphasis has been placed on technological research to give people a sense of immersion in VR. However,

© ICST Institute for Computer Sciences, Social Informatics and Telecommunications Engineering 2024
Published by Springer Nature Switzerland AG 2024. All Rights Reserved
A. L. Brooks (Ed.): ArtsIT 2023, LNICST 565, pp. 290–304, 2024.
https://doi.org/10.1007/978-3-031-55312-7_22

more research needs to be conducted on designing and constructing immersive spaces that combine art and VR and their evaluation.

We designed and constructed an immersive space suitable for art content to evaluate how art content affects the human mind. The video art of Naoko Tosa (hereafter "Tosa art"), one of the authors, will be used as art content. As described later, Tosa art uses technology to extract the beauty hidden in natural phenomena and transform it into video art characterized by its abstract and organic forms. Many people who have viewed Tosa art have commented that they feel like they are in outer space or feel a sense of floating. Therefore, the characteristics of Tosa art are best expressed when viewed in a vast space. To give viewers the feeling of being in an infinite space, we conceived the idea of constructing a space surrounded by mirror displays that function as both a mirror and a display and having viewers appreciate Tosa art in that space.

This paper describes the design and construction method of an immersive space constructed using a mirror display. In addition, we evaluated how combining the immersive space and Tosa art would affect people's minds through a psychological experiment.

2 Related Studies and Activities

2.1 Research on Immersive Spaces in VR

The purpose of VR is to create a space different from reality and to give people an immersive feeling as if it were reality [3, 17]. VR space can be constructed by projecting images into an actual space using a projector or displaying images on an HMD (Head Mounted Display). In both cases, there is much research on adding the senses of touch, taste, and smell to increase the sense of presence. As these are primitive human senses, however, there is a problem that research progress takes time [4].

2.2 Fusion of VR and Art

Attempts to fuse art and VR occurred with the advent of VR and have continued until now. For example, William Latham of Goldsmiths, University of London, has been actively creating an art-expressed artificial life form called "Mutator VR" [5]. In the 1990s, there were many attempts to create an immersive space (CAVE is a typical example of such an immersive space [6]) using projectors, etc., and to display art in the space.

2.3 Construction of Immersive Space Using Mirrors

Mirrors are often used in art expression because it is relatively easy to create a seemingly endless space by using mirrors. One well-known example is Yayoi Kusama's "Infinity Mirror Room," in which she installed her art in a mirrored space [7].

3 Digital Art "Sound of Ikebana"

3.1 Concept of "Sound of Ikebana"

One of the authors, Naoko Tosa, discovered that by applying sound vibrations to a fluid such as paint and photographing it with a high-speed camera, the fluid creates a shape similar to that of a flower arrangement. Tosa further edited the resulting video to match the colors of the Japanese seasons and created a digital artwork called "Sound of Ikebana." Fig. 1 shows a scene from the work. For the details of the art creation process, please refer to [8, 9, 13]. Although there have been various research on the visualization of sound, called "Cymatics," (for example [18]), this is another way of sound visualization.

Fig. 1. A scene from "Sound of Ikebana."

3.2 Effects of "Sound of Ikebana" on Human

When Tosa exhibited her digital art around the world with a focus on the "Sound of Ikebana," many overseas art professionals pointed out that "Tosa's digital art, which expresses beauty latent in physical phenomena in an abstract form, expresses beauty previously unnoticed by Westerners, and this is Japan's consciousness and sensitivity." Since then, "Sound of Ikebana" has taken on challenges in new directions, such as attempting to create new shapes by using the birth cries of newborn babies and the voices of Olympic athletes as sound sources and attempting to create art in the space age by creating works under microgravity [9]. Many people who have viewed Tosa art have commented that they feel their creativity is enhanced. A stimulating new art effect can be found if art appreciation enhances the viewer's creativity. Such effects are apparent in a space with infinite expansion. This also led us to design and construct a space that gives a sense of infinite expansion and have visitors view Tosa art in that space to see if it improves creativity.

4 Design and Construction of Immersive Spaces Using Mirror Displays

4.1 Mirror Display

As mentioned in Sect. 2.3, using mirrors is appropriate for constructing a system that gives the impression of being in an infinite space. Here, we decided to use a "mirror display" with the functions of both a mirror and a display.

Several companies have commercialized mirror displays that combine the functions of a mirror and a display. We used a mirror display developed by AGC Corporation and commercialized under the name "Mirroria" [10]. The feature of this display is that it achieves a half-mirror reflectance of approximately 65%, the same level of reflectance as that of an ordinary mirror, by utilizing the company's glass manufacturing technology.

4.2 Design and Construction of Immersive Spaces

Several psychological experiments we have conducted have confirmed that art content positively affects the human mind [11, 12]. Art content was displayed on large LED and mirror displays in these experiments. To take this further and confirm whether art content is effective in improving people's creativity, placing people in a more immersive environment would be effective.

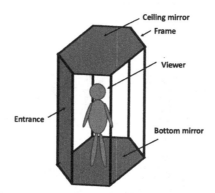

Fig. 2. Conceptual diagram of immersed space.

Therefore, we decided to construct an environment in which the space is surrounded by mirrors, and part of the mirrors are used as mirror displays, in which art contents are displayed. First, a hexagonal space surrounded by rectangular mirrors is constructed. The concept is shown in Fig. 2, where the hexagonal space comprises three sets of two mirrors facing each other. It is well known that mirrors create an infinite number of images by mutually reflecting each other. By having three sets of mirrors, the person inside feels as if he/she is surrounded by countless mirror images of himself/herself. Furthermore, by using the ceiling and floor as mirrors, one feels as if one is surrounded by an infinite number of images of oneself, both above and below.

The six mirrors that make up this hexagonal space are mirror displays and can display images. Since the vertical length of the mirrors is longer than the vertical length of the display, the display on which the images are shown forms part of the mirrors. At the same time, the position of the display is variable in the vertical direction (Fig. 3). This makes it possible to shift the position where the six mirrors display the image. Suppose the mirrors facing each other have the same position for displaying the images. In that case, the respective images will interfere with each other, reducing the sense of an endless series of images. Thus, by shifting the position of the image display, it is possible to create the effect of an endless series of images without having each image interfere with the other.

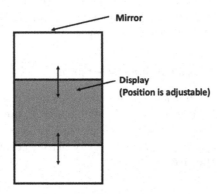

Fig. 3. Configuration of individual mirror displays.

Fig. 4. Exterior view of the immersed space (left: exterior view, right: door open).

The appearance of the constructed immersive space is shown in Fig. 4. Inside this device, even simple shapes can generate an environment of beauty by continuing back and forth, left and right, and up and down indefinitely (Fig. 5). Figure 6 shows several scenes where Tosa art is shown as an example of art content.

In this immersive space, preliminary experiments have confirmed that people can experience a sense of floating and liberation. Since a sense of liberation and floating

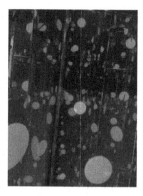

Fig. 5. Geometric figure (circle) displayed in the immersed space.

Fig. 6. Example of displaying Tosa Art in the immersive space.

are linked to creativity, people's creativity is expected to be aroused when art content is displayed in this immersive space. Then we can set the following hypothesis.

Hypothesis: The combination of the immersive space and Tosa art increases human creativity.

Next, we conduct psychological experiments to check this hypothesis.

5 Evaluation of Immersive Space by Psychological Experiments

5.1 Evaluation Concept

We evaluated the constructed immersive space. This immersive space gives people inside it the feeling of being in an infinitely expanding space. Therefore, when evaluating art in it, it is considered a good match with contemporary art, such as video art and media art. As mentioned earlier, there have been many attempts to combine VR and art, but art content can only demonstrate its actual value if viewed in a space suitable for it.

To evaluate this immersive space, we conducted a psychological experiment to compare and evaluate the impression subjects receive when art content and content to be compared are displayed. As for the art content, we decided to use Tosa art. The reasons for this are as follows.

(1) As mentioned, Tosa art is created by filming fluid phenomena with a high-speed camera. This indicates that Tosa art is based on physical phenomena. Therefore, it is more compatible with the scientific evaluation method than compared to art created by the artist's own hands.
(2) Since it is based on fluid phenomena, various variations can be created by changing parameters, such as the type of fluid and the type of sound. In this respect, it is also compatible with the scientific evaluation method under different conditions.

5.2 Contents Used in the Experiments

(1) Art Content
    For the reasons stated above, we decided to use Tosa art. Specifically, we used a 3-min video with the "Sound of Ikebana" as its primary content.
(2) Comparative Content
    When conducting evaluation experiments using art content, preparing content for comparison with the art content is essential. We used simple geometric figures such as circles and squares as comparison contents. We conducted a preliminary experiment and evaluated several geometric figures through psychological experiments to determine the geometric figures to be compared with the art contents. The following three types of geometric figures were used in the preliminary experiment.

Geometric Fig. 1: The shape is a circle, and only the color changes with time.
Geometric Fig. 2: The shapes change to circles and squares in sequence along with the colors.
Geometric Fig. 3: The shape is a square, and the square rotates. The colors change with time, as in Geometric Figs. 1 and 2.

Here, the colors were set to be the same as the representative color of the art content, in synchronization with the time variation of the color of the art content, to create a similar impression as the art content. Preliminary experiments showed no significant differences among the three types of geometric shapes. As the degree of change for Geometric Fig. 2 is in the middle among the three types, we decided to use Geometric Fig. 2 (hereafter referred to as "Figure") for comparison with the art content. The details of the preliminary experiments are described in the literature [13] and can be found there.

5.3 Evaluation Items

Regarding the evaluation items, first, an evaluation item, "Impression factor," was established to determine what impression the subjects had. This has been used in several psychological experiments such as [14–16] and used by us for art evaluation.

In addition, since one of the purposes of this evaluation is to assess whether the combination of "immersive space + art content" arouses people's creativity, we decided to add an evaluation item regarding how it affects people's minds. As a result of discussions led by one of the authors, Michio Nomura, who specializes in psychology, we decided to evaluate the content in terms of whether it relaxes people's minds ("Relaxation factor"), whether it inspires people's minds ("Motivation factor"), and whether it arouses people's creativity ("Creativity factor"). Specific evaluation items are shown in Table 1 below. Overall, there are 24 evaluation items, which is done on a 7-point scale. The difference in meaning between "immersed," one of the Motivation items, and "immersive," one of the Creative items, is subtle, but "immersed" corresponds to logical brain processing, such as "immersed oneself in studying." In contrast, "immersive" corresponds to sensory brain processing, such as "listening to music makes me immersive."

Table 1. Evaluation Items

1. Impression factor (9 items)	3. Motivation factor (5 items)
Comfortable - Uncomfortable	Enthusiastic – Not enthusiastic
Friendly - unfriendly	Immersed – Not immersed
Beautiful - Not beautiful	Curious – Not curious
Calm - Restless	Motivated – Not motivated
Interesting - Boring	Aroused – Not aroused
Warm - Cold	**4. Creativity factor (5 items)**
Changeable - Not changeable	Associate – Do not associate
Luxury - Sober	Immersive – Not immersive
Individual – Ordinary	Activated – Not activated
2. Relaxation factor (5 items)	Inspired – Not inspired
At ease – Not at ease	In the zone – Not in the zone
Secure – Not secure	
Pleasant – Not pleasant	
Relaxed – Not relaxed	
Healed – Not healed	

5.4 Subjects

Forty male and female students (32 males and eight females) in their first through fourth year at Kyoto University were used as subjects.

5.5 Experimental Procedure

Below is the process of the experiment.

(1) First, after briefly explaining the purpose and content of the experiment, a subject signed a consent form.
(2) The subject moved into the immersive space.
(3) Then the subject performed an initial evaluation before viewing Content 1 ("No content" condition). The subject brought his/her own smartphone into the space, and the evaluation was done using Google Forms.
(4) Before Content 1 was displayed, a resting period (3 min) was taken to reset the subject's psychological state. During this time, the display was kept black.
(5) Contents 1 was displayed (3 min).

(6) After viewing Content 1, the subject was asked to complete a second evaluation.

(7) Before Content 2 was displayed, a resting period (3 min) was taken to reset the subject's psychological state. During this time, the display was kept black.

(8) Contents 2 was displayed (3 min).

(9) After viewing Content 2, the subject was asked to complete a third evaluation.

(10) Then, the subject exited the immersive space.

Regarding the order of presentation of art and geometric figures, to ensure that order effects did not affect the results, the order was set randomly for each subject so that the total order of art → geometric figures and geometric figures → art was 20.

6 Evaluation Results

6.1 Results for Each Evaluation Factor

The average evaluation scores of 40 subjects for each Impression, Relaxation, Motivation, and Creativity factor are shown in Figs. 7, 8, 9, and 10. In these figures, the graphs show the differences in the evaluation scores for three different contents: while the display was kept black ("No content"), after viewing the geometric figures ("Figure"), and after viewing the art content ("Art"). Also, the results of the analysis of variance (ANOVA), which will be described later, are overlapped on these figures.

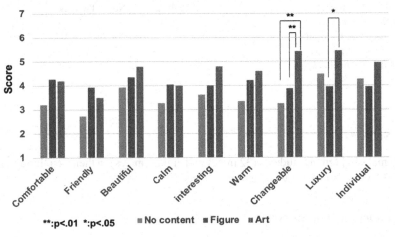

Fig. 7. Evaluation results for "Impression factor."

In the individual evaluation results, for many of the evaluation items, the results were higher in the order of "No content < Figure < Art," indicating the effectiveness of the "immersive space + Tosa art" approach.

6.2 Analysis of Variance (ANOVA)

In order to verify the significance of the differences in evaluation scores between each content in Sect. 6.1, an analysis of variance with the two factors (two-way ANOVA)

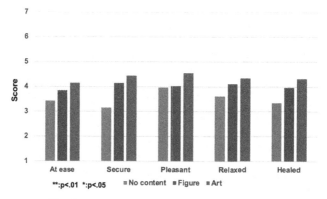

Fig. 8. Evaluation results for "Relaxation factor."

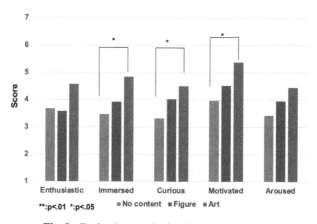

Fig. 9. Evaluation results for "Motivation factor."

was conducted for each of the Impression factor, Relaxation factor, Motivation factor, and Creativity factor. The two factors involved in the two-way ANOVA are "content" and "evaluation item." The results are shown in Fig. 11 (Impression factor), Fig. 12 (Relaxation factor), Fig. 13 (Motivational factor), and Fig. 14 (Creativity factor).

Also, multiple comparison was conducted on the individual assessment items for a more detailed analysis. The results are overlapped in Fig. 7 (Impression factor), Fig. 8 (Relaxation factor), Fig. 9 (Motivation factor), and Fig. 10 (Creativity factor).

6.3 Considerations

(1) ANOVA results by factor

A two-way ANOVA was conducted on the overall Impression, Relaxation, Motivation, and Creativity factors. The results showed that, except for the Relaxation factor, there were significant differences among "No content," "Figure," and "Art" (Figs. 11 through 14). Specifically, for the Impression factor, there was a significant difference at the 1% level among each combination of "No content," "Figure," and

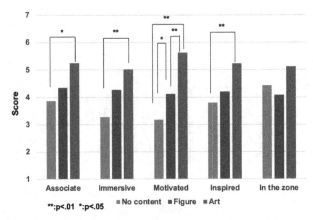

Fig. 10. Evaluation results for "Creativity factor."

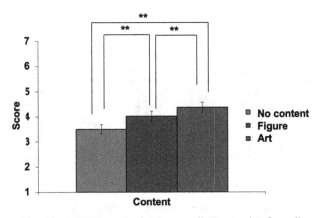

Fig. 11. ANOVA results for the overall "Impression factor."

"Art" (Fig. 11). As for the Motivation and Creativity factors, there was a significant difference at the 1% level between "No content" and "Art" and between "Figure" and "Art." Also, there was a significant difference between "No content" and "Figure" at the 5% level (Figs. 13 and 14).

For the Relaxation factor, we found a significant difference at the 1% level between "No content" and "Art" but no significant difference for the other combinations (Fig. 12).

(2) Multiple comparison results for individual evaluation items

Results of the multiple comparisons for individual assessment items showed the following.

For the nine items on the Impression factor, for the item "Changeable," there were significant differences between "No content" and "Art," and "Figure" and "Art" at the 1% level. For the item "Luxury," there was a significant difference between "Figure" and "Art" at the 5% level.

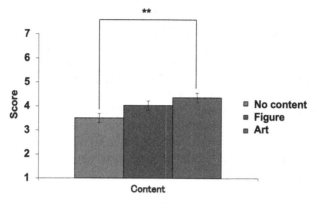

Fig.12. ANOVA results for the overall "Relaxation factor."

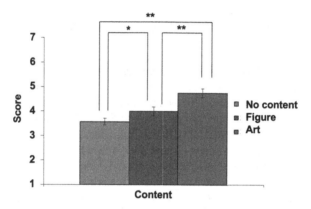

Fig. 13. ANOVA results for the overall "Motivation factor"

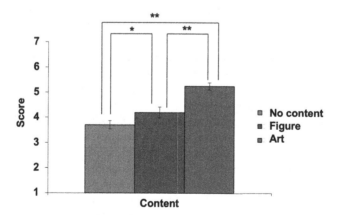

Fig. 14. Analysis of variance results for the overall "Creativity factor."

There were no significant differences among "No content," "Figure," and "Art" on any of the five items of the Relaxation factor.

There was a significant difference for three of the five items on the Motivation factor. For the items "Immersed," "Curious," and "Motivated," there were significant differences between "No content" and "Art" at the 5% level.

There were significant differences for four of the five items on the Creativity factor. For the item "Motivated," there were significant differences between "No content" and "Art" and between "Figure" and "Art" at the 1% level. Also, there was a significant difference between "No content" and "Figure" at the 5% level. For the items "Immersive" and "Inspired," there were significant differences between "No content" and "Art" at the 1% level. Also, for the item "Associate," there was a significant difference between "No content" and "Art" at the 5% level.

These results indicate that Tosa art effectively motivates people and improves their creativity. Therefore, the hypothesis set at the beginning of this psychological experiment was supported.

7 Conclusion

Previous studies have shown that art has a calming and inspirational effect on the human mind [1, 2]. Through several psychological experiments, we have also found that art has a relaxing and inspiring effect on the human mind [12, 13]. In addition to this, we hypothesize that art has the effect of increasing a person's creativity. This is because when we exhibited video art created by Naoko Tosa, one of the authors, in various parts of the world, many people commented that they "felt a life force" and "felt Japanese beauty." In contrast, others said they "felt a sense of levitation" and "their creativity was aroused." If art appreciation increases people's creativity, then a new benefit of art can be found. This study was conducted to confirm this through a psychological experiment.

In this paper, we first described designing and constructing an "immersive art space" suitable for art appreciation. The fact that projection mapping using art images is used in many situations means that displaying art in a vast space increases the sense of immersion. In this study, we proposed and constructed a hexagonal immersive space using a mirror display that functions as both a mirror and a display to create a sense of being in a vast space. In this space, three sets of mirrors facing each other create the impression of an infinite space. By displaying art images in the mirrors, it is possible to give people the feeling of being in an infinite space surrounded by art images.

In the latter half of this paper, we described the results of an experiment to confirm whether displaying art in the constructed space enhances creativity through psychological evaluation. The art used for the evaluation was the video art by Naoko Tosa described above. Geometric figures were used as the content to be compared with the art content. Based on the results of an experiment to compare multiple types of geometric figures with different shapes and movements [13], one of them was selected and used in this study.

We asked 40 subjects to rate on a 7-point scale how they felt when viewing the three types of content, "No content," "Figure," and "Art," using 24 items in four groups related to "Impression," "Relaxation," "Motivation," and "Creativity." The results revealed the following.

First, a two-way ANOVA was used to test whether there were statistically significant differences between the ratings of the three types of content for the four groups of "Impression," "Relaxation," "Motivation," and "Creativity." The results showed that the main effects for "Impression," "Motivation," and "Creativity" were significant among all combinations of "No Content," "Shapes," and "Art." For "Relaxation," the main effect was significant only between "No content" and "Art."

Multiple comparison was conducted for a more detailed analysis. The results showed significant differences between the contents for two of the nine "Impression" items. Regarding "Motivation," three out of five items showed significant differences among the contents. In addition, significant differences were found in 4 out of 5 items for "Creativity." Among the "Creativity" items, significant differences were found for "Motivated" in all combinations of "No content," "Figure," and "Art" among the contents. These results indicate combining "immersive art space" and "Tosa art" contributes to creativity.

There are several possible directions for future research. One is to see if the results of this study are generalizable by using art other than Tosa art as content. Another is to verify whether or not similar results can be obtained by measuring physiological data and psychological evaluation. We want to continue these studies in the future.

References

1. Winner, E.: How Art Works: A Psychological Exploration. Oxford University Press (2018)
2. Beard, R.L.: Art therapies and dementia care: a systematic review. Dementia **11**, 633–656 (2012)
3. Greensgard, S.: Virtual Reality. The MIT Press (2019)
4. Jones, L.: Haptics. The MIT Press (2018)
5. Latham, W.: https://en.wikipedia.org/wiki/William_Latham_(computer_scientist)
6. Cave Automatic Virtual Environment. https://ja.wikipedia.org/wiki/Cave_automatic_virtual_environment
7. Kusama, Y.: Infinity Mirror Rooms. https://www.tate.org.uk/whats-on/tate-modern/yayoi-kusama-infinity-mirror-rooms
8. Pang, Y., Zhao, L., Nakatsu, R., Tosa, N.: A study of variable control of sound vibration form (SVF) for media art creation. In: 2017 International Conference on Culture and Computing (2017)
9. Tosa, N., et al.: Creation of Fluid Art 'Sound of Ikebana' under Microgravity Using Parabolic Flight. Leonard/ISAST, MIT Press (2022). https://doi.org/10.1162/leon_a_02360
10. Mirroria. https://www.asahiglassplaza.net/products/mirroria/
11. Nakatsu, R., Tosa, N., Niiyama, S., Kusumi, T.: Evaluation of the effect of art content on human psychology using mirror display with AR function. Nicograph International 2021, pp. 54–61 (2021)
12. Nakatsu, R., Tosa, N., Takada, H., Kusumi, T.: Psychological evaluation of image and video display by large-screen LED display and projection. Trans. Soc. Art Sci. **20**(1), 45–54 (2021)
13. Nakatsu, R., et al.: Development of Immersive Art Space Using Mirror Display and Its Preliminary Evaluation (WMSCI 2023) (2023)
14. Flynn, J.E., Hendrick, C., Spencer, T., Martyniuk, O.: A guide to methodology procedures for measuring subjective impressions in lighting. J. Illuminat. Eng. Soc. **8**(2), 95–110 (1979). https://doi.org/10.1080/00994480.1979.10748577

15. Loe, L., Mansfield, K.P., Rowlands, E.: Appearance of lit environment and its relevance in lighting design: experimental study. Light. Res. Technol. **26**, 119–133 (1994)
16. Oi, N.: The difference among generations in evaluating interior lighting environment. J. Physiol. Anthropol. Appl. Hum. Sci. **24**(1), 87–91 (2005)
17. Bowman, D.A., McMahan, R.P.: Virtual reality: how much immersion is enough. Computer **40**(7), 36–43 (2007)
18. Misseroni, D., Colquitt, D.J., Movchan, A.B., Movchan, N.V., Jones, I.S.: Cymatics for the cloaking of flexural vibrations in a structured plate. Sci. Rep **6**(1), 23929 (2016)

I've Gut Something to Tell You: A Speculative Biofeedback Wearable Art Installation on the Gut-Brain Connection

Line Krogh Sommer, Johanna Møberg Lauritzen, Alberte Spork, Louise Biller, Mathilde Merete Jensen, and Brian Bemman[✉][iD]

Aalborg University, Rendsburggade 14, 9000 Aalborg, Denmark
`bb@create.aau.dk`

Abstract. The gut-brain connection is an increasing area of focus in many research domains. In the field of human-computer interaction (HCI), numerous biofeedback technologies provide detailed information on various bodily functions, however, those designed for the bowels are limited. Additionally, arts-based research methods are increasingly being utilized in HCI for investigating lived experiences with technology. In this paper, we present an interactive art installation and wearable artifact called *I've Gut Something to Tell You*, which provides a speculative way of somatically connecting with one's bowels through technology, namely, through 'translations' of bowel sounds into written language derived from the researchers' own investigations into experiences with their bowels. Participants' experiences with the installation were evaluated through a thematic analysis of written narratives and interviews while poetic inquiry was used to communicate the prominent themes that emerged. Our findings highlight a recognition of the importance of the gut-brain connection but suggest further that one's subjective experience of this connection remains complicated due to societal taboos surrounding the topic. More broadly, our work challenges the notion of what constitutes biofeedback and our findings provide interesting reflections from the perspective of soma design regarding how we view our bowels as a separate but vocal entity which cannot be controlled in contrast to some other organs.

Keywords: Interactive Art · Wearable · Biofeedback · Soma Design · Speculative Design · Narrative Inquiry · Poetic Inquiry · Gut-Brain Connection

1 Introduction

A person's relationship to their body and, in particular, their bowels can be complex for many reasons, not least of which concerns norms in much of the Western world aimed at often suppressing bodily functions in public discourse and experience. However, research efforts in studying the nature of the connection between the gut, brain, and mental states are steadily increasing with

A. L. Brooks (Ed.): ArtsIT 2023, LNICST 565, pp. 305–324, 2024.
https://doi.org/10.1007/978-3-031-55312-7_23

wide-ranging and notable contributions from the medical community, medical humanities researchers, and artists alike [10]. Additionally, related contributions from the field of human-computer interaction (HCI), have resulted in countless biofeedback devices and technologies (e.g., smart watches) that provide users across a variety of contexts (e.g., games and sports) with intimate information and near continual awareness of various bodily functions from heart rate variability [33], blood pressure [17], and respiratory response [32] to neurological [18], thermal [22], and galvanic responses [8]. However, comparatively fewer devices and technologies relate to the gut [19,35] while what such biofeedback would mean for controlling the functioning of one's bowels and a general understanding of the long-term effects that such information and use of technologies have on the relationship between one's mind/psyche and body remain unclear. Fortunately, there are increasing calls in HCI for the need to consider the body not merely as an object for interaction with technology but as lived experience [24,31], which may help alleviate this concern. Artistic research methods, such as speculative design [9] and soma design [13], offer compelling ways in which to answer this call while at the same time challenging the ways in which we relate to the inner workings of our bodies. Other related methods, such as poetic inquiry [7], can possibly provide further means for better communicating the emotional meaning of any insights that may emerge.

1.1 Contribution

This paper presents an interactive art experience and wearable artifact called *I've Gut Something to Tell You* intended as a speculative- and soma-design inspired installation that provides an alternative way of connecting with one's bowels through technology. Specifically, the intensity of a person's bowel sounds are artistically interpreted as written words and phrases derived from the researchers' investigations into their own relationships with their bowels. The primary aim of this research is to investigate the use of art and artistic methods of inquiry for probing the feelings and reflections we have about our bodies through a speculative device that quite literally gives 'voice' to one's voiceless bowels and further challenges the notion of what constitutes biofeedback among devices that ordinarily monitor and inform users of their bodily functions. In Sect. 2, we provide an introduction to arts-based research and the methods of narrative and poetic inquiry as well as overviews of speculative- and soma- design. In Sect. 3 we present the design and implementation details of *I've Gut Something to Tell You*, which utilized soma-design practices and narrative inquiry. In Sect. 4, we introduce the experiment we carried out with participants who experienced the installation/wearable and present the themes that emerged from a thematic analysis of their narratives and interviews as well as a collection of poems the researchers made with respect to these themes using poetic inquiry. In Sect. 5, we conclude with a brief summary of our findings and directions for possible future work.

2   Related Work

Artistic practice has been argued to be a useful part of research insofar as artists can ask important questions while also responding to the unexpected in ways that traditional researchers are not necessarily so inclined [29]. Importantly, visual and sensory explorations in artistic practice serve in research as the "basis for compiling thematic patterns of evidence from which meaning is made vivid" [29, p. 56]. Both artists and their audiences alike can make meaning by seeing and sensing in the ways that art affords, namely, with one's entire body, as a means for discovering new ways of interacting with and understanding the world. This understanding is facilitated by a process of qualitative data collection and analysis which typically centers on "creating rich literary word portraits...that reflect the insight of the [artist or other] insider and the critical focus of the dispassionate observer" [29, p. 56].

2.1   Narrative and Poetic Inquiry

Narrative inquiry is a qualitative data analysis methodology for collecting, interpreting, and organizing spoken and written data in the form of personal anecdotes or narratives [7]. Narratives are particularly useful in documenting personal experiences because they are "a distinctive way of thinking and understanding that is unique and embodied, that is,...[they] integrate...the physical and psychological dimensions of knowing" [7, p. 73]. It is perhaps not surprising then that narrative inquiry has been employed in various humanities disciplines and appears particularly well suited to investigating necessarily embodied activities, such as artistic expression and experience.

A related mode of inquiry and arts-based research methodology known as poetic inquiry has been described as a particularly artful way of doing research [7] which, like narrative inquiry, has been employed in different social science disciplines, such as psychology and education [23], as a means for collecting, interpreting, and organizing qualitative data. Central to this approach is the creation of poetry as inquiry and the recognition that poetry itself can be an effective means for expressing and learning within research. The goals of poetic inquiry may vary from improving researcher reflexivity [12] and preserving an individual's voice [34] to providing more emotionally impactful, engaging, and accessible research [11,21]. One type of outcome with poetic inquiry is found poetry, which is the "rearrangement of words, phrases, and sometimes whole passages that are taken from other sources and reframed as poetry" [7, p. 97]. Found poems can be created by changing the phrasing and order of words in a set of data, resulting in a so-called treated poem, or by allowing the data's original phrasing to remain mostly intact, resulting in an untreated poem. Poetry clusters are collections of such found poems that center on a singular theme often corresponding to a participant's experience and object of study, that investigates the data on a deeper level than only one poem could manage.

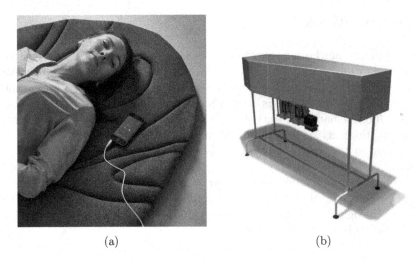

(a)   (b)

Fig. 1. The Soma Mat [27] in (a) and Afterlife (2009) [4] in (b).

2.2   Soma Design

Soma design is a methodology for helping to ensure that designers create artifacts that better support the sensations, emotions, and subjective experiences of their users while encouraging deeper aesthetic appreciation for living a better life [13]. One's goal with soma design might be the relatively basic pleasure and appreciation of exploring one's soma, or the more complex task of unlearning damaging habits, such as bad posture [13]. When designing with and for the soma, designers place considerable emphasis on their own first-person lived experience by actually engaging with their own somas, for instance, through various somaesthetic practices [25] or working thoroughly with materials before or during the actual design process [3]. It is argued that such an approach, which prioritizes care and understanding for one's own soma, improves one's ability to tend to and empathize with the somas of others. One key component of effective soma design is slow-storming, which – in contrast to the frequently fast-paced formation of ideas typically encountered in industrial brainstorming design sessions – requires that designers intentionally slow themselves in order to better articulate bodily experiences, be more thoughtful and reflective, and possibly produce more sustainable outcomes [3]. Not unlike traditional cyclic design processes, however, soma designers also iteratively test their design but with the goal of better engaging with and immersing themselves in their experience [13]. Figure 1(a) shows an example of a commercially developed product, Soma Mat [27], that was made in collaboration with IKEA and Boris Design using soma design [13,15]. The aim of the Soma Mat is to support a person's somaesthetic abilities [25] by directing their attention to different parts of the body through heat, e.g., under a heel or hip, along with audible instructions (taken from a form of meditation known as a Feldenkrais lesson) to focus on the same body part.

Over the course of its design, the creators of the Soma Mat engaged in various bodily practices at least once a week – describing their feelings before and after each exercise to better understand each other's progress – and participated in somaesthetic workshops and Feldenkrais training. The Soma Mat was evaluated by its principal designer through an inquiry into her first-person lived experience using narratives written about her own experience using it.

2.3   Speculative Design

Speculative design is a design practice which arguably stands in stark contrast to design thinking [9]. In its more frequent usage, design is concerned with solving problems. However, in the face of particularly complex problems and world-wide crises, it has been argued that simply designing our way out of these may not always be sufficient, as this may lead to a false understanding of their severity and shift focus away from the importance of ideas and attitudes that can potentially have just as much impact towards being able to solve them than the problem-solving designs themselves [9]. Speculative design is thus more concerned with asking questions – it provokes, offers alternative viewpoints and ways of being to the status quo, and encourages imagination, reflection, and critical thinking, particularly with respect to technological advancements and serious global concerns [9,30].

The value of speculative design lies in how it makes people feel, and there are several instructive criteria to consider. Often times, speculative designs work with improbable scenarios that could emerge within the bounds of science but are not likely to, and straddling this border between the real and fictional is what makes speculative design impactful. Such scenarios are frequently placed within the realm of some possible future or alternative present as tangible and often humorous or absurd objects of study intended to spark thoughts of not necessarily how things should be but how things could be – leaving it up to the viewer to decide if such visions are preferable or not. Figure 1(b) shows one such speculative artwork called Afterlife (2009) [4] that envisions a fictional future technological device in which the chemical potential of a corpse is converted into electrical energy and contained within a battery. Viewers of the artwork were asked to consider how they might use an afterlife battery charged either by themselves or a loved one and in so doing were provided with an alternative and humorous way of thinking about death that made an ordinarily difficult subject to discuss more accessible.

3   Design and Implementation of I've Gut Something to Tell You

In this section, we provide an overview of the design process and implementation details for the two primary components of I've Gut Something to Tell You: (1) a wearable artifact and (2) an interpretation screen for visualizing bowel sounds and their intensities through text – both of which were developed using practices inspired by speculative- and soma-design under the guiding premise and speculative problem of "Listening to what my bowels are saying".

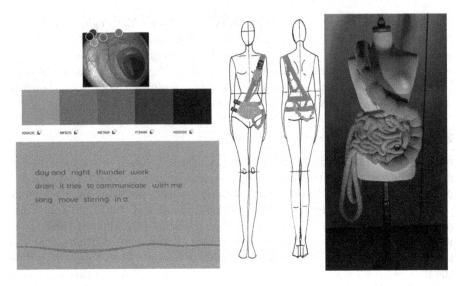

Fig. 2. Design details of *I've Gut Something to Tell You* with chosen color palette (top left), visualization screen of bowel sounds through textual interpretation and intensity (bottom left), early conceptual sketch of the wearable bowels on a figurine (middle), and final implemented version of the wearable bowels (right).

3.1 Design Overview

Figure 2 shows the various design details of the wearable artifact and visualization screen that make up *I've Gut Something to Tell You*. One will first note that the color palette (top left) used in both the visualization screen (bottom left) and the wearable itself (middle, right) consists of colors inspired by those present in the human digestive tract. The wearable artifact consists of approximately anatomically correct dimensions of the human intestines (small and large) that have been crotcheted using patterns taken from [28] having a visual and textual appeal that invokes empathy and comfort (rather than disgust), as emphasized in soma design. The intestines lie fixed to a harness worn about the waist such that they resemble their actual internal placement of the wearer while a shoulder strap keeps the electronic components in place which are further obscured by a portion of the large intestine. Iterative prototyping and simple majority decision voting were used to improve upon the materials used, the fit and form of the underlying harness, and the overall visual aesthetic of the wearable and interpretation screen.

Interpreting Bowel Sounds as Words and Phrases. As a central component of *I've Gut Something to Tell You*, the process of gathering the reflections and feelings which arise when experiencing bowel sounds, the way in which these sounds are interpreted into words, as well as how their interpretation is presented to the wearer, has a significant impact on the overall experience. The process of interpreting bowel sounds into words followed four main steps carried out by the researchers: (1) a logging period in which all the researchers used a stethoscope to listen to the sounds of their bowels over the course of a day and recorded various information of notable events (2) first-person narrative writing about the experience of listening to their bowels inspired by narrative inquiry, (3) a collective discussion of the researchers' reflections and feelings concerning their experience, and (4) a collective slow storming session borrowed from soma design for aligning the words and phrases taken from the logging period, narratives, and discussions on the researchers' respective experiences.

The logging period allowed for an understanding of the first-person lived experience of the researchers' relationship with their bowels, as emphasized in soma design. The information collected concerned basic physiological reports regarding e.g., what times during the day any sounds were most prominent, how these sounds changed due to bodily processes, such as eating, drinking, and bowel movements, as well as how they responded to different body positions e.g., sitting, standing, and lying down, which helped to inform the set up of the evaluation (discussed in Sect. 4). Additional information was collected, such as which of the four abdominal quadrants (upper and lower-right and left) resulted in the most sound as well as a description of what the noises sounded like and specific words and phrases that came to mind while listening, which served to inform the construction of the wearable and design of the interpretation of bowel sounds, respectively. Many of the words that emerged during the logging period were associated with water or animals, some examples being 'diving', 'bubbles' and 'a roaring frog'. In communicating our experiences to one another, different perspectives emerged. For example, one researcher imagined how bowel sounds could be thought of as notifications that alert us to various bodily needs. Several researchers experienced listening to the sounds of their bowels as a meditative practice involving getting lost in the sounds and losing awareness of thoughts. Listening to our bowel sounds was often amusing as well and inspired different visualizations of the bowels – for example, thinking of them as an animal or an alien, as a home to several beings, or as a friend in need of compassion.

The information collected from the logging period, narratives, and collective discussions served as the text from which we drew on during our slow storming phase. During this phase, we selected words and phrases that most aligned with our experiences over the course of several weeks of relating to our bowels (e.g., through meditation and bodily awareness sessions) during the design process and internal testing of the wearable. Towards the middle of this phase, we categorized the words and phrases that had so far emerged through a thematic analysis and then in the remaining time of this phase added additional words or phrases within these themes. The final set of words and phrases that emerged within these themes as part of this process became the text that would be selected from and displayed on the interpretation screen and represented a variety of experiences, reflections, and feelings from the researchers e.g., animals, movement, music, feelings, the weather and traveling, as well as questions and song lyrics.

3.2 Implementation Overview

Figure 3 shows an overview of the implementation details of *I've Gut Something to Tell You* which consist of its two main components of the wearable and the interpretation screen (top) and how the electronic and technical features were connected (bottom). A stethoscope is placed within the lower left quadrant of the small intestines of the wearable and the sounds detected with it are passed to a simple wireless audio transmitter which sends these sounds to a sound mixing board and later to the computer. The software, Processing [1], is used to interpret the intensities of bowel sounds into text that appears on a visualization window which is then projected using a projector. Two speakers connected to the sound mixing board allow participants to hear the sounds of their bowels directly from the stethoscope while the projected interpretations allow participants to read the text and imagine any voice of their bowels themselves. In order to further reinforce the sometimes imperceptible physical feeling of bowel activity, the amplification of bowel sounds were visually represented at the bottom of the visualization as a waveform whose amplitude fluctuated in accordance with the intensity of the sounds detected by the stethoscope.

Visualizing Interpreted Bowel Sounds as Words and Phrases. In implementing the text-based interpretation of bowel sounds, we decided on an approach in which words and phrases are selected and presented in a (mostly) nonsensical and non-grammatically correct "stream-of-consciousness" way, which we believed encouraged imagination in line with speculative design while remaining relatable insofar as they reflect the thoughts and feelings of the researchers' own experiences with their bowels. The textual interpretation of bowel sounds are thus mapped according to three conditions: (1) low intensity bowel sounds result in a randomly chosen word or phrase (e.g., 'day and night'), (2) high intensity bowel sounds result in a randomly chosen exclamatory word or phrase (e.g., 'fantastic!'), and (3) five seconds of bowel sound inactivity results in a randomly chosen pause word (e.g., 'umm', 'hmm' or 'like'). As shown in Fig. 2, the randomly chosen words or phrases from any given condition are displayed on the visualization screen in sets of four available slots one row at a time, beginning from the bottom row and 'scrolling' upwards to the next row in type-writer fashion following each new condition and disappearing after three rows.

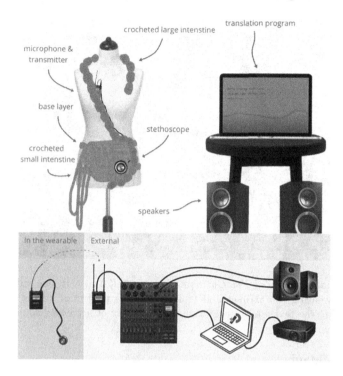

microphone & transmitter

crocheted large intenstine

translation program

base layer

stethoscope

crocheted small intenstine

speakers

In the wearable   External

Fig. 3. Overview of the implementation details of *I've Gut Something to Tell You.*

4   Evaluation

In this section, we detail our evaluation process carried out with participants who experienced *I've Gut Something to Tell You* both with and without textual interpretation of their bowel sounds and were asked to detail their experience in the form of written narratives and interviews. We conclude with an analysis of our participants' experiences that consists of the common themes that emerged through thematic analysis of their responses and poems created by the researchers for these themes using poetic inquiry.

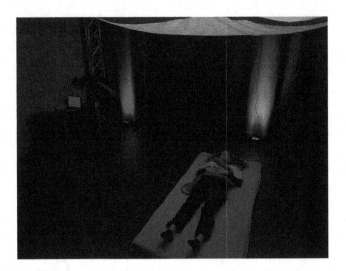

Fig. 4. Evaluation space of *I've Gut Something to Tell You* with one individual lying down while wearing the artifact and looking up at a projection of the visualization screen showing the textual interpretation and intensity of her bowel sounds.

4.1 Participants

We collected data from five, volunteer participants (4 female and 1 male) with an average age of 24.2 ± 1.7 years. All participants reported having similar levels of education and socio-economic standing, as well as having the same ethnicity and culture background. Of these participants, one reported current or prior formal training in the health care sector while another reported prior formal experience with interactive art. All other participants reported either a basic level or no prior experience with either interactive art or health. All participants were informed of what would occur during their experience with the installation as well as how their data would be used. Consent was obtained in accordance with the participating university's ethical guidelines for conducting non-medical experiments with human participants.

4.2 Procedure

We began by inviting each of our five participants to experience *I've Gut Something to Tell You* alone in a relatively quiet, dimly lit, and comfortable room, free of external stimuli which could detract from the experience of getting in touch with their bowels. After obtaining consent, participants were introduced to the wearable and informed of what would happen over the course of the session and approximately how long it would last. Afterwards they were offered something to drink as a means for encouraging any bowel activity and subsequent sounds. Participants were then assisted with putting on the wearable and then asked to lie face-up (with eyes open) on a thin mattress placed in the center of the room. As shown in Fig. 4, approximately two meters above each participant was a piece of light-colored fabric onto which the visualization screen was projected from behind by a projector fixed to the ceiling of the room. For each participant, the visualization screen first displayed the interpreted text and waveform intensity of their bowel sounds for five minutes, after which the text vanished and they saw only the waveform intensity for an additional five minutes. The text appearing in the first condition served as a visual cue for priming the participant to attend more closely to the sounds of their bowels (similar to the heat spots in the Soma Mat [27]) while the following period without text allowed them to more freely and directly connect with their bowels in a somatic manner more akin to their ordinary daily experience. Two speakers placed in opposite corners of the room provided the amplified sounds of each participant's bowels during both conditions of the session. Following this 10-minute period, one researcher entered the room and asked the participant to provide a written narrative (in any form) on a laptop concerning their initial thoughts and feelings about what they experienced. Additionally, they were provided with the text of their bowel-sound interpretations as a memory recall aid. Participants were given an additional 10 minutes alone to complete their narratives after which two researchers returned to the room to carry out a final semi-structured interview. The total experience for each participant lasted anywhere from 45 to 60 min depending on how long the interview lasted.

Interview Questions. The semi-structured interview consisted of 11 main questions with 6 follow-up sub-questions to several of the main questions. Questions such as "Do you think about your gut in your daily life?" and "How would you normally interpret your bowel sounds?" were aimed at understanding the participants' existing somatic relationship with their bowels. Other questions such as "How did it feel to have the wearable on?", "Did the experience make you particularly aware of something inside yourself?", "What was it like to listen to the bowel sounds with the written interpretation?", and "What did the words mean to you?" were intended to understand the participants' experiences with the installation and how it may have promoted a somatic connection with their bowels. The question, "Do you feel that the experience gave you either a new or different understanding, or greater awareness of your bowels?" was aimed at determining whether their experience provided novel insight into their relationship with their bowels or possibly expanded their somatic awareness and understanding. The question, "If you had a way to always talk to your intestines or for your intestines to talk to you, e.g., via an app, how do you think that would affect your daily life?" was meant as a speculative question designed to provoke (1) imaginative reflections from participants regarding what such a relationship with their bowels, if possible in the future, would mean for them and the somatic relationship to their bowels, and (2) critical assessments of biofeedback devices that already provide continual monitoring of our bodies.

Table 1. The four themes and final set of codes for each that emerged from a bottom-up thematic analysis of participants' narratives and interview responses after having experienced *I've Gut Something to Tell You*.

	Theme	Codes
1	Speculating on the pros and cons of listening to one's body	Hyper-awareness of own body, medical/anatomical reflections, overthinking, contradictions, knowing if something is good or bad for you
2	Thinking of the gut as an independent actor	Memories, personification of bowels, overthinking, contradictions, knowing if something is good or bad for you
3	Pondering how the body and mind are connected	Positive feelings, medical/anatomical reflections, whole body and mind is connected, feeling understood, personification of bowels
4	Reflecting on the taboos surrounding one's bowels	Taboos, negative feelings, overthinking regarding what others think, medical/anatomical reflections, hyper-awareness of bodily sounds

4.3 Analysis Methods

In our analysis of the responses provided by participants in their narratives and transcribed interviews, we used a bottom-up approach to thematic analysis [5] wherein five of the researchers were assigned to five distinct coder pairs such that each researcher would code the responses from two different participants and all participant responses would be coded. Each researcher in a pair independently read one of the participant's responses and then generated a set of codes which were then checked against those found by their respective coding partner in a coder-pair consensus period. These five sets of codes were then further narrowed down to a final, single set of codes following a consensus period with all five researchers. Following this consensus period, the researchers worked together to identify a set of themes that best encapsulated the experiences of participants by grouping the final set of codes into distinct subgroups or themes on the basis of their similarity. These themes were then collectively reviewed and named accordingly. Finally, the method of poetic inquiry known as found poetry was used to create a poetry cluster [7] consisting of a single treated found poem from one researcher for each of the themes identified by our thematic analysis described above and using the coded responses from all participants in the given theme.

4.4 Results

There were four significant themes shown in Table 1 (and discussed below) that emerged from the participants' experiences with *I've Gut Something to Tell You*.

Speculating on the Pros and Cons of Listening to One's Body. It was apparent in most of the participants' responses that the experience created a hyper-awareness of their own body that was at times strange, confusing, or uncomfortable, with participants stating that "I became very aware of the feeling in my stomach and it was a little uncomfortable", "I became very aware of how it feels inside my stomach and how I was feeling and what they [the bowels] were saying", and "I don't think I have ever been faced with how uncomfortable it [bowel sounds] actually makes me feel". However, other responses nonetheless indicate that the experience provoked positive feelings and reflections of amazement, amusement, or contentment, with one participant stating that "the sound became kind of oddly relaxing to me" and another expressing that "I started giggling at one point" and that "It was very fun, very eye opening. Way more intense than I had anticipated". Moreover, for many of the participants, whether the experience was negative or positive, it encouraged reflections with respect to their own health, with one participant remarking, "I also thought about if it's a good or bad thing that it makes a lot of noise or not, and how important it actually is that you think about it." This same participant reflected in the following way on whether it would be useful to have such a device that provided feedback about one's bowels: "I think that it could be a good idea, then it could say if it's unwell or if you should eat something or drink something or, do something differently. So, I think that it could be pretty nice, very healthy… as long as one doesn't become obsessed with it." While another further cautioned that "I think there needs to be some balance. Listening to your body can get very extreme, and not listening can be very dangerous, too. So, I think there needs to be some kind of balance between knowledge and listening to your body and your own experiences". Despite rather contradictory perspectives taken by some participants, most acknowledged the importance of listening to one's body, with one participant remarking that "I want them [bowel sounds] to be heard, but I do not want anybody else to hear!"

Thinking of the Gut as an Independent Actor. The combination of text and amplified bowel sounds also led many of the participants to personify their bowels as something which acts in accordance with its own autonomy. One participant recalled a thought he had during the experience in the following way: "Okay, this is my bowels talking or am I talking to the machine and the bowels talk back to me or something like that? I thought some of those noises represent feelings, like 'hmmm' means you wonder. And I thought, what do your bowels sound like when you wonder upon something? And it got me to think maybe my bowels, they have thoughts, too, of themselves somehow." Another participant expressed a similar sentiment, "Well, with the words that came, I was wondering what it meant, and how to interpret that. Because it [the gut] felt like it was trying to say something but couldn't get the sentence done". In some of the participants, certain memories were triggered that suggested the bowels as an independently acting entity. For example, one participant recalled, "It [the experience] reminded me of when I was little and had to sleep and I could hear my parents who were outside my room. I could hear them walking around and doing different things".

Pondering How the Body and Mind Are Connected. Many of the participants were provoked to reflect broadly on their relationship to their body as a whole. For some participants, this included medical or anatomical reflections, for example, with one stating, "The interpretation of the words that was on the screen, like, how would the mind communicate with the bowels? Cause I read the words, my mind read the words, but that's not how it works, it's just like...I don't know, stuff in your body that does stuff like that, cells and electrodes and I don't know. So, I guess I thought about, how do they actually communicate?". Other reflections spoke more to the mind as being distinct from the body with this same participant stating, "I guess you experience the extremes, the highs and lows of whatever your bowels do. But all the things in between, they're not that noticeable because you're just going on inside your head", while another participant acknowledged a similar distinction but reached a unified conclusion with the following: "Maybe you tend to think that your top part, that your head is the only part that's you, but that it [the experience] became so personal it was reassuring that my whole body is me. Even the gross intestines also experience with you what happens throughout your life".

Reflecting on the Taboos Surrounding One's Bowels. The overall negative feelings and reflections expressed by participants during their experience possibly speak to the general taboos and negative perceptions in society surrounding our bowels and our relationship to them. For example, one participant stated, "I feel like the digestive process is kind of taboo. So, we've as a society been conditioned to believe that excrement...is disgusting. And while I don't believe it's a topic we should be discussing as openly as the weather, it is quite interesting how we almost blush just at the thought of mentioning something like that. Even if it's to a medical professional. It's a completely natural process, so we shouldn't have to be ashamed of it", and another acknowledged that "I know in my everyday life just if I am feeling hunger and my stomach starts to growl because of that I get so embarrassed and I will start to blush". Notably, many of the female participants drew parallels to similar stigmas in society regarding menstruation, with one participant remarking in response to the words 'blood' and 'lava' that "I kind of associated it with the whole menstruation liberation-type situation that's been going on in terms of making that [sic] okay to speak about your period, and how I mean the issue is kind of the same in terms of your bowel movements, but it's not talked about at all".

Fig. 5. A poetic cluster consisting of one treated poem created by a different researcher for each of the four themes identified in the participants' narratives and interview responses after having experienced *I've Gut Something to Tell You.*

Poetry Cluster Based on Identified Themes. Figure 5 shows a poetic cluster consisting of one treated poem created by a different researcher for each of the four themes identified in the participants' narratives and interview responses after having experienced *I've Gut Something to Tell You.* All four poems provide the reader with an internal monologue from the perspective of the narrator concerning their experience of listening to their bowels. The poem for our first theme concerns speculating on the pros and cons of doing so while simultaneously being hyper-aware and acknowledging a proper balance must be struck. The poem for our second theme captures the experience of coming to view the bowels as an autonomous entity that can perhaps think, feel, make decisions, and speak on its own, rather than as just something one has and that can be somewhat controlled. Despite varying degrees of agency and independence attributed to the narrator's bowels throughout as well as the tension that inevitably arises as a consequence, the narrator ultimately and reluctantly acknowledges a shared journey of two, independent equals. The poem for our third theme reinterprets this shared journey of independent actors through the narrator's expressions of surprise, wonder, and the ultimate realization and acceptance that the whole of their body and mind are in fact connected as one. The poem for our fourth and final theme encapsulates the narrator's experience of a general sentiment evident across all themes, namely, of the complexity that exists in the relationship one has with their bowels by virtue of them being viewed negatively by both the individual and society.

4.5   Discussion

Recall that the overall aim of *I've Gut Something to Tell You* was to investigate the use of art and artistic methods of inquiry for probing the feelings and reflections we have about our bodies and challenge the notion of what constitutes biofeedback through a speculative device that interpreted bowel sounds as written words and phrases derived from soma design practices. While the combination of the particular methodologies used here is not novel [3], the outcomes presented here with respect to our stated aims above are worth noting. The design process for selecting words and phrases used in the installation based on the researchers' own experiences was intended to better empathize with the participants in service towards strengthening the unification of body and mind in alignment with soma design principles while the speculative component of pairing these words and phrases with the participants' bowel sounds was intended to provoke deeper feelings and reflections. It was clear for most participants, however, that the insecurity participants expressed regarding the amplified sound of their bowels overrode much of the effect that the speculative feature of words and phrases may have had. Nonetheless, this experience certainly provided an opportunity for participants to confront an uncomfortable topic or problem in line with the aims of speculative design [9,30]. The participants' negative reactions – echoed in the well-acknowledged shame within much of Western culture surrounding the topic of our bowels – raise interesting questions regarding the general nature of biofeedback. Proper bowel functioning is arguably as essential as proper heart functioning yet why should one be celebrated and another shunned? If a biofeedback device could be created to provide users with continual and detailed information concerning their bowel functioning, would they want such a device or would they equally shun it? What kind of feedback could be provided that would be informative without being embarrassing? Could or should this feedback be provided with the aim to control the bowels in the same way a smart watch can provide limited control over the heart? In the experiences of some participants, bowel functioning appeared uncontrollable as its perhaps erratic and unpredictable behavior (unlike the heart and lungs) led to its personification as an independent actor. Despite the stigma experienced by many of the participants regarding the sound of their bowels during the experience, participants largely reflected nonetheless on the importance of listening to one's body but stressed also that care should be taken to strike a proper balance in support of previous work on the long-term effects and usage of other biofeedback devices [26]. Additionally, the third theme indicated that while one participant maintained a dualist perspective with respect to their mind and body, most others attested that the words and phrases helped to better connect their bowels with their mind, possibly indicating further support for the non-dualist approach to design purported by soma design practices [14]. Moreover, being mentally invested in one's body while recognizing how it might be connected to the mind speaks to a more holistic viewpoint which is now more widely adopted in medicine, particularly with respect to e.g., the relationship one's personality traits and emotional patterns have with treatments for irritable bowel syndrome [2,20].

Finally, the poetry cluster we presented in accordance with our four identified themes is perhaps more challenging to assess. From the perspective of poetic inquiry, however, simply using a thematic analysis to identify these themes as being important to human experience and the relationship between the mind and body in our participants arguably falls short in being able to communicate this importance in an emotionally meaningful way [11,21,23]. While we leave it up to the reader to decide how emotionally meaningful these poems are and how well they capture the underlying themes, we believe the use of actual coded responses from participants serves as an arguably more authentic way to represent our participants' lived experiences with the installation by better preserving their voice, in support of [34], while also better grounding the reader's emotional meaning derived from each poem in its respective underlying theme – much in line with the goal of poetic inquiry which seeks to "synthesize experience in a direct and affective way" [23, p. xxii]. Particularly with respect to our main area of focus in the gut-brain connection, it could be further argued that the poems align well with a "more-than-human" approach to health literacy proposed in previous work with poetic inquiry [6,16], and that they provide a more directly accessible form of communicating complex and speculative topics, particularly for those of our participants who reported no formal health training.

Limitations. There are several limiting factors which may have hindered our research. First, it is clear that the relatively small sample size means that the feelings and reflections about the experience would likely not represent those expressed in the wider population – despite the care taken by the researchers to somatically empathize with eachother's own experiences during the design. Similarly, working with more culturally and ethnically diverse participants spanning a broader age range could yield more interesting insights into how artistic interpretations of bodily functions are received and related to. Second, it was evident that the sounds of participants' bowels overpowered some experiences with the written interpretations, limiting possible insights into their effect on how participants related to their bowels during their experience. Finally, it could be argued that presenting participants with the same order of bowel sounds with interpreted text followed by only bowel sounds could have biased the participants' experiences in some way, akin to an order effect of some kind. However, as each of the two scenarios were not treated as testable conditions, we do not believe this was a significant factor.

5   Conclusion and Future Work

In this paper, we presented the interactive art experience and wearable artifact, *I've Gut Something to Tell You*, which artistically interpreted the sounds of a person's bowels as written words and phrases derived from the researchers' investigations into their own relationships with their bowels. We believe our work serves as inspiration for other medical humanities researchers and artists wishing to stimulate meaningful conversations on awareness of one's body while the themes we identified help elucidate the relationship people have with the inner workings of their body through the words and phrases they use to speak about it. More broadly, we believe the combination of multiple artistic research methodologies, namely, speculative- and soma-design practices with narrative and poetic inquiry, as applied here, offer an interesting and effective means for more empathetically probing the feelings and reflections we have about our bodies, communicating these findings in a more emotionally meaningful way, and challenging the notion of what constitutes biofeedback, particularly with respect to taboo or difficult topics. In possible future work, it would be worthwhile investigating deeper phenomenological investigations into the participants' experiences as well as how to more reliably detect various bowel sounds, possibly through the use of additional stethoscopes. It would also be interesting to reconfigure the installation as a "feedback loop" in which participant responses are gradually added to the software, resulting in an ever-growing database encapsulating the collective 'voice' of all our bowels.

References

1. Processing (2023). https://processing.org/, Computer Software (version 4.2)
2. Agostini, A., Scaioli, E., Belluzzi, A., Campieri, M.: Attachment and mentalizing abilities in patients with inflammatory bowel disease. Gastroenterology Research and Practice 2019 (2019). https://doi.org/10.1155/2019/7847123
3. Asgeirsdottir, T., Comber, R.: Making energy matter: Soma design for ethical relations in energy systems. In: Proceedings of the 2023 CHI Conference on Human Factors in Computing Systems, pp. 1–14 (2023)
4. Auger, J., Loizeau, J.: Afterlife (2009) Artwork
5. Braun, V., Clarke, V.: Using thematic analysis in psychology. Qual. Res. Psychol. **3**(2), 77–101 (2006). https://doi.org/10.1191/1478088706qp063oa
6. Brown, M.E., Kelly, M., Finn, G.M.: Thoughts that breathe, and words that burn: poetic inquiry within health professions education. Persp. Med. Educ. **10**(5), 257–264 (2021)
7. Butler-Kisber, L.: Qualitative inquiry: thematic, narrative and arts-based perspectives. SAGE Publications Ltd (2018). https://doi.org/10.4135/9781526417978
8. Dillon, A., Kelly, M., Robertson, I.H., Robertson, D.A.: Smartphone applications utilizing biofeedback can aid stress reduction. Front. Psychol. **7**, 832 (2016)
9. Dunne, A., Raby, F.: Speculative Everything: Design, Fiction, and Social Dreaming. The MIT Press (2014)
10. Enders, G.: Gut: The inside story of our body's most underrated organ; trans. by David Shaw. Greystone Books (2018)
11. Fitzpatrick, E., Fitzpatrick, K.: What poetry does for us in education and research. Poetry, method and education research: doing critical, decolonising and political inquiry. Oxon: Routledge (2020)
12. Freeman, M.: "Between eye and eye stretches an interminable landscape": The challenge of philosophical hermeneutics. Qual. Inq. **7**(5), 646–658 (2001)

13. Höök, K.: Designing with the body: Somaesthetic interaction design. The MIT Press (2018)
14. Höök, K., et al.: Unpacking non-dualistic design: the soma design case. ACM Trans. Comput.-Human Interact. (TOCHI) **28**(6), 1–36 (2021)
15. Höök, K., Ståhl, A., Jonsson, M., Mercurio, J., Karlsson, A., Johnson, E.C.B.: Cover story somaesthetic design. Interactions **22**(4), 26–33 (2015)
16. Lupton, D.: 'Things that matter': poetic inquiry and more-than-human health literacy. Qual. Res. Sport, Exercise Health **13**(2), 267–282 (2021)
17. Ma, H., Bian, Y., Wang, Y., Zhou, C., Geng, W., Zhang, F., Liu, J., Yang, C.: Exploring the effect of virtual reality relaxation environment on white coat hypertension in blood pressure measurement. J. Biomed. Inform. **116**, 103721 (2021)
18. Marin-Pardo, O., Vourvopoulos, A., Neureither, M., Saldana, D., Jahng, E., Liew, S.-L.: Electromyography as a suitable input for virtual reality-based biofeedback in stroke rehabilitation. In: Stephanidis, C. (ed.) HCII 2019. CCIS, vol. 1032, pp. 274–281. Springer, Cham (2019). https://doi.org/10.1007/978-3-030-23522-2_35
19. Mladenović, J.: Considering gut biofeedback for emotion regulation. In: Proceedings of the 2018 ACM International Joint Conference and 2018 International Symposium on Pervasive and Ubiquitous Computing and Wearable Computers, pp. 942–945 (2018)
20. Muscatello, M.R.A., Bruno, A., Mento, C., Pandolfo, G., Zoccali, R.A.: Personality traits and emotional patterns in irritable bowel syndrome. World J. Gastroenterol. **22**(28), 6402 (2016)
21. Nichols, T.R., Biederman, D.J., Gringle, M.R.: Using research poetics "responsibly": Applications for health promotion research. Int. Q. Community Health Educ. **35**(1), 5–20 (2014)
22. Peper, E., Schmid, A.: The use of electrodermal biofeedback for peak performance training. Somatics **4**(3), 16–18 (1983)
23. Prendergast, M.: The phenomenon of poetry in research."Poem is what?" Poetic inquiry in qualitative social science. In: Prendergast, M., Leggo, C., Sameshima, P. (eds.) Poetic inquiry: Vibrant voices in the social sciences, p. xix. Sense Rotterdam (2009)
24. Rapp, A., Boldi, A.: Exploring the lived experience of behavior change technologies: Towards an existential model of behavior change for HCI. ACM Transactions on Computer-Human Interaction (2023)
25. Shusterman, R.: Thinking through the body essays in Somaesthetics. Cambridge University Press (2012)
26. Siepmann, C., Kowalczuk, P.: Understanding continued smartwatch usage: the role of emotional as well as health and fitness factors. Electron. Mark. **31**(4), 795–809 (2021)
27. Ståhl, A., Jonsson, M., Mercurio, J., Karlsson, A., Höök, K., Banka Johnson, E.C.: The Soma Mat and breathing light. In: Proceedings of the 2016 CHI Conference Extended Abstracts on Human Factors in Computing Systems, pp. 305–308 (2016)
28. Stendys, G.M.: Hæklet Anatomi. Turbine (2020)
29. Sullivan, G.: Art practice as research: Inquiry in visual arts. Sage Publications (2010)
30. Sustar, H., Mladenović, M.N., Givoni, M.: The landscape of envisioning and speculative design methods for sustainable mobility futures. Sustainability **12**(6), 2447 (2020)
31. Svanæs, D.: Interaction design for and with the lived body: Some implications of merleau-ponty's phenomenology. ACM Trans. Comput.-Hum. Interact. **20**(1) (apr 2013). https://doi.org/10.1145/2442106.2442114

32. Tu, L., Bi, C., Hao, T., Xing, G.: Breathcoach: A smart in-home breathing training system with bio-feedback via VR game. In: Proceedings of the 2018 ACM International Joint Conference and 2018 International Symposium on Pervasive and Ubiquitous Computing and Wearable Computers, pp. 468–471 (2018)
33. Umair, M., Chalabianloo, N., Sas, C., Ersoy, C.: HRV and stress: a mixed-methods approach for comparison of wearable heart rate sensors for biofeedback. IEEE Access **9**, 14005–14024 (2021)
34. Vincent, A.: Is there a definition? Ruminating on poetic inquiry, strawberries and the continued growth of the field. Art/Res. Int.: Transdiscip. J. **3**(2), 48–76 (2018)
35. Vujic, A., Tong, S., Picard, R., Maes, P.: Going with our guts: potentials of wearable electrogastrography (EGG) for affect detection. In: Proceedings of the 2020 International Conference on Multimodal Interaction, pp. 260–268 (2020)

MappEMG: Enhancing Music Pedagogy by Mapping Electromyography to Multimodal Feedback

Ziyue Piao[1] , Marcelo M. Wanderley[1] , and Felipe Verdugo[1,2(✉)]

[1] IDMIL, CIRMMT, McGill University, Montreal, Canada
ziyue.piao@mail.mcgill.ca, marcelo.wanderley@mcgill.ca
[2] Faculté de musique and S2M laboratory, Université de Montreal, Montreal, Canada
felipe.verdugo.ulloa@umontreal.ca

Abstract. Music learning and practice may be enhanced by the use of biofeedback based on both learners' and teachers' muscle activity, an essential component of music performance typically unavailable to listeners. By incorporating haptic vibrations, MappEMG enables the audience to experience the performers' muscle effort. This paper updates the MappEMG system to make muscle effort explicit in music lessons. We integrated a low-cost EMG system (BITalino MuscleBIT) and modified processing, communication, and mobile application modules. We conducted a series of experimental teaching workshops where a piano professor guided beginner and intermediate piano students with the updated MappEMG. Four interaction scenarios with MappEMG were identified from these workshops, and we gathered feedback on the initial effectiveness of using MappEMG in music pedagogy.

Keywords: Music Pedagogy · Electromyography · STEAM · Multimodal Interaction · Digital Musical Interfaces

1 Introduction

Muscle activity plays a vital role in music performance. It allows body motion for sound-producing and expression-facilitating purposes, thereby impacting the overall quality of the performance [19]. However, the hidden or non-visible nature of muscle activity poses a significant obstacle to fully understanding its role in music performance. To address this, the MappEMG system was developed to enable individuals to experience performers' muscle activity, leading to new opportunities for understanding performance intentions and playing techniques [3], and potentially enhancing knowledge transfer among musicians, listeners, and music students.

Supported by Pôle lavallois d'enseignement supérieur en arts numériques et économie créative, Partnership Development program of Social Sciences and Humanities Research Council of Canada, Natural Sciences and Engineering Research Council of Canada Discovery grant, and CIRMMT.

A. L. Brooks (Ed.): ArtsIT 2023, LNICST 565, pp. 325–341, 2024.
https://doi.org/10.1007/978-3-031-55312-7_24

The MappEMG system offers interactive opportunities designed to enrich the classical concert experience by giving audiences access to musicians' bodily engagement [3,4]. The system captures musicians' muscle activity using electromyography (EMG) and translates it into haptic vibrations on a mobile phone through the hAPPtiks application. Initial use of MappEMG during a 45-minute immersive classical piano performance yielded diverse listener responses, as perceptions of the vibrations influenced their experiences [3]. Our goal was to extend the use of the MappEMG system to music lessons to enhance the interaction between music teachers and students. Several system limitations had to be addressed for its application in pedagogical settings. First, the original MappEMG was only interfaced with the Delsys Trigno^{TM} Wireless EMG system, which, while reliable, proved less suitable for teaching due to its high cost. Second, while audiences in performances experience the performer's muscle activity, a pedagogical set-up requires interaction flexibility (e.g., recurrent changes of the person/muscle targeted, the system calibration, etc.) to allow a deep understanding, interpretation, and reproduction of the aimed muscle activity [19].

To address these limitations, we updated the MappEMG system. We incorporated the low-cost EMG system BITalino MuscleBIT; enhanced processing and communication capabilities for acquiring, processing, streaming, and transmitting EMG data to multiple phones; we also enhanced the hAPPtiks application to manage both haptic vibration and color changes on mobile phones. Our main contribution lies in applying the updated MappEMG system in exploratory piano teaching workshops and identifying four interaction scenarios for students to learn performance techniques using MappEMG. We found that different scenarios provided varying potential benefits to learners at different levels of musical proficiency.

2   Related Works

2.1   Electromyography in Music Pedagogy

In music education, biofeedback enhances tactile and auditory feedback, promoting "physiological self-regulation" for behavior adjustment. EMG stands out as a primary acquisition tool to develop biofeedback systems based on muscle activity for rehabilitation and learning purposes [23]. Its popularity is attributed to its straightforward and non-invasive nature, making it a valuable tool for directly measuring muscle activity levels and patterns [17]. Surface EMG (sEMG) has diverse applications, including medical research for diagnosing neurophysiological conditions [35] and prosthetic control [14], as well as in music biomechanics research, where it aids in investigating muscle load associated with different playing strategies (e.g., [3,4]).

EMG has been used for music pedagogy since the 1990 s, enriching instructional practices and improving musical performance [42]. Many studies focus on supplying students with feedback and using EMG as a teaching analysis tool, which involves recording and analyzing specific muscle signals within a controlled laboratory environment [42]. For instance, EMG biofeedback effectively addressed left-hand muscle tension and reduced unwanted muscle tension in violin and viola pedagogy [29]. EMG has also been used to analyze muscle activity during techniques like trills and vibrato, providing ergonomic considerations in instrument design, such as the impact of shoulder rests on muscle engagement [30,43,44]. Some real-time EMG analysis and biofeedback interfaces have been developed for music learning [31,33], however, few interfaces are used in lessons to support dynamic music instruction.

2.2 Haptic Augmenting Music Pedagogy

Researchers have investigated multi-modal detection and haptic feedback as means to improve instrumental learning, highlighting haptics' benefits in rhythm and posture guidance, as well as creating immersive experiences. A study [21] on snare drum learning found that haptic feedback accelerates the learning process, making music pedagogy more effective. Haptic feedback has been applied to posture learning for string instruments like the violin [22], enhancing musical immersion due to touch and acoustic signal correlations [37]. However, there is a gap in applying and evaluating haptic feedback in music lessons [2]. Haptic devices may have varying outcomes for students of different skill levels, making it essential to explore haptic feedback's potential and impact on music instruction to benefit students of all levels.

Using Music Information Retrieval technologies, computer-aided music learning can facilitate self-practice scenarios by analyzing performance elements [11]. The Piano Tutor [9] is one of the smartest piano teaching systems with expert assessment and feedback, but most of such intelligent systems lack bodily interaction essential for sensorimotor engagement during instrumental playing [7]. This highlights the irreplaceable role of music teachers in providing physical guidance and personalized instruction.

2.3 Motion in Piano Performance

While many studies concentrate on functional aspects of music-related motion analysis, muscle activity, which is within the intrinsic level of motion, remains relatively underexplored [24,34]. Various methods are used to capture motion data in piano performance, including optoelectronic systems and electromyography (EMG) [26]. Optoelectronic systems have been broadly employed for analyzing and visualizing pianists' movements, enabling the development of performer stimulations [27], real-time hand gesture following and learning in mixed reality [28], and analysis of performers' sound-producing [5] and ancillary whole body gestures [32]. In contrast, EMG provides data on the activity of individual muscles and muscle groups during movement, which is an essential tool for understanding muscle control and movement organization [26]. In piano performance, muscles of the upper limbs, particularly at the forearm and hand segments, have been traditionally associated with sound production and control needs [15]. However, recent research and research-creation works have shown that trunk and upper limb segments (and therefore muscles) might have inter-related functions for sound-production [39,41], expression-facilitating [3], and ergonomic purposes [40].

3   MappEMG

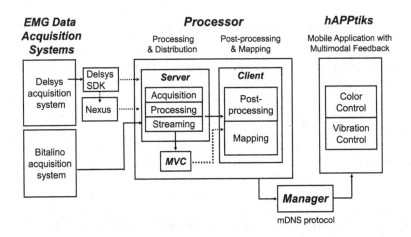

Fig. 1. The diagram framework of MappEMG. The data streaming procedure involves the following steps: a. EMG Data Acquisition Systems: EMG signals are acquired using Delsys or BITalino systems via a Bluetooth connection; b. Processor: The server receives live stream data, the MVC software collects MVC data for processing, and the client transfers the data to hAPPtiks; c. Manager: Transmits color and vibration changes to hAPPtiks; d. hAPPtiks: Provides real-time visual and haptic feedback updates.

The MappEMG system (Fig. 1) features a modular live-streaming structure that acquires EMG data from Delsys and BITalino acquisition systems, processes, streams, and emits the processed data to iPhones, offering real-time vibration feedback with multiple mobiles.

The initial version of MappEMG used in musical performance faced challenges, including costly and limited portable EMG acquisition and the need to manually insert the IP addresses of all the phones used to produce the haptic feedback in the pipeline code [3]. To address these challenges and meet the demands for using MappEMG in music lessons, subsequent designs and improvements were made. We present here the new MappEMG framework, including the acquisition interface, the Python-based Processor, and the hAPPtiks mobile application, with key goals of:

– Enabling both Delsys and Bitalino EMG devices to acquire EMG data.

- Incorporating the latest biosiglive[1] Python library to support better data acquisition, processing, and streaming [1].
- Refactoring of the communication protocol between the processor and the hAPPtiks application to allow automatic discoverability of the used phones (in the previous version, IP addresses of each phone had to be integrated manually in an emitter module coded in Max).
- Using the hAPPtiks application to control not only vibration but also the screen color of mobile phones.

3.1 Electromyography Acquisition Systems

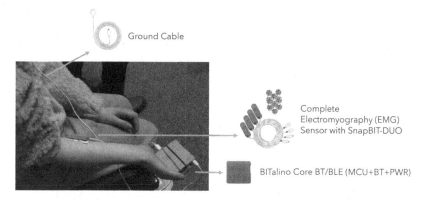

Fig. 2. The Bitalino MuscleBIT bundle, designed for EMG signal acquisition, consists of the following components: a. A ground cable: This serves as a reference for recorded electrical signals (in the photo, it's connected to the participant's clavicular notch); b. EMG Sensors: These Assembled Electromyography (EMG) Sensors with SnapBIT-DUO allow for fast and precise measurements, ensuring accuracy in repeated measurements; c. BITalino Core: The Core includes multiple connection ports, allowing for the simultaneous connection of one to four sensors.

The Delsys Trigno^{TM} System is a high-reliability commercial EMG acquisition system, serving as the primary acquisition system in the first generation of MappEMG [3]. However, low-cost alternatives with similar functionalities, such as OpenEMG, BITalino, and SparkFun Muscle Sensor, are available. We chose the BITalino MuscleBIT due to its standalone and easy-to-carry/wear design, enabling convenient use on performers and wireless data streaming via low-latency and low-energy Bluetooth (BLE) communication. The system comprises four bipolar electrodes and one reference electrode, allowing EMG data acquisition up to 1000 Hz (Fig. 2). Despite its advantages, the BITalino system faced difficulties establishing Bluetooth connections with macOS during

---

[1] Biosiglive is a Python library that aims to provide a simple and efficient way to access and process biomechanical data in real-time. https://github.com/aceglia/biosiglive.

our testing, which could potentially be linked to the absence of recent updates to its Python API. As a solution, the MappEMG Processor was operated on Windows. To connect the BITalino MucleBIT to MappEMG, we developed a BITalino Interface in biosiglive, including the BITalino device as an input data acquisition system for our Python Processor.

3.2   The Processor

The Python Processor uses the biosiglive package [1] to implement real-time data processing and sharing modules. It consists of three separate modules: the Server, the Client, and the MVC modules[2].

The Server. The Server allows users to connect to the BITalino MucleBIT system by inserting its Bluetooth address to define the EMG acquisition channels (number of electrodes used), to process and plot in real-time EMG data, and to create a TCP/IP connection for data streaming. Users can adjust the EMG acquisition sampling rate, window size for processing raw EMG data, and server acquisition rate for smoothing and downsampling purposes. The server utilizes Python's multiprocessing for concurrent data acquisition, processing, and streaming. The default EMG processing function in biosiglive involves a bandpass filter with adjustable frequencies. EMG signals are collected at 1000 Hz and rectified using a bandpass filter.

The MVC. The MVC module connects to the TCP/IP created by the server to receive the real-time collected data. The module is used to perform Maximum Voluntary Contraction (MVC) trials for each of the targeted muscles and to compute and store MVC values for each muscle. The saved MVC values are used for normalization purposes in an ulterior stage.

The Client. The Client can receive raw and processed EMG data from the server through the TCP/IP connection. This module applies a secondary set of processing procedures: it normalizes the EMG rectified signals using the previously stored MVC values; it determines the mapping curve from processed EMG data to vibration amplitude/frequency values and to screen brightness/color values. The Client initiates data streaming (OSC messages) with an 'emg' command message. A 'close' command message is sent to stop data streaming when the connection is closed.

3.3   hAPPtiks

We integrated the mDNS communication protocol in the hAPPtiks application to allow automatic mobile phones' discoverability and communication. A module

---

[2] https://github.com/IDMIL/MappEMG.

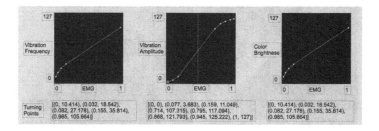

Fig. 3. The relationships between processed EMG signals and feedback (vibration frequency, vibration amplitude, and color brightness) are initially configured in the MaxMsp software [3]. These configurations are then stored as a list of points on the Processor. Parameters related to vibration are fine-tuned for tactile comfort, and brightness follows the same mapping function as frequency.

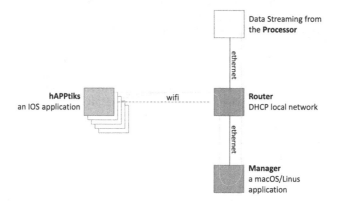

Fig. 4. The network connection between devices

named Manager was implemented (macOS and Linux) to manage the mDNS communication protocol. All computer and mobile devices must be connected to the same WIFI network to allow successful communication between the different modules. Therefore, we created a local Network (DHCP) (see Fig. 4).

hAPPtiks is an iOS application designed to receive vibration amplitude/frequency and screen brightness/color messages from the Processor, providing real-time multimodal feedback with a simple interface without additional widgets.

Fig. 5. During the MappEMG workshop and piano master class, some photos were captured: a. A comprehensive demonstration of the entire system in action. b. The teacher is providing guidance to a student, utilizing feedback from MappEMG. c. A student performing a piece while connected to MappEMG. d. Enthusiastic beginner students volunteering to experience MappEMG.

4   Piano Workshops

MappEMG aims to enhance teacher-student interaction during real piano lessons through body interaction, addressing aspects challenging to convey through language or touch. Workshops were conducted in different Quebec Conservatories (Canada) involving piano students of different levels as preliminary work for exploring MappEMG's potential use in music lessons. Led by Prof. A., a pianist and professor with a kinesiology background, the use and application of MappEMG was adapted to the level of students' and to the specific needs of each workshop. It is important to note that the workshops were exploratory lessons [36]; they were not based on a formal experimental design and evaluation, and they focused on addressing teacher A.'s pedagogical strategies and preferences.

Table 1. MappEMG Workshops

No.	Age	Number	Piano level	Interaction with MappEMG
1	5–15	13	Beginner	Scales + struck/pressed touch
2	12–18	11	Intermediate	Whole pieces
3	6–13	12	Beginner	Scales + struck/pressed touch
4	16–21	8	Intermediate	Whole pieces

4.1   Participants

Table 1 overviews the participants' number and level of four workshops, including beginner and intermediate-level piano students. The piano teachers from the hosting conservatories invited their students to participate in the workshop. The students were grouped according to their piano level (assessed by their respective teachers) and participated in the workshop corresponding to their piano level. The beginner group comprised participants with less than three years of piano experience, while the intermediate group included those with around 4 to 7 years of practice. Workshop attendees voluntarily joined the sessions to receive guidance from Prof. A. The analysis excludes other workshop attendees, such as piano teachers, parents, and students of different levels.

4.2   Procedure

Each workshop included four sessions to enhance piano technique and understanding of basic body and mechanic principles of piano playing. Sessions 1 and 2 focused on fundamental piano techniques and essential body movements in piano playing. Session 3 was the core MappEMG session, exploring the potential of MappEMG benefits for students and teachers. Session 4 involved a master class where 2–3 piano students received guidance from Prof. A.

Session 1 and 2: Fundamental Technique and Body Movement. Session 1 differentiated "pressed touch" and "struck touch" using key velocity, hammer velocity, force, and position data from research papers [18,25]. Session 2 covered vital body movements in the trunk, shoulders, elbows, wrists, and fingers when playing the piano [5]. These sessions provided beginners (40 minutes) and intermediate-level (20 min) piano students with fundamental insights into body movement in piano playing.

Session 3: MappEMG. Session 3 introduced biofeedback, EMG, and the MappEMG system. Participants used iPhones with the hAPPtips application to connect to a testing interface for personalized feedback. A selected group of 5–10 piano students wore MappEMG consecutively while playing the piano or receiving instructions from both Prof. A. and a researcher from our team.

Session 4: Master Class. Session 4 featured 2-3 piano students performing prepared pieces for a master class, receiving guidance from Prof. A. Beginner-level and intermediate-level students wore MappEMG for corresponding guidance using hAPPtiks. The master class lasted about 30 min, focusing on refining piano-playing skills and technique.

4.3 Feedback

We gathered anonymous feedback from student participants by submitting notes at the end of the MappEMG session, and Prof. A. provided his feedback after each workshop.

Teacher's Feedback. Prof. A. explained, "In this workshop series, I aimed to explore the use of MappEMG while adapting to each student level. I dynamically adjusted and refined teaching methods, especially after the first two workshops." Then Prof. A. gave feedback on different levels of students (in Table 2).

Students' Feedback. The most commonly mentioned word for beginner-level students was "interesting". One beginner student stated, "I think this helps us understand our gestures when playing and prevents bad postures."

Other three intermediate students commented: "A very useful educational tool that benefits students at all levels and is not limited to just piano players." "It helps learn from mistakes, optimize techniques, and eliminate unnecessary and inefficient movements." "I appreciated the connection between music and physics; it adds an interesting and more scientific dimension to learning."

Some intermediate students provided valuable suggestions to improve MappEMG. Another two students mentioned that the system could benefit from more visual interaction and pre-recorded performances. Additionally, they suggested that it would be more helpful for their self-practice using hAPPtiks to recognize and evaluate muscle gestures for specific techniques.

5 Discussion

During the workshops, Prof. A. focused on guiding and improving students' piano techniques. Developing from the ecological dynamics theory of motor learning [10], which emphasizes the interaction between the individual and the environment, we specified interaction scenarios in music pedagogy with MappEMG. Moreover, we garnered insights from feedback provided by both Prof. A. and the participating students. These insights shed light on the potential efficacy of MappEMG in improving the teaching and learning processes, further contributing to our understanding of motor skill acquisition in the domain of music.

Table 2. Teacher's Feedback for Different Levels of Piano Students

For Beginners	"I focused on using MappEMG to teach and let them experience the biomechanical differences between 'pressed touch' and 'struck touch', introduced in the previous session. After connecting them to MappEMG, I let them repeatedly play a note or a scale with a different touch and feel the vibrotactile feedback. Feeling their muscle activity through MappEMG helped them establish a connection between the playing goal and its corresponding muscle effort. They were pleasantly surprised to feel their muscle contractions and they were willing to play with MappEMG... Beginner students, especially younger ones, were more attracted to MappEMG. They gathered closer to me during the MappEMG session, displaying excitement to participate, possibly because the system can be seen as a game interface. They also engaged in warm discussions with their classmates"
For Intermediate Students	"With MappEMG, I could address more complex playing contexts. For example, during workshop No. 4, an intermediate student, M., performed a piano etude which included fast scale and arpeggio passages at the right hand. Although M. displayed playing proficiency, there was room for refining the technique, particularly in terms of minimizing unnecessary finger-lifting movements (i.e., overuse of extension movement at the metacarpophalangeal joints). I placed an EMG electrode on the corresponding extensor muscle, situated at the forearm, to address M.'s issue. I first guided M. to adjust the playing gesture until I could feel slight variations of the vibration based on the targeted muscle. Then, I handed my iPhone to M.'s left hand. Firstly, I asked M. to feel the vibrations with the left hand while playing the excerpt of the etude with the right hand using the playing technique she normally used. Secondly, I asked M. to adapt the playing technique to reduce as much as possible the vibrations while maintaining the same musical goal. After a few attempts, M. was able to introduce greater hand pronation/supination movements, which helped reduce finger extension movements to release the keys. This is a playing technique called 'rotation' in the piano community, and it is widely taught to reduce both finger muscle activity and exposure to risk factors of injuries, as finger extensors can be more affected by muscle fatigue than finger flexor muscles (e.g., [20]). ... Surprisingly, several intermediate students demonstrated a similar phenomenon. When they held the iPhone with one hand, and I asked them to play with the other hand, they could generally make gestural adjustments based on their previous knowledge or the content taught in previous sessions"

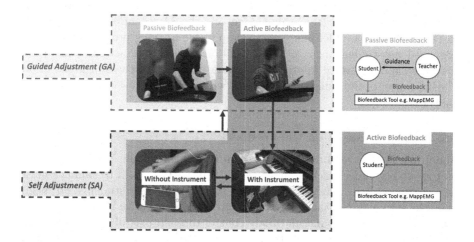

Fig. 6. Interaction scenarios for teacher-student interaction in music lessons with feedback tools

5.1   Interaction Scenarios

After conducting the workshops, we identified four types of interactions based on Prof. A.'s exploration of teaching using MappEMG:

- Guided Adjustment (GA): student adjustments with teacher guidance based on passive (1) and active (2) biofeedback during the lesson.
- Self Adjustment (SA): student self-made adjustments based on biofeedback during practice with the instrument (3) and without the instrument (4) between lessons.

As shown in Fig. 6, we categorize feedback from the student's perspective into passive and active biofeedback. Passive biofeedback refers to the case where the teacher feels the MappEMG feedback based on the student's muscle activity and provides guidance based on the teacher's perceived feedback. Conversely, active biofeedback entails students actively feeling their biofeedback and adjusting their gestures according to the student's perceived feedback. The two scenarios were used as organized pedagogical steps. In the initial step (passive GA), students' learning of new techniques was purely based on guidance from the teacher. Moving to the active GA scenario, students could practice the acquired techniques by actively feeling their muscle activity through biofeedback. They could make real-time adjustments based predominantly on their own biofeedback perception and gestural proprioception while benefiting from reduced guidance from the teacher. The next two scenarios could be introduced during practice between lessons. In the SA scenarios (both with and without an instrument), students could fine-tune their gestures by interpreting haptic feedback themselves (active biofeedback). During our workshop, we used the SA with and without instrument scenarios as a demo session, allowing students to sense their motion with MappEMG.

Following Gallahue's motor skill learning model [16], which was derived from the Fitts and Posner as well as Gentile models [13], the GA scenarios fall within the beginning/novice level learning stage, and the SA scenarios target the intermediate/practice level learning stage. Despite the learning processes outlined in advanced/fine-tuning music piece acquisition [8], our interactions aim to help students explore diverse approaches to gain a basic awareness of gestural and playing concepts and skills. The scenarios will serve as the basis for our following analysis.

5.2   Analysis Based on the Feedback

Teaching Beginner Level Piano Students. During our workshops, based on the subjective feedback from learners, MappEMG seemed to enhance beginners' learning interest and attention more than intermediate students. For beginners, the active biofeedback interaction proved most effective in establishing the relationship between sound and body control through playful engagement. Due probably to their young age, these students demonstrated a greater interest in the system and capacity to adapt their playing while actively feeling the biofeedback [6].

Teaching Intermediate Level Piano Students. Intermediate students benefited the most from optimizing their playing techniques using MappEMG in passive biofeedback interaction (passive GA), which aligns with Gallahue's motor learning model, emphasizing skill enhancement and effective feedback at the intermediate level [38]. The teacher's ability to dynamically switch teaching methods with MappEMG maximized efficiency in the intermediate students' groups compared to the beginner group. Intermediate students showed a better capacity to move from passive GA to active GA to understand and experience the gestural modifications demanded by Prof. A. This outcome of our preliminary work should, however, be confirmed by a formal evaluation of intermediate-level students' use of MappEMG in teaching and learning contexts.

For both Levels of Piano Students. During and after the use of MappEMG, students demonstrated improved body awareness and the ability to adjust their gestures according to the perceived biofeedback, which suggests that muscle biofeedback might be a promising complementary tool to traditional teaching methods. Students' focus shifted from imitating the teacher's demonstration to actually experiencing the link between sound production and muscle effort, particularly evident in the active GA scenario, as noted in the teacher's feedback. Based on our workshop experience, we speculate that a combination of passive and active biofeedback scenarios can effectively help teach new techniques and reinforce playing adjustments related to body posture or movements. The active feedback scenarios show significant potential in music learning as they could complement students' proprioception and self-inform required adjustments [12, 23].

5.3   Limitation and Future Work

We are currently developing the Android version of hAPPtiks to improve system accessibility. Our current partnership project, "Science at the Service of Music Performance" (three university music departments and four music conservatories around Quebec), will help refine the workshop procedure, with biannual workshops planned in the next two years. To further evaluate the results, we intend to conduct formal user experiments, assessing students' improvement and learning efficiency in controlled passive and active learning scenarios. Additionally, we plan to explore whether beginner students sustain their learning interest beyond the initial novelty, potentially requiring a longitudinal experiment with MappEMG. Ultimately, we aim to broaden MappEMG's music lessons and performance application through other learning, performance, or artistic activities.

6   Conclusion

This paper introduced the updated MappEMG system and its initial application in music pedagogy through exploratory piano workshops. The updated

MappEMG system enables the acquisition, processing, and low-latency streaming of muscle data with a low-cost EMG system. It maps and emits processed EMG data to the hAPPtiks iOS application, allowing multi-users to feel the performer's muscle activity through color and vibration changes. We conducted four workshops with beginner and intermediate piano learners, exploring various interactions between teacher and students using MappEMG. We identified four interaction scenarios: Guided Adjustment with Passive and Active Biofeedback, and Self Adjustment with and without Instrument. The user feedback showed that MappEMG could be used to teach playing techniques to beginner and intermediate piano students, increasing interest and attention in beginners and optimizing playing techniques in intermediate students. Moreover, MappEMG enhanced students' focus on gesture-sound relations and might be used to prevent poor ergonomic movements and postures. The four interaction scenarios have great potential for music learning, augmenting traditional teaching and practice contexts.

Acknowledgment. The work is funded by Pôle lavallois d'enseignement supérieur en arts numériques et économie créative (call for projects 2021–2022), the Partnership Development program of Social Sciences and Humanities Research Council of Canada (SSHRC-890-2021-0072), a Natural Sciences and Engineering Research Council of Canada Discovery grant to the second author, and CIRMMT. We thank Amedeo Ceglia for the support in updating the pipeline to the new biosiglive version, former interns Karl Koerich and Noa Kemp for their work in the pipeline processing refactoring, Alex Burton for its work on the implementation of the mDNS protocol and on the new version of the hAPPtiks application, and Sylvie Gibet for discussions on the previous version of the MappEMG system. We also thank all IDMIL lab members' suggestions and comments (especially Travis West, Bavo Van Kerrebroeck, Pierrick Uro, Paul Buser, and Erivan Duarte). Finally, we warmly thank the piano teachers and students of Quebec conservatories who participated in the workshops and provided feedback on our work.

References

1. Ceglia, A., Verdugo, F., Begon, M.: Biosiglive: an open-source Python package for real-time biosignal processing. J. Open Source Softw. **8**(83), 5091 (2023)
2. Tom, A., Singh, A., Daigle, M., Marandola, F., Wanderley, M.M.: Haptic tutor-a haptics-based music education tool for beginners. In: Proceedings of International Workshop on Haptic and Audio Interaction Design (2020)
3. Verdugo, F., et al.: Feeling the effort of classical musicians-a pipeline from electromyography to smartphone vibration for live music performance. In: Proceedings of the International Conference on New Interfaces for Musical Expression. PubPub (2022). https://doi.org/10.21428%2F92fbeb44.3ce22588
4. Verdugo, F.,: MappEMG: supporting musical expression with vibrotactile feedback by capturing gestural features through electromyography. In: Proceedings of International Workshop on Haptic and Audio Interaction Design (2020)

5. Verdugo, F., Pelletier, J., Michaud, B., Traube, C., Begon, M: Effects of trunk motion, touch, and articulation on upper-limb velocities and on joint contribution to endpoint velocities during the production of loud piano tones. Front. Psychol. **11**, 1159 (2020)

6. Arifin, A., Mashuri, M.T., Lestari, N.C., Satria, E., Dewantara, R.: Application of interactive learning games in stimulating knowledge about object recognition in early childhood. Educenter: Jurnal Ilmiah Pendidikan **2**(1) (2023)

7. Bremmer, M., Nijs, L.: The role of the body in instrumental and vocal music pedagogy: a dynamical systems theory perspective on the music Teacher's bodily engagement in teaching and learning. Front. Educ. **5**, 79. Frontiers Media SA (2020)

8. Chaffin, R., Imreh, G., Lemieux, A.F., Chen, C.: seeing the big picture: piano practice as expert problem solving. Music. Percept. **20**(4), 465–490 (2003)

9. Dannenberg, R.B., Sanchez, M., Joseph, A., Joseph, R., Saul, R., Capell, P.: Results from the piano tutor project. In: Proceedings of the Fourth Biennial Arts and Technology Symposium, pp. 143–150 (1993)

10. Davids, K., Araújo, D., Hristovski, R., Passos, P., Chow, J.Y.: Ecological dynamics and motor learning design in sport. Skill acquisition in sport: Research, theory and practice, pp. 112–130 (2012)

11. Dittmar, C., Cano, E., Abeßer, J., Grollmisch, S.: Music information retrieval meets music education. In: Dagstuhl Follow-Ups. vol. 3. Schloss Dagstuhl-Leibniz-Zentrum fuer Informatik (2012)

12. Dupee, M., Forneris, T., Werthner, P.: Perceived outcomes of a biofeedback and neurofeedback training intervention for optimal performance: Learning to enhance self-awareness and self-regulation with olympic athletes. Sport Psychol. **30**(4), 339–349 (2016)

13. Fitts, P.M., Posner, M.I.: Human performance (1967)

14. Fougner, A., Stavdahl, y., Kyberd, P.J., Losier, Y.G., Parker, P.A.: Control of upper limb prostheses: terminology and proportional myoelectric control-a review. IEEE Trans. Neural Syst. Rehab. Eng. **20**(5), 663–677 (2012)

15. Furuya, S., Altenmüller, E.: Flexibility of movement organization in piano performance. Front. Hum. Neurosci. **7**, 173 (2013)

16. Gallahue, D.L., Donnelly, F.C.: Developmental physical education for all children. Human Kinetics (2007)

17. Gazzoni, M., Afsharipour, B., Merletti, R.: Surface EMG in ergonomics and occupational medicine. Surface electromyography: physiology, engineering, and applications, pp. 361–391 (2016), publisher: Wiley Online Library

18. Goebl, W., Bresin, R., Galembo, A.: Touch and temporal behavior of grand piano actions. J. Acoust. Society America **118**(2), 1154–1165 (2005)

19. Gonzalez-Sanchez, V., Dahl, S., Hatfield, J.L., Godøy, R.I.: Characterizing movement fluency in musical performance: toward a generic measure for technology enhanced learning. Front. Psychol. **10**, 84 (2019)

20. Goubault, E., Verdugo, F., Pelletier, J., Traube, C., Begon, M., Dal Maso, F.: Exhausting repetitive piano tasks lead to local forearm manifestation of muscle fatigue and negatively affect musical parameters. Sci. Rep. **11**(1), 8117 (2021)

21. Grindlay, G.: Haptic guidance benefits musical motor learning. In: Proceedings of symposium on haptic interfaces for virtual environment and teleoperator systems, pp. 397–404. IEEE (2008)

22. Grosshauser, T., Hermann, T.: Augmented haptics – an interactive feedback system for musicians. In: Altinsoy, M.E., Jekosch, U., Brewster, S. (eds.) HAID 2009. LNCS, vol. 5763, pp. 100–108. Springer, Heidelberg (2009). https://doi.org/10.1007/978-3-642-04076-4_11

23. Gruzelier, J.H., Egner, T.: Physiological self-regulation: biofeedback and neuro-feedback. Musical excellence: strategies and techniques to enhance performance, pp. 197–219 (2004), publisher: Oxford University Press, Oxford, UK

24. Jensenius, A.R.: Action-Sound: Developing Methods and Tools to Study Music-related Body Movement. Ph.D. thesis, University of Oslo (2007)

25. Kinoshita, H., Furuya, S., Aoki, T., Altenmüller, E.: Loudness control in pianists as exemplified in keystroke force measurements on different touches. J. Acoust. Society America **121**(5), 2959–2969 (2007)

26. Kotov-Smolenskiy, A.M., Khizhnikova, A.E., Klochkov, A.S., Suponeva, N.A., Piradov, M.A.: Surface EMG: applicability in the motion analysis and opportunities for practical rehabilitation. Human Physiol. **47**(2), 237–247 (2021), iSBN: 0362-1197 Publisher: Springer

27. Kugimoto, N., et al.: CG animation for piano performance. In: SIGGRAPH'09: Posters, pp. 1–1 (2009)

28. Labrou, K., Zaman, C.H., Turkyasar, A., Davis, R.: Following the Master's Hands: Capturing Piano Performances for Mixed Reality Piano Learning Applications. In: Extended Abstracts of the 2023 CHI Conference on Human Factors in Computing Systems, pp. 1–8. ACM, Hamburg Germany (Apr 2023). https://doi.org/10.1145/3544549.3585838, https://dl.acm.org/doi/10.1145/3544549.3585838

29. LeVine, W.R., Irvine, J.K.: In vivo EMG biofeedback in violin and viola pedagogy. Biofeedback Self Regul. **9**, 161–168 (1984)

30. Levy, C.E., Lee, W.A., Brandfonbrener, A.G., Press, J., Levy, A.E.: Electromyographic analysis of muscular activity in the upper extremity generated by supporting a violin with and without a shoulder rest. Med. Probl. Perform. Artist. **7**(4), 103–109 (1992)

31. Mani, S., Vinay, C.K., Deepika, P., Rao, M.: Surface EMG signal classification for unsupervised musical keyboard learning application. In: 2020 IEEE SENSORS, pp. 1–4. IEEE (2020)

32. Massie-Laberge, C., Cossette, I., Wanderley, M.M.: Kinematic analysis of pianists' expressive performances of romantic excerpts: applications for enhanced pedagogical approaches. Front. Psychol. **9**, 2725 (2019)

33. Oku, T., Furuya, S.: A novel vibrotactile biofeedback device for optimizing neuromuscular control in piano playing. In: 2019 IEEE Conference on Virtual Reality and 3D User Interfaces (VR), pp. 1554–1555. IEEE (2019)

34. Ramstein, C.: Analyse, représentation et traitement du geste instrumental: application aux instruments à clavier. Ph.D. thesis, Institut National Polytechnique de Grenoble-INPG (1991)

35. Reaz, M.B.I., Hussain, M.S., Mohd-Yasin, F.: Techniques of EMG signal analysis: detection, processing, classification and applications. Biol. Proc. online **8**, 11–35 (2006)

36. Reimer, P.C., Wanderley, M.M.: Embracing less common evaluation strategies for studying user experience in NIME. In: NIME 2021. PubPub (2021)

37. Remache-Vinueza, B., Trujillo-León, A., Zapata, M., Sarmiento-Ortiz, F., Vidal-Verdú, F.: Audio-tactile rendering: a review on technology and methods to convey musical information through the sense of touch. Sensors **21**(19), 6575 (2021)

38. Salehi, S.K., Tahmasebi, F., Talebrokni, F.S.: A different look at featured motor learning models: comparison exam of gallahue's, fitts and posner's and ann gentile's motor learning models. Movement Sport Sci. **2**, 53–63 (2021)

39. Turner, C., Goubault, E., Dal Maso, F., Begon, M., Verdugo, F.: The influence of proximal motor strategies on pianists' upper-limb movement variability. Hum. Mov. Sci. **90**, 103110 (2023)

40. Turner, C., Visentin, P., Oye, D., Rathwell, S., Shan, G.: An examination of trunk and right-hand coordination in piano performance: a case comparison of three pianists. Front. Psychol. **13**, 838554 (2022)
41. Verdugo, F., Begon, M., Gibet, S., Wanderley, M.M.: Proximal-to-distal sequences of attack and release movements of expert pianists during pressed-staccato keystrokes. J. Mot. Behav. **54**(3), 316–326 (2022)
42. Visentin, P., Shan, G.: Applications of EMG pertaining to music performance-A review. Arts BioMechanics **1**(1), 15 (2011)
43. Ziane, C., Goubault, E., Michaud, B., Begon, M., Dal Maso, F.: Muscle fatigue during assisted violin performance. Ergonomics (just-accepted), pp. 1–19 (2023)
44. Ziane, C., Michaud, B., Begon, M., Dal Maso, F.: How do violinists adapt to dynamic assistive support? a study focusing on kinematics, muscle activity, and musical performance. Human Factors (2021)

Exploring Perception and Preference in Public Human-Agent Interaction: Virtual Human Vs. Social Robot

Christian Felix Purps[1]([envelope]), Wladimir Hettmann[1], Thorsten Zylowski[1,2], Nathalia Sautchuk-Patrício[1], Daniel Hepperle[1,2], and Matthias Wölfel[1,2]

[1] Karlsruhe University of Applied Sciences, Karlsruhe, Germany
{christian_felix.purps,matthias.woelfel}@h-ka.de
[2] University of Hohenheim, Stuttgart, Germany

Abstract. This paper delves into a comparison between virtual and physical agent embodiments with the aim to gain a better understanding of how these are perceived and interacted with in a public setting. For our experiment we developed two agent embodiments: a virtual human and a mechanical looking social robot, that encourage passerby in public space to exercise squats through speech and non-verbal cues. We analyzed user behavior during the interaction with one of the distinct systems that differ in representation but share the same purpose and intent. We recorded 450 encounters in which a passerby listened fully to the agent's instructions and used body tracking to analyze their exercise engagement. At least one squat was performed in each of 145 encounters, which generally indicates fairly high system acceptance. Additional feedback came from 61 individuals (aged 13 to 74, 41 males, 20 females) through a questionnaire on perception of competence, autonomy, trust, and rapport. There was no significant difference found between the virtual human and the social robot concerning these factors. However, responses to single questions indicate that interactions with the social robot were perceived as significantly more responsive, and gender differences in perceived interaction pressure emerged, with women reporting significantly higher values compared to men. Despite public space challenges, the agent systems prove reliable. Complexity-reducing technical and methodological simplifications and possible sampling biases must be taken into account. This work provides a glimpse into public interactions with virtually and physically embodied agents, and discusses opportunities and limitations for future development of such systems.

Keywords: human-agent interaction · virtual humans · social robotics · public space · physical exercise

1 Introduction

Recent advances in artificial intelligence (AI) have ushered a new era of human-computer interaction (HCI) in which virtual humans and social robots controlled

© ICST Institute for Computer Sciences, Social Informatics and Telecommunications Engineering 2024
Published by Springer Nature Switzerland AG 2024. All Rights Reserved
A. L. Brooks (Ed.): ArtsIT 2023, LNICST 565, pp. 342–358, 2024.
https://doi.org/10.1007/978-3-031-55312-7_25

by intelligent agents could play an increasing role in further shaping our socio-technical landscape [1]. Voice-based agents (e.g. Alexa, Siri, Cortana etc.) have already become socially acceptable and ubiquitous [5]. In contrast, applications in which these software agents are embodied, e.g., by virtual characters, are still emerging or are, in the case of predominantly robotic embodiment, only a niche [4,10]. However, these systems have the potential to significantly improve interaction quality by leveraging the multi-modality of communication channels, such as combining speech with non-verbal cues. This enables a more fluent and natural humanlike interaction, as embodied agents can convey meaning, emotions, and intentions more effectively through gestures, facial expressions, body language, and vocalizations and thus foster social acceptance [21].

Both, virtually embodied agents (e.g. virtual humans) and physically embodied agents (e.g. social robots) can take advantage of these features and serve similar application purposes. However, they differ in how users perceive these entities, even if they share a similar appearance, visual and behavioral fidelity, level of humanlikeness and other aspects (e.g. equivalent voice and speech communication). This may affect e.g. the generation of rapport and relation, trust, and perceived competence, which are all crucial facets of human communication [8]. For a better understanding of these distinctions, researchers have conducted studies and experiments in the field of HCI, comparing different embodiments for software agents. In addition to numerous applications in private usage, the potential utilization of these technologies in public spaces is also becoming increasingly conceivable [11]. However, there has been limited focus on conducting field tests involving comparisons between different types of embodied agents in public spaces. Thus, further study and research is needed to understand the implications and opportunities of using embodied agents in public settings, especially at a time when expectations for intelligent and accessible interactive systems are rising and, in many parts, remain unmet [3]. Conducting these types of experiments in public spaces can identify causal effects through randomization while studying people and groups in their natural settings, and brings a decidedly sociological perspective to the practice of experimentation by treating differences between people and places as a research opportunity rather than unwelcome threats to experimental control [2].

In our research, the primary objective is to compare two types of agent representations: one that is virtual and another that is physical. However, comparing these two types of embodiment poses challenges, especially when dealing with virtual humanlike characters and their physical counterparts (humanoid social robots). Constructing a high fidelity humanoid robot with the same appearance and movement quality as a virtual counterpart is difficult. Consequently, such robots often fall into the "uncanny valley", displaying unnatural traits, causing unease. Conversely, mechanical looking social robots lack humanlike features (and thus avoid eeriness), but also make a direct comparison with realistic virtual humans problematic. However, while the proposed entities differ in appearance, their underlying behavior and intent may remain consistent. This justifies a comparison between them, despite the inevitable variation in the natural expression of non-verbal cues depending on their respective embodiment. Rather than focusing on minor differences, our aim is to gain a general understanding of how people interact with these agents and how they perceive these very different entities in public space. We are particularly interested in exploring which type of embodiment, virtual or physical is preferred by the participants when interacting with these agents. Our investigation will deal with the participant's perception of the agent's competence, autonomy, and trustworthiness based on its embodiment. Furthermore, we will measure how well rapport is established and how long participants are willing to engage with these agents. By exploring these aspects, we aim to uncover insights into the attractiveness and effectiveness of both virtual and physical agent representations.

To conduct our study, we designed a simple interaction scenario to encourage participants to engage in physical activity with the agent in public space. The agent uses verbal and non-verbal cues and motivates participants to perform physical exercises (squats). Participants are greeted, instructed, and then asked to perform the exercises. The system tracks performance and provides verbal and non-verbal feedback. The agent cycles through different states depending on participants behavior and interactions aiming to provide a rewarding and motivating training experience.

2   Related Work

Social robotics has gained a valuable role in assisting, influencing and motivating human behavior in many HCI contexts [19]. Virtually embodied agents such as virtual humans may serve similar purpose and application while not requiring a physical representation. Similarities and differences in interacting with the distinct entities have been studied for different applications such as movie recommendation [16], socio-emotional interactions for children [6], and human decision-making in general [19].

Thellmann et al. compared virtual and physical agent embodiment (Nao robot and it's virtual representation) and it's effect on social interaction. Their investigations consider the relationship between physical and social presence. The results suggest that social presence of an artificial agent is important for interaction with people, and that the extent to which it is perceived as socially present might be unaffected by whether it is physically or virtually present [20].

Schneider and Kummert investigated the effect of an agent's embodiment type (humanoid social robot vs. virtual humans in three levels of humanlikeness) on motivation during the performance of versatile sport exercises. They figured out that participants tend to exercise significantly longer when interacting with a social robot than with a virtual embodied training partner. Additionally the participants found the robotic partner more likable than the virtual representation [18].

As likability, other factors, such as trust play a vital role in HCI research. One of the most commonly used definitions is that of Mayer et al. according to whom trust is *"The willingness of a party to be vulnerable to the actions of another party*

based on the expectation that the other will perform a particular action impor-
tant to the trustor, irrespective of the ability to monitor or control that other
party" [12]. Trust is established between two parties, and when designing reliable systems, attention is given to the qualities of both parties involved. On the
computer side, considerations such as fairness, accuracy, and transparency come
into play. As for the human aspect, traditional personal traits like introversion,
extraversion, honesty, and affinity for technology become relevant. These human
characteristics often remain outside the scope of technological systems. Van der
Werff et al. present a model that links the motivation to trust with attributes
derived from the Self-Determination Theory (SDT) [22]. By addressing these
attributes through interface design, the human side of trust can be taken into
account. SDT, originating from the work of Ryan and Deci, revolves around fulfilling psychological needs and encompasses competence, autonomy, and social
relatedness as its key dimensions and is used, among other things, to explain
motivation [17]. This motivation to trust can give an explanation why a person
is willing to trust a system initially. As perceived competence, autonomy, and
social relatedness increase, trust can be expected to increase as well.

3   Human-Agent Interaction System

We designed two distinct technical entities: A social robot, based on a one-arm
robot with a smartphone attached, and a virtual human, displayed in life-size on
a vertically arranged screen (see Fig. 4). To compare and evaluate both systems
we created a trivial agent behavioral use case: a squat trainer application. Both
systems are based on the same perceptive system (interlocutor localization and
recognition of a squat physical exercise repetition) and rule based behavioral
state machine and generate identical verbal output. Both forms of representation
are capable of expressing a set of non-verbal social signals that are intended to
be similar in meaning but differ in actual performance depending on whether
the embodiment is virtual or physical.

3.1   Perception

To enable the agent interaction system to sense and interpret its environment,
with a particular focus on observing and decoding the non-verbal signals of its
interlocutors, optical sensors were used. In the context of a use case in public
spaces, we were looking for a mobile and easily deployable perceptive system.
Therefore, and since depth sensors have shown promising results in recognizing
the shape of people, matching skeletons, and to recognize gestures, we decided
to use Microsoft's Azure Kinect. The Kinect comes with a full body-tracking
SDK[1] that allows extraction of joint information in real-time from the depth

---

[1] Body Tracking SDK for Azure Kinect enables segmentation of exposed instances
and both observed and estimated 3D joints and landmarks for fully articulated,
uniquely identified body tracking of skeletons. (http://www.azure.microsoft.com/
en-us/services/kinect-dk).

Fig. 1. Virtual Human and Social Robot. The image depicts the life size virtual human (left) while idling and the one-arm robot based social robot (right) during the "Seek Attention" animation.

image. To access data from an Azure Kinect, we use a self-developed middleware that seamlessly integrates the Kinetic Space[2] gesture-recognition module. Using Kinetic Space, we recorded squat exercise performances of different people to be used later for recognition. Kinetic Space can learn and recognize gestures from just few examples while the normalization of skeleton data ensures recognition accuracy by removing the influence of individual features, body orientation or localization. The middleware transmits social signals and spatial coordinates to control the agent's behavior for further processing.

3.2 Embodiment

For the embodiment of the two agents, we opted for fundamentally different representations. For the virtually embodied agent, we chose a virtual double of a real-life person, displayed life-size on a vertically arranged large screen. For the physically embodied agent, we created a very simple mechanical representation consisting of a robotic arm and a cell phone with abstract eyes and mouth.

Virtual Human. The virtual embodied agent's representation in high fidelity (see Fig. 4) was created using Blender[3]. We chose largely neutral clothing for the agent instead of, for example, a sports outfit, so as not to create an additional bias in comparison with the social robot, where it is difficult to represent it in

---

[2] Kinetic Space is an open-source tool that enables training, analysis, and recognition of individual gestures with a depth camera like Microsoft's Kinect family [23].

[3] https://www.blender.org/.

one way or another in an athletic manner. The avatar's body and rigging was extracted from a Mixamo[4] character and merged with a head, created based on a 3D scan and post-processing of a real person's head augmented with a fine-tuned facial rig for realistic facial expressions based on the proposed method of Purps et al. [15]. The avatar uses blend shape based lip synchronization for realistic mouth movements during speech. We used Mel Frequency Cepstral Coefficients (MFCC) to extract the phoneme profiles and assign the corresponding blend shape values. The virtual human is rendered in real-time using Unity[5] runtime application that connects with the agent behavioral system via localhost network connection. Vocal utterances and speech are played by the locally connected speaker.

Social Robot. We created a very mechanical appearing social robot based on a robotic arm and a mobile phone. As the robotic arm base we used Elephant Robotics mechArm 270-Pi[6] and attached a mobile phone car holder to it. The movements of the robotic arm (rotation of it's six "joints") are controlled by a python application (that interfaces via C code to the serial port of the arm) and receives commands by the agent behavioral system (see Sect. 3.5) via TCP network connection. As "robotic face" serves a Xiaomi Mi 4 smartphone horizontally attached to the phone holder. The smartphone runs a 3D application created with Unity that displays two abstractly stylized eyes. These eyes can morph into different shapes depicting an abstraction of different facial expressions respectively emotions. If the robot performs vocal utterances or speaks to the interlocutor it plays a sine wave as a stylized mouth. The application receives commands via TCP Wi-Fi connection by the agent behavioral system, too. The mobile phone is connected via Bluetooth to a speaker to play vocal utterances and speech.

3.3 Non-verbal Cues

While both representation, virtual human and social robot are congruent in their scripted behavior and speech, the performance of certain behavior significantly differs based on the possibilities provided by their embodiment. This does concern the appearance but mainly the possibilities of eliciting other non-verbal social signals to the conversation partner. The intention is for these distinct signals to be interpreted unequivocally by the agent's interlocutor, despite the marked differences between them (see Fig. 2). In states of idleness or focused attention on an interlocutor, the virtual human employs subtle idling behaviors, while the social robot tilts its cell phone slightly back and forth. Additionally,

---

[4] https://www.mixamo.com/.

[5] https://www.unity.com/.

[6] MechArm 270-Pi is a lightweight and compact 6-axis robotic arm manufactured by Elephant Robotics using Raspberry Pi as controller, with a payload of 250g, which is sufficient to lift an average mobile phone. (www.shop.elephantrobotics.com/ende/collections/mecharm/products/mecharm).

the virtual human breaths and blinks frequently while the social robot simulates that through pulsing its abstract eyes. To encourage the potential participant to take action and join the interactive session for the exercise procedure, the virtual human greets with a friendly wave of the hand while the social robot performs a potentially similar associated gesture through arm movement and tilting/turning the mobile phone. To provide non-verbal feedback during the joint exercise execution, the virtual human simulates a squat, while the social robot moves its arm (and thus the attached mobile phone) up and down. Thus, the participant receives feedback if the squat was detected and executed correctly. In active conversation, the virtual human's lip movements simulate talking, while the social robot displays an audio sine wave on its cell phone. Smiling is depicted as a lip smile for the virtual human and a "happy" facial expression using its eyes and mouth for the social robot. In cheerful states, the virtual human shows a cheering/dancing animation, and the social robot engages in a side-to-side bouncing movement. In this way, the two representations lead to unique manifestations of identical behaviors.

3.4   Speech

To give the agents speech capabilities, we used the internal Windows speech synthesis API (SAPI[7]), which transcribes the assigned text content into an audio file using text-to-speech. In the embodied agent variant, we extended the speech synthesis with the use of the Unity Plugin uLipSync[8]. The playback of the audio file and the lip visemes for the pronounced vowels activate the required blend shapes in sync. Due to technical constraints for the social robot the text content to be used has been generated as an audio file with speech synthesis, so that the same voice, pronunciation, speed and pitch are used in both variants. The only difference is that in one case it is played back for each of the different states, and in the other it is transcribed and synchronized in real time. In addition, the pronunciation of the social robot is supported by the simulation of audio signal waves on the display. For each call and situation, the agent has a number of different phrases and utterances to choose from randomly.

3.5   Behavior

The core of our human-agent interaction system is the business logic for agent behavior. The business logic was implemented as Node.js application (social robot), respectively python application (virtual human) using the same finite state machine (FSM) architecture. The FSM is connected to the perception unit (see Sect. 3.1) via a local network and switches internal states based on

---

[7] The Speech Application Programming Interface or SAPI is an API developed by Microsoft to allow the use of speech recognition and speech synthesis within Windows applications.

[8] https://github.com/hecomi/uLipSync.

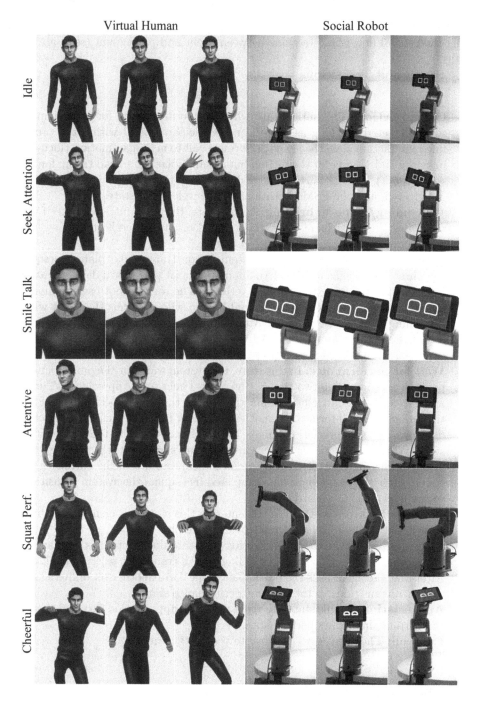

Fig. 2. Non-verbal cues of virtual human and social robot. Depicted are the social cues that describe the generic non-verbal behavior and how it is interpreted/performed by the two entities virtual human and social robot.

detections. The proposed state machine consists of eight main states representing the general flow of interactions between agent and participant (see Fig. 3).

States and Transitions

- **Call to action:** The initial state where the system waits for a participant to be detected. As long as there is no participant found within the predefined interaction zone the agent performs a call to action animation (focusing passerby and waving for the virtual human and focusing a head tilting for the social robot). Once a participant is detected by the perceptive system, they are greeted, and the FSM transitions to the welcoming state.
- **Welcoming/General briefing:** In this state, the agent introduces itself and motivates the participant to start an exercise. After the short briefing, the FSM moves to the briefing for exercise state.
- **Briefing for exercise:** Here, the system demonstrates and explains the squat exercises. This means for the virtual human that it performs a demonstration squat while explaining the exercise verbally. The social robot uses its mobile phone to display a video of the virtual human who performs the squat while explaining the exercise verbally. Afterwards, the participant is asked to perform five squats as an exercise. A five-second countdown starts and the FSM transitions to the "Wait for performance" state.
- **Wait for performance:** In this state, the system waits for the participant to perform a squat while two timers, the Wait-Timer and Appreciation-Timer, are started. If the participant is inactive during the 5-s Wait-Timer, the FSM transitions to the intermediate motivation state. If the participant is inactive during the 15-s Appreciation-Timer, the FSM transitions to the appreciation state. Whenever a squat is detected, these timers are reset, the agent gives a visual animation feedback (squat performance) and utters the current squat count. After the participant has completed five squats the system transitions to the intermediate reward state.
- **Intermediate motivation:** When entering this state the agent holds a randomly (out of three) selected motivational speech.
- **Intermediate reward:** When entering this state the agent plays a rewarding animation (fixed) and holds a rewarding speech (randomly selected of three). After that the agent asks the participant to perform another set of five squats and then transitions to the wait for performance state.
- **Appreciation:** In this state the agent utters an appreciation text and thanks the participants for its performance.
- **Farewell:** The agent says goodbye to the participant.

4 Method

For our research we propose a method that aims to generally study passersby's perceptions, behaviors, and interactions with the virtual human and social robot

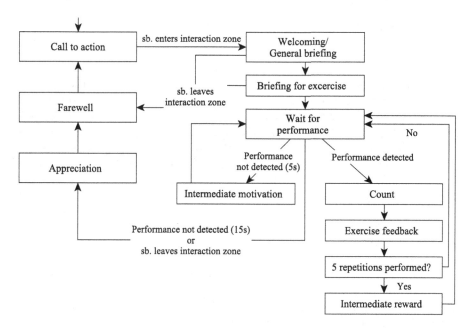

Fig. 3. System State Machine/Agent Behavioral Pattern. The agent initiates engagement by welcoming participants entering the interaction zone, guides squat exercises with feedback, offers motivation if squats aren't done, rewards every five repetitions, appreciates those who finish exercising and farewells those who leave.

in public space using the above stated human-agent interaction system. We aim to achieve this by combining quantitative performance metrics based on system tracking data with qualitative insights from questionnaires and behavioral observations.

4.1 Experiment Setup

The two installations have been placed side by side. The interaction zone was set at a distance of 1.3 m from the embodied agent display and has a radius of 0.5 m. The interaction zone for the social robot was centered around it with a radius of 2 m. These dimensions were determined to reduce distraction or interference when using the installation simultaneously in a 5 × 5 m room. Meanwhile, the social robot was placed on the far right, facing the entrance. Next to it, to the left of the social robot, approximately in the middle of the left side, facing the robot, was the embodied agent installation. To provide orientation and visual feedback, the two interaction zones were marked on the floor. A lateral boundary was also marked in the form of an open triangle leading from the agent to the sweet spot of the interaction zone. In addition, 0.5 m diameter spaces were prepared to the right and centre of the entrance for participants to complete the questionnaires. These were also used for the final interviews at the end of the experiment.

4.2   Participants and Procedure

All participants in our study were passerby recruited ad hoc in public space at the "Effekte Festival" in Karlsruhe, Germany, a public event for universities and research institutions to present their scientific work. Passerby had the opportunity to engage with one of the showcased systems when approaching close enough to the social robot or virtual human. Upon their proximity, they were automatically registered as participants and the agent starts it's training procedure (see Fig. 3). Participants could interact with the systems for as long as they wished. The system reset when a participant left the interaction zone. In total, 450 times (214 social robot, 236 virtual human) passerby started an interaction with one of the distinct systems and remained in the tracking area long enough to listen fully to the exercise briefing. Noteworthy, our system did not identify individuals (e.g., through face identification), so the interaction count may not reflect the exact number of distinct individuals involved. From the 450 started interactions, 145 resulted in at least one squat execution that was detected by the system (78 social robot; 67 virtual human). After the interaction, participants were asked to fill out a post-experiment questionnaire. Overall, 66 participants completed the questionnaire. However, four participants did not fill out this correctly and for one participant the system logging was corrupted resulting in 61 cases for analysis (38 social robot, 23 virtual human). 41 participants identified themselves as male, 20 as female. The average age of the participants was 41 (13 youngest, 74 oldest).

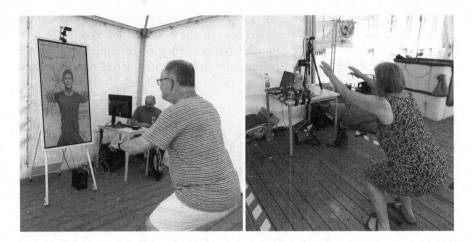

Fig. 4. Participants during Interaction. The participants perform a squat with the virtual human (left) and social robot (right).

4.3   Data Collection

For each participant, the system logged the interaction duration, the performed squat repetitions and the triggered number of rewards/motivations to evaluate performance numbers. Additionally, the participant's behavior was observed during participation for qualitative analysis. We created a questionnaire consisting of 17 questions (5 point Likert scale, 1=strongly disagree to 5=strongly agree) and assessed factors considering perception of the agent's **competence** (3 questions, "When using this agent I feel as a competent person", "If using this agent I feel effective", "I am convinced to be able to interact with this agent", $\alpha = 0.84$), **autonomy** (3 questions, "I can use my interactions freely when interaction with the agent", "I'm fully in control when interacting with the agent", "I can express my intentions when interacting with the agent", $\alpha = 0.74$), **rapport** (6 questions, "I perceive the agent as an independent person", "I feel attached to the agent", "I feel respected when using this agent", "It's fun to interact with the agent", "I will feel empathy with the agent", "I try to treat the agent as a human being", $\alpha = 0.79$) and **trust** (4 questions, "the agent is trustable", "I have a good feeling when relying on the agent", "I could trust information provided by the agent", "I trust this agent", $\alpha = 0.87$). Additionally, we asked "Do you feel under pressure while interaction with the agent?". The questionnaire is based on the Need Satisfaction Scale [7] and the Rapport-Expectation with a Robot Scale (RERS) [14].

5   Results

We calculated the arithmetic mean values for all data records and questionnaires. The respective results can be found in Table 1. A between-subjects one-way ANOVA was calculated to compare the perception of social robot and virtual human. There were no significant differences found between the virtual human and social robot for all factors assessed from the questionnaire nor system data (squat repetitions, number of rewards or motivations or interaction duration). However, the mean values for competence, autonomy, relationship, and trust are slightly higher (better) for the social robot. Moreover, we found significance with moderate effect size for single item ("I am convinced that I am able to interact with the assistant.", $F = 4.43$, $p = 0.04$, $\eta_p^2 = 0.07$). We also found a significant difference with large effect size between male and female gender in another item ("I feel pressured to behave in a certain way.", $F = 5.35$, $p = 0.007$, $\eta_p^2 = 0.15$). Eight participants performed squats in a way that was not recognized by our system. A between-subjects one-way ANOVA showed that rapport ($F = 3.36$, $p = 0.03$, $\eta_p^2 = 0.08$) and trust ($F = 5.67$, $p = 0.02$, $\eta_p^2 = 0.09$) were significantly lower rated in these cases.

6   Discussion

This research contribution investigated the public audience's preference for interaction partners, comparing virtual and physical embodiment. The study explored

Table 1. Results of data evaluation (virtual human vs. social robot). The first group shows results (arithmetic mean, standard deviation, maximum values) based on our system logging. The second group shows the evaluated results of the questionnaire (arithmetic mean, standard deviation).

	Virtual human ($N = 23$)			Social robot ($N = 38$)		
	$\bar{x}$	σ	max.	$\bar{x}$	σ	max.
Interaction Duration (min.)	1:40	0:43	4:33	1:36	0:35	3:50
Squat Repetitions	5.48	5.34	20	6.76	5.70	30
Motivation tiggers	2.43	1.99	6	2.79	1.65	7
Reward triggers	1.04	1.07	4	1.13	1.26	6
Competence	3.16	1.19	-	3.54	0.91	-
Autonomy	2.17	0.79	-	2.54	1.07	-
Rapport	2.67	0.84	-	2.93	0.87	-
Trust	2.98	1.17	-	3.32	0.96	-

how perceived competence, autonomy, rapport and trust in the agent varied based on its embodiment, while also measuring the strength of generated rapport and engagement into the exercise task.

Although no significant distinctions emerged between the virtual human and the social robot across the assessed questionnaire factors or system data (squat repetitions, rewards, motivations, and interaction duration etc.), we generally measured slightly higher mean values in terms of competence, autonomy, rapport and trust for the social robot. This slightly better rating could possibly be attributed to the participants' increased confidence in their ability to effectively engage with the agent. This observation would be supported by significant differences in responses to the question "I am convinced that I am able to interact with the assistant", which has a moderate effect size. The fact that participants perceived responsiveness differently when interacting with the agent could be due to the robot's movements taking place in the dimensions of the physical world and thus being more discernible in nuances. This may play a role especially in public spaces, where uncontrolled conditions such as incident stray light, reflections, etc. make it harder to perceive small changes on an albeit large display. However, it is also possible that the strong degree of abstraction of the social robot is decisive here. At this point, further investigations (among others with comparisons a virtual embodiment of the same social robot) are interesting and necessary. We found a gender difference in the item "I feel pressured to behave in a certain way" with a large effect size indicating that women feel significantly more pressured during the interaction. This could be caused by the circumstance that women report higher subjective stress levels and arousal of emotional experience in HCI tasks [9]. This however, requires careful further investigation.

Our agent soft- and hardware and the perceptive system proved to be reliable and robust during the whole experiment duration even under the uncontrolled conditions in public space. However, some technological limitations and problems with the local setup were identified. The ambient noise in the public space occasionally disrupted the exercise instructions, accentuated by the lack of acoustic isolation between the assistants, which ended up causing confusion among the participants when the interaction with the two assistants took place simultaneously. The lighting at the study site was not optimal for viewing the virtual human and changing over course of the day. Another technical problem caused the virtual human's cheering animation to not trigger regularly. The agents occasionally faced challenges in accurately detecting exercises, which were often influenced by the participants' squat execution, including factors like arm angles and squat depth. Additionally, instances of rapid squats occasionally led to detection hiccups, impacting the accuracy of counting. Furthermore, accurately identifying an individual's exercise among multiple people present within the detection area (camera field of view) occasionally posed a challenge for the perceptive system. Generally, instances where squats went unrecognized by the system (due to bad execution or sensor failure) or tracking was disturbed correlated with lower rapport and trust ratings, which is explainable by an increased frustration in these cases what could also be observed during the performance. Children preferred the social robot, displaying joyful behavior trying to figure out the boundaries and possibilities of the robot movements and body tracking. Unpleasantly, our system was neither trained nor calibrated to interact with persons below a certain body size. Thus, we cannot report reliable numbers considering actual interactions of children with the robot.

Overall, in approximately ⅓ (145 of 450) encounters in which passersby listened fully to the exercise briefing, the interaction was continued at least until the completion of the first squat. This shows a rather high acceptance rate for both systems, given the public space scenario [13]. It should be noted that the lack of face identification may cause the number of encounters to not reflect the exact number of distinct individuals involved. Participants interacted naturally with the agents, engaging in conversations, greetings, and waving. Initial movement detection problems frustrated some, causing them to leave early. Others, citing issues like heat or mobility, directly told the agents they wanted to stop. In interviews, participants consistently wanted agents to provide exercise feedback, especially for squatting techniques. Some were curious about the technology and expressed a desire for the agents to act as fitness instructors and homework assistants. It is important to note that the participants were not a random sample from the general population; rather, it is important to be aware that the individuals who participated most likely included many who had technical or scientific interests in one way or another due to the kind of event where they were recruited. There may also be bias in the sample because most participants were from Western cultures, particularly Germany. Interactions with agents were conducted in German, with observers translating for non-fluent participants. The questionnaire, solely in German, unintentionally acted as a participation criterion in this step. We also acknowledge biases in our study, particularly within qualitative observations. These biases, unintended and often subconscious, can inadvertently influence the observed phenomenon. Among the four observers, three were men, one was a woman and all were Western-educated, which leads to interpretations influenced by this worldview.

7   Conclusion

In this paper, we presented a simple interaction scenario for public spaces in which a virtual human and a mechanical looking social robot encouraged participants to perform squats using verbal and non-verbal cues. Our system tracked their training performance and provided various incentives and feedback to motivate users. In our study, we investigated how the agent's embodiment type influences participants' perceived competence, autonomy, trust, rapport, and interaction engagement. The results indicated a relatively high acceptance rate for both systems for a public space scenario. No significant differences emerged between the virtual human and the social robot across the assessed factors competence, autonomy, trust, rapport and system data. However, interactions with the social robot felt significantly more responsive to the participants. Notably, gender differences were observed in perceived pressure during interaction, with women reporting significantly higher levels. Our agent software, hardware, and perceptive system demonstrated reliability and robustness despite the challenges of an uncontrolled public environment. However, limitations related to technology, methodology, and sample bias were identified. Addressing the challenge of comparing vastly distinct embodiments (virtual and physical) demands innovative methodologies. Future research endeavors should encompass a repetition of the experiment under controlled conditions, effectively mitigating biases caused by the turbulent and unregulated public space environment. If discrepancies arise, it would be worthwhile to examine the extent to which these can be attributed to either verbal or non-verbal feedback. In addition, studies are needed that examine participants' understanding of the intended non-verbal cues of agents' embodiments, especially when these are not primarily human but rather resemble parts of animal bodily communication. Additionally, comparative investigations contrasting a virtual human with a virtual social robot hold promise for minimizing the influence of physical embodiment bias. Further research into how to generate easily interpretable artificial non-verbal signals for trivial physical embodiments, such as the one presented, is essential to ensure that this technology is suitable for future studies and applications. An increasing use of embodied agents in public space is quite conceivable and part of the zeitgeist. Depending on the area of application and requirements, both virtual and physical agent embodiments (be they humanoid or mechanical) are potentially useful. Therefore, further efforts should be made to understand the mode of action of the different types of agent embodiments and their respective communication capabilities, which have a crucial impact on natural interaction.

References

1. Aljaroodi, H.M., Adam, M.T., Chiong, R., Teubner, T., et al.: Avatars and embodied agents in experimental information systems research: a systematic review and conceptual framework. Australasian J. Inform. Syst. **23** (2019)
2. Baldassarri, D., Abascal, M.: Field experiments across the social sciences. Ann. Rev. Sociol. **43**, 41–73 (2017)
3. Cho, M., Lee, S.s., Lee, K.P.: Once a kind friend is now a thing: understanding how conversational agents at home are forgotten. In: Proceedings of the 2019 on Designing Interactive Systems Conference, pp. 1557–1569 (2019)
4. Dereshev, D., Kirk, D., Matsumura, K., Maeda, T.: Long-term value of social robots through the eyes of expert users. In: Proceedings of the 2019 CHI Conference on Human Factors in Computing Systems, pp. 1–12 (2019)
5. Hoy, M.B.: Alexa, siri, cortana, and more: an introduction to voice assistants. Med. Ref. Serv. Q. **37**(1), 81–88 (2018)
6. Jeong, S., Breazeal, C., Logan, D., Weinstock, P.: Huggable: the impact of embodiment on promoting socio-emotional interactions for young pediatric inpatients. In: Proceedings of the 2018 CHI Conference on Human Factors in Computing Systems, pp. 1–13 (2018)
7. La Guardia, J.G., Ryan, R.M., Couchman, C.E., Deci, E.L.: Within-person variation in security of attachment: a self-determination theory perspective on attachment, need fulfillment, and well-being. J. Pers. Soc. Psychol. **79**(3), 367 (2000)
8. Li, J.: The benefit of being physically present: A survey of experimental works comparing copresent robots, telepresent robots and virtual agents. Int. J. Hum Comput Stud. **77**, 23–37 (2015)
9. Liapis, A., Katsanos, C., Sotiropoulos, D., Xenos, M., Karousos, N.: Stress recognition in human-computer interaction using physiological and self-reported data: a study of gender differences. In: Proceedings of the 19th Panhellenic Conference on Informatics, pp. 323–328 (2015)
10. Ling, E.C., Tussyadiah, I., Tuomi, A., Stienmetz, J., Ioannou, A.: Factors influencing users' adoption and use of conversational agents: A systematic review. Psychol. Market. **38**(7), 1031–1051 (2021)
11. Luria, M., Reig, S., Tan, X.Z., Steinfeld, A., Forlizzi, J., Zimmerman, J.: Re-embodiment and co-embodiment: Exploration of social presence for robots and conversational agents. In: Proceedings of the 2019 on Designing Interactive Systems Conference, pp. 633–644 (2019)
12. Mayer, R.C., Davis, J.H., Schoorman, F.D.: An integrative model of organizational trust. Acad. Manag. Rev. **20**(3), 709–734 (1995)
13. Narumi, T., Yabe, H., Yoshida, S., Tanikawa, T., Hirose, M.: Encouraging people to interact with interactive systems in public spaces by managing lines of participants. In: Yamamoto, S. (ed.) HIMI 2016. LNCS, vol. 9735, pp. 290–299. Springer, Cham (2016). https://doi.org/10.1007/978-3-319-40397-7_28
14. Nomura, T., Kanda, T.: Rapport-expectation with a robot scale. Int. J. Soc. Robot. **8**, 21–30 (2016)
15. Purps, C.F., Janzer, S., Wölfel, M.: Reconstructing facial expressions of hmd users for avatars in vr. In: International Conference on ArtsIT, Interactivity and Game Creation. pp. 61–76. Springer (2021). https://doi.org/10.1007/978-3-030-95531-1_5
16. Rossi, S., Staffa, M., Tamburro, A.: Socially assistive robot for providing recommendations: comparing a humanoid robot with a mobile application. Int. J. Soc. Robot. **10**, 265–278 (2018)

17. Ryan, R., Deci, E.: Self-determination theory and the facilitation of intrinsic motivation, social development, and well-being. Am. Psychol. **55**, 68–78 (2000)

18. Schneider, S., Kummert, F.: Comparing the effects of social robots and virtual agents on exercising motivation. In: Ge, S.S., Cabibihan, J.-J., Salichs, M.A., Broadbent, E., He, H., Wagner, A.R., Castro-González, Á. (eds.) ICSR 2018. LNCS (LNAI), vol. 11357, pp. 451–461. Springer, Cham (2018). https://doi.org/10.1007/978-3-030-05204-1_44

19. Shinozawa, K., Naya, F., Yamato, J., Kogure, K.: Differences in effect of robot and screen agent recommendations on human decision-making. Int. J. Hum Comput Stud. **62**(2), 267–279 (2005)

20. Thellman, S., Silvervarg, A., Gulz, A., Ziemke, T.: Physical vs. virtual agent embodiment and effects on social interaction. In: Traum, D., Swartout, W., Khooshabeh, P., Kopp, S., Scherer, S., Leuski, A. (eds.) IVA 2016. LNCS (LNAI), vol. 10011, pp. 412–415. Springer, Cham (2016). https://doi.org/10.1007/978-3-319-47665-0_44

21. Turk, M.: Multimodal interaction: a review. Pattern Recogn. Lett. **36**, 189–195 (2014)

22. van der Werff, L., Legood, A., Buckley, F., Weibel, A., de Cremer, D.: Trust motivation: the self-regulatory processes underlying trust decisions. Organ. Psychol. Rev. **9**(2–3), 99–123 (2019)

23. Wölfel, M.: Kinetic Space - 3D Gestenerkennung für Dich und Mich. Konturen 32 (2012)

A Historical Perspective of the Biofeedback Art: Pioneering Artists and Contributions

Hosana Celeste Oliveira[✉] [iD]

Federal University of Pará, Belém, PA 66075-110, Brazil
hosanaceleste@ufpa.br

Abstract. The biofeedback interfaces were initially developed for medical reasons and revolutionized the way we represent the body and its functions due to the possibilities they brought to expand, conceptualize, investigate, and present body information through sounds and images. Biofeedback interfaces outside the medical context demand new interaction paradigms that radicalize the relationship between the body and the machine. The appropriation of the biofeedback technology by artists resulted in an artistic-scientific research program named biofeedback art or biosignal art, which involves, above all, transdisciplinary practices. Such a transdisciplinary aspect is explained: biofeedback art is inspired by applied techno-science research while contributing to and informing part of them since it deals with interoception, perception, and cognitive processes in general. Biofeedback art should not be seen only as a specialized technical operation field but also as an area of inventiveness and cultural criticism. The paper presents an overview of biofeedback art, focusing on its emergence in the 1960s and 1970s, and its transdisciplinary contributions. The historical research methodology and a narrative and systematic literature review guide the study.

Keywords: Biofeedback Art · Art and Science · Art and Neuroscience

1 Biofeedback Interfaces

This article presents a historical perspective of artistic practices with biofeedback, highlighting their emergence, pioneer artists, and their contributions. The methodology used to perform the study is hybrid and merges historical research methodology and narrative and systematic literature review. In future in-depth studies, considering the current results, this methodology will be improved to include the participation of women, both artists and scientists. The procedures used to survey, document, and categorize the mentioned artistic practices outline a historiography of art and technology converging with neuroscience. The official history of art and technology has rarely documented biofeedback art in its beginnings; therefore, the history of medicine, mainly linked to the development of medical devices, can provide a complementary source to deepen this research in the future. The primary sources to investigate biofeedback art were anthologies, proceedings, and forums on art, science and technology, human-computer interface

A. L. Brooks (Ed.): ArtsIT 2023, LNICST 565, pp. 359–371, 2024.
https://doi.org/10.1007/978-3-031-55312-7_26

design, and experimental electronic music; art catalogs and records from museums and electronic art festivals; and also papers, documents, and writings by the first generation of artists who dedicated themselves to the artistic practice with biofeedback. All these sources were considered through techniques of crossing information and snowballing. Because there is no exclusive denomination for "biofeedback art", several keywords as search criteria were used. The Table 1 presents the criteria summary for surveying the biofeedback art.

Table 1. General summary of the criteria for surveying the biofeedback art.

| Keywords | Biofeedback art, pioneers of biofeedback art, artistic practice with biofeedback, artistic practice with medical devices, enactive interface, biosensitive interface, affective interface, affective computing, physiological interface, embodied interaction interface, brain-computer interface, brain-controlled interfaces, neural interface, EEG interface, electroencephalographic signals in art, biosignal art, encephalographic creations, biosignal-driven art, cognitive computing and other keywords combined from these previously mentioned | Sources: electronic art festivals documents in general, such as Ars Eletrônica, Artech, #ART, SIIMI, FACTO; anthologies (such as published by Springer and Routledge), proceedings, forums and repositories on art, science and technology, human-computer interface design and experimental electronic music (such as LEONARDO) |

1.1 New Interaction Paradigms

"Biofeedback art" [1] or "biosignal art" [2], in the context of this paper, refers to incorporating biofeedback technology into artistic practices. Biofeedback is a technology initially developed for medical purposes and allows the integration of motor and physiological data from the user to a device. This integration inspired the development of a new interface genre that brought about a paradigm shift concerning conventional interfaces, designed since the 1970s predominantly from graphic references to conduct interaction [3, 4].

While in the traditional interface, interaction takes place through mediation based on symbolic knowledge (words, mathematical symbols, or other systems of symbols) or iconic (visual form of images, such as diagrams and illustrations, accompanied or not by verbal information), in the biofeedback interface the biological body is used and the "interaction" takes place via motor (natural and intuitive), physiological and organic responses (conscious and unconscious) constituted in the act of interaction, and required for performing tasks [3].

As an emerging and experimental media, biofeedback interfaces outside the medical context have few conventions to follow and allow free exploration precisely because of it. Many questions remain open, such as the type of language this interface supports and how to guide the user's attention. Because it bases on controlling psychophysiological states and dealing with the focus of attention, new cognitive skills are required [3]. In human-computer interaction, the biofeedback interface receives different names: enactive interface, biosensitive interface, affective interface, physiological interface, or embodied interaction interface. Regardless of the terminology, they all refer to a specific interface based on problems related to the body-mind relationship, which considers the biological body as a literal interface with the machine.

1.2 Art, Body and Biofeedback Interfaces

Throughout the history of art, we have observed different levels of involvement of the body, both the artist and the audience, with the artwork, even though the body as a whole is always required for the execution or fruition of the artwork, once there is no separation between thoughts, sensations, actions and the entire body itself [5]. However, art paradigms changed in the 20th century when the body came to be used as a canvas, brush, support, and platform [6]. With the advent of the use of computers in the art field, the relationship between body and artwork changed due to the possibilities of producing interactive artworks – the body becomes required and essential, and the public acquired a new status, which is "user" or "interactor" [5–7]. But with the use of biofeedback technologies in artistic practices, the body-artwork relationship radicalized, given the incorporation of the biological level of the body. In this case, there is no more discontinuity between the body and the artwork, making them interconnected via structural coupling [8]. A peculiar type of artwork begins to be produced that depends exclusively on the body for its existence and functioning. The body has no longer just the role of a user; it is also part, in the strict sense, of the machinery that builds the artwork.

2 Artistic Practices with Biofeedback

2.1 History: Pioneer Artists

Although biofeedback has been used in artistic practices for approximately six decades [9, 10], only some artists have explored this technology. The reason is that the manipulation of biosignals is complex, involving many techniques and knowledge from many areas, such as the medical, biological, and computational. Artworks employing biofeedback are mainly developed in the context of interdisciplinary research, and unfortunately, they are not well-known and documented and can not be easily found gathered in publications that provide an overview of them [1, 2].

Although the art nourished by the biological body goes back to the ideas of "meat machines" in cyberculture in the 1980s [11], the first experience with a biofeedback device documented dates back to the 1960s. The pioneer was the composer Alvin Lucier (USA, 1931–2021) when he used his alpha brain waves to create *Music for solo performer* (1964–1965), a soundscape presented at Brandeis University (Massachusetts, USA), and considered the first artistic experience with biofeedback [2, 12].

Also, in the 1960s, other artists pioneered biofeedback technologies, developing artworks that contributed to art and science by investigating correlations between human physiology and sound sensory stimuli. Like Lucier, the other precursors are also from the field of experimental electroacoustic music, collaborated with scientists, and pavimented biofeedback art as an artistic-scientific research program in the following decades.

Alvin Lucier. Alvin Lucier was one of the experimental electronic music forerunners and sound installations that explored acoustic phenomena, auditory perception, and physical properties of sound, and much of his work was influenced by science. In this regard, Lucier approached body sciences around 1960, when he began collaborating with physicist Edmond Dewan (USA), who was particularly interested in the interrelationship between music, nature, and science. Dewan was the first to have a human being controlling an external machine from his brain waves. At Cambridge's air force research laboratories, Lucier and Dewan developed a brain wave reading device (see Fig. 1) that gained notoriety and was used to create *Music for solo performer* [12]. Bart Lutters and Peter Koehler [13] highlight the importance of the device created by Lucier and Dewan and how it helped to improve research on electroencephalogram signals conversion into sounds. According to the authors, this technique has been used since the 1930s, but it was only in the 1960s that it became "popular", becoming a tool for use in the field of experimental music.

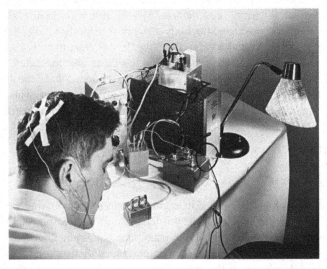

Fig. 1. Edmond Dewan and the acoustic wave control system developed in 1964. Source: Lutters e Koehler (2016, p. 3).

In *Music for solo performer* (see Fig. 2), Lucier sits alone in a chair in the stage center with EEG electrodes tied to his head. The electrodes explore the rhythmic cerebral modulations of alpha bands: "[…] through direct audification with the addition of percussion instruments: cymbals, drums, and gongs that coupled to large speakers; […] bursts of alpha activity cause the loudspeakers to excite the acoustic instruments which, in turn, activate a disembodied percussion ensemble" [2]. Alpha rhythms, used by Lucier, are brain waves with a frequency between 8 and 13 Hz related to relaxed awareness, lack of focus of specific attention, and a Zen state of relaxation and awareness. The similarity and connection between the mental state achieved during Zen meditation and that required to produce waves of alpha frequencies was established in the early stages of biofeedback research [10, 14].

Fig. 2. *Music for solo performer* (1965 version), by the composer Alvin Lucier. Source: University of Calgary (http://syneme.ucalgary.ca/tiki/tiki-view_blog_post.php?postId=584).

Thomas B. Holmes [12] observes that, except for Lucier's facial expressions and the opening and closing of his eyes, there was no visible index between the performer and the concert of sounds. However, he notes that the alpha waves became more intense when Lucier closed his eyes or became rarer when he opened them – the intensity of the sounds in the speakers could distinguish such observation. Closing the eyes helped to reach the relaxation state and intensify the production of alpha waves in the brain, and vice versa. Lucien's self-control is guided by sound and, at the same time, by his mental life (see Fig. 3). The artist proposes a correlation for the work as following [12]: increased relaxed awareness/decreased attentional focus/increased relaxation state - increased alpha wave intensity - increased sound loudness (persistent tinnitus).

Fig. 3. *Music for solo performer detail* (1965 version), by the composer Alvin Lucier. Source: University of Calgary (http://syneme.ucalgary.ca/tiki/tiki-view_blog_post.php?postId=584).

The Lucier's work conception evokes two themes, brain waves and resonance since the central idea of his work was not just to generate sounds through the amplification of brain waves but to set the vibrating surfaces of the speakers in contact with percussion instruments that, in turn, would create sounds of their own. Drums, gongs, and other small objects were placed under, on top of, or against the speakers [12].

David Rosenboom. David Rosenboom (USA, 1947), another pioneering artist of the 1960s, was a composer-scientist and director of the Laboratory of Experimental Aesthetics at the University of York and the Electronic Music Studios, in Ontario, in addition to having produced several works using biofeedback, also organized *Biofeedback and the arts: results of early experiments Aesthetic Research Center* (1976), which presents the research carried out during the years 1966–1974 by Rosenboom and colleagues, among them Lloyd Gilden and Richard Teitelbaum. According to Thom Blum [10], the book is relevant for the arts, especially for experimental music, because it exposes the concepts, techniques, scientific methods, and research and production processes of the biofeedback art pioneering artists, besides elucidating the thinking that guided their production.

Rosenboom's early experiments used electroencephalographic, electrocardiographic, and respiratory activity measurements to control sound generation modules. The work *Ecology of the skin* (1970–1971) (see Fig. 4) allowed up to ten participants to use electrodes that detect alpha brain waves in connection with a digital logic module. The device designed by Rosenboom for *Ecology of the skin* (see Fig. 5) allowed tracking the amount of time it took the participant to generate alpha waves – the more time he spent producing alpha waves, the better his control over the synthesis modules of sound connected to the circuit. To maintain the alpha state, the participant had to get positive feedback in the form of greater control over the resulting sound output. Usually, Rosenboom's works requested two or more participants who received visual or aural feedback when they managed to generate alpha waves synchronously – the idea here was to provide, through musical improvisation, a sort of collective behavior [10].

Fig. 4. David Rosenboom and colleagues preparing for the performance *Ecology of the Skin* (1970). Source: http://davidrosenboom.com.

Rosenboom's work offers a unique insight into the blending of Eastern disciplines related to spirituality and meditation practice with Western research methods and scientific physiological monitoring. This combination of knowledge, according to Blum, is the thinking foundation of artists at that time.

Richard Teitelbaum. American composer Richard Teitelbaum (USA, 1939), also a pioneer in biofeedback art, created *Spacecraft* (1967) using biosignals from his body and other participants. Teitelbaum built the artwork with electroencephalography, electrocardiography electrodes, and contact microphones with amplifiers to capture sounds from the heart, thoracic cavities, and breathing, all used as musical biological material

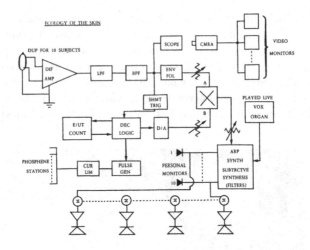

Ecology of the Skin (1970)

Fig. 5. Biofeedback system setup diagram. The electronic setup for this artwork included the ability to adjust the degree of brainwave control over the sound, for each of the 10 participants, according to a simple statistical measure, the amount of time spent per minute producing alpha waves [1].

in real time. The experience combined the participants' conscious musical actions integrated with Teitelbaum's biological actions. In this work, the biosignals operated in an automatic loop system or manually adjusted in the output control for a Moog synthesizer (see Fig. 6).

Fig. 6. Richard Teitelbaum, avant-garde figure and synthesizer pioneer, working with a Moog. Source: https://ra.co/news/72420.

Later works by Teitelbaum, such as the *Alpha bean lima brain* (1972) and *Tai chi alpha tala* (1974), were similarly controlled by brain waves but extended the experience beyond sonification and included "visual display such as strobe lights, sculptural objects, Paik-Abe Video Synthesizer, and the Dan Sandin Image Processor to produce pieces that extend to other media such as martial arts, telecommunications and television" [15]. *Alpha bean lime brain* includes an unusual manipulation of brain waves. These were telephoned from California to New York to make a pot of beans jump during Charlotte Moorman's Avant Garde Festival.

Lloyd Gilden. Lloyd Gilden (USA), another pioneering artist and experimental psychologist, worked with biofeedback, music, and qualitative research in the 1960s and 1970s. Gilden used a questionnaire with questions about the feelings aroused in the participants during the experience with alpha wave measurement electrodes, something

Fig. 7. *Heart beating recording and light* (1970), by Teresa Burga.

very similar to current neuropsychology tests. The artist took information from the questionnaire to remodel the work or theorize about it. Recurring statements were usually associated with feelings of "letting go", "freeing yourself", "avoiding becoming emotionally involved in the production of feedback signals", "blank mind", "not repeating refrains of some activity desired/unwanted", "serenity" and "peace" [10].

Teresa Burga. It is essential to observe that biofeedback art was, for many years, a male artistic practice, a situation recently changed. However, there is Teresa Burga (Peru, 1935–2021), one of the precursors of Latin American art technology who created works in different supports focused on the female body.

She created *Heart beating recording and light* (1970) (see Fig. 7), which was later part of a session in the installation *Self-portrait* (1972) – the artist's most complex work that documented her body through drawings, photographs, medical records, and visual and auditory data, such as the artist's heartbeat recorded for one day. The installation was divided into three parts: "face reports," "heart reports," and "blood reports" and aimed to demonstrate the standardization of the subject in graphic and numerical formats, also a critique of medical images. For Burga, using the heartbeat as a self-portrait is a way of escaping the aesthetic restrictions of art and pointing out another way of relating human beings, information, and technology and suggesting how this type of representation alters the strategies of control. Burga's biofeedback artwork completely differs from pioneering American artists closely linked to experimental music [16].

3   Conclusion

The pioneers in the biofeedback art have something in common, except for Burga: their works are the result of collaboration with scientists and mainly explore the biosignals sonification in the context of electroacoustic music, which uses a device, usually a loudspeaker, that transduces electrical energy to acoustic energy [12]. The pioneering artists' preference for this type of production is explained by Rosenboom [9] who comments that music is a privileged resource to deal with paradigms of apprehension by offering freedom "to work, interact and explore interrelationships within speculative realities" (in this case, to explore the subject's psychophysiology). This forwarding given by pioneering artists to their production will nourish the art practices that followed in the 1990s associated with computational technology.

Still, regarding the correlation between biosignals and music using biofeedback, Rosenboom says, "[…] Events in this psychophysical environment can become symbolic of almost everything that the mind can conceive. Acoustic waves remain relatively abstract through the primary processing stages of auditory perception in the nervous system. Acoustic feature detectors in the auditory nervous system closely resemble sound's physical nature. Consequently, music is an ideal medium for exploring how we hear, process, store, associate, and retrieve sonic images. Within music, it is relatively easy to identify levels of abstraction and explore the processes through which we apprehend and assign meaning to psychophysical events" [9]. Nevertheless, the author notes that although music uses acoustic waves and auditory perception to create psychophysical models, we can explore other possibilities.

Experiments with brain waves and sounds by pioneering artists showed that fruition in artworks produced with biofeedback requires skill (to reach and maintain alpha states, for example), training, and agility. The literature on biocontrol techniques argues that it can be achieved with discipline and practice. However, it is known that the alpha stage is easily disturbed by minor visual and mental distractions and minimal motor effort [2]. As such, the experience with biofeedback devices can be frustrating for some. Being aware of this problem, Richard Teitelbaum [10] even trained the participants of his performances in body awareness disciplines such as yoga and Zen meditation, aiming to improve their skills during the performances.

According to Blum [10], artistic practice with music/sounds and biofeedback was discontinued by the end of the 1970s, but it revived in the 1990s with the emergence of digital technologies. The tension of the Vietnam War in the 1970s, according to this author, reflected a lot on the biofeedback research scene, undermining psychedelia movements and the hope of social application of this type of technology. The following cultural scenario created a different thought dominated by science fiction fantasies and the cyberpunk universe, Blum points out. The artists abandoned using bio-electrodes but carried with them the experiences they had acquired with their use. Some of them even, decades later, resumed their use, as is the case of David Rosenboom, with the interactive works *On being invisible II* (*Hypatia Speaks to Jefferson in a Dream*) (1994), *Ringing minds* (2014) and *Portable gold and philosophers' stones* (*Deviant resonances*) (2015).

The fascination with neural network representations of the brain in the 1960s and early 1970s differs from current motivations, partly because of the political and historical context in which biofeedback technologies emerged and were used. In the 1960s, research in this field and its applications were funded by the United States Department of Defense and did not have the current commercial and entertainment aspect [10]. For Rosenboom [10], at that time, the application of biofeedback technologies in understanding the neurological, psychological, and creative processes of the human being should follow an orientation with social purposes and be used, above all, to improve the "ability to experiment with and bring conscious self-control to the hitherto unconscious neural processes on which mental life is founded."

While pioneers prioritized music to create psychophysiological representations, other artists later invested in different languages and material elements facilitated by the computational medium. Particularly Stelarc, a very emblematic artist, adopts a series of resources beyond the musical ones, although these are present in his works, to create psychophysical models of himself and, like the Brazilian artist Tânia Fraga, both produced biofeedback systems that include robotics elements; on the other hand, Pia Tikka is inspired by the cinematographic language to create what she claims enactive cinema. In general, the musical language is still present in contemporary production, such as works created by artists such as Stelarc, Laura Guerra, Lisa Park, among others. Interest in electroencephalography by artists continues, and it has been widely used to develop audiovisual pieces and artistic practice with emotion induction procedures. In the last ten years, thanks to the popularization of biosensors, which have become economically accessible and easier to integrate into computers, work with biofeedback devices in the art field has increased.

In a few words, biofeedback art is the result of the transdisciplinarity between art and neuroscience and finds parallels with the history of interoception, the body-mind relationship research, and the techniques of detection, evaluation, auscultation, and body percussion in medicine. It is a field of artistic-scientific research in which artists explore new ways of combining technology and the human body physiology that lead us to reflect on our personal experience and physiological and cognitive-emotional responses. There is the hypothesis that biofeedback art is relevant for the embodied mind research program because it introduces complementary arguments for the scientific models of the mind by subverting medical technologies and exploring aspects outside the usual framework of their use. When associated with sounds, images, robotic elements, or cinematic language, biofeedback art throws light and reinforces concepts or philosophical matters of the embodied mind thesis. In this sense, art plays a role in converting concepts into visual and sound forms by creating poetics that connect body, mind, and environment. Furthermore, biofeedback art might be a source of alternative data visualization and a basis for interdisciplinary research on the history of ideas, philosophical psychologies and medical, and visual culture.

The artworks produced by the pioneering artists or those who came later, regardless of the biofeedback poetic implementation, can be discussed employing notions such as systems, competent stimuli, affordances, and self to foster dialogues with neuroscience. As a type of process art, biofeedback art exposes the dynamics between the mind and body and how environmental stimuli influence both; the artwork gains competence throughout the fruition process, and emotionally efficient stimuli cause physiological changes in the interactor's body and alter themselves by these body changes. Because it only exists in an interdependent relationship between the body and the externalization modes of the mind made possible by the biofeedback device, the biofeedback art is a radically embodied system (Anthony Chemero). The structural coupling concept between the interactor/device and the environment (Tom Ziemke) explores a similar idea. As for the "self"-based approach, biofeedback art focuses on self-perception enhanced through competent stimuli or affordances (James Gibson) and enactions that induce the encounter with the self. One of the functions of biofeedback art is precisely this: activating the self through how the interactor coordinates the set of experiences constructed from the body's perceptual and proprioceptive signals induced by the artwork. This coordinated activity opens the involucre of the self. These conceptual approaches illustrate how one could investigate biofeedback art and inform the model we are developing to study it.

The research does not end here. The historical overview was built mainly from the literature available in the leading repositories of scholarly communication. Our study needs to be deepened in art, science, and technology archives and museums to verify women artists' participation in the construction of biofeedback art history. We may find other "Teresas Burgas" utterly invisible in official documents.

References

1. Oliveira, H.: Arte de biofeedback: uma proposta epistemológica para a compreensão da mente corporificada [doctoral dissertation]. Universidade Estadual Paulista Júlio de Mesquita Filho, São Paulo (2019)
2. Ortiz, M., Coghlan, N., Jaimovich, J., Knapp, B.: Biosignal-driven art: beyond biofeedback. Sonic Ideas **3**(2), 1–27 (2011)
3. Paraguai, L.: Interfaces multisensoriais: espacialidades híbridas do corpoespaço. Revista FAMECOS **15**(37), 54–60 (2009)
4. Aprile, W., et al.: Enaction and Enactive Interfaces: A Handbook of Terms. Enactive Systems Books, Grenoble (2023)
5. Sogabe, M.: O corpo do observador nas artes visuais. In: Anais do 16° Encontro Nacional da Associação Nacional de Pesquisadores de Artes Plásticas: Dinâmicas Epistemológicas em Artes Visuais, 1582–1588. ANPAP, Florianópolis (2007)
6. Warr, T., Jones, A.: The Artist's Body. Phaïdon Press, London (2000)
7. Couchot, E.: A Tecnologia na Arte: Da Fotografia à Realidade Virtual. Editora da UFRGS, Porto Alegre (2003)
8. Varela, F., Thompson, E., Rosch, E.: Embodied Mind: Cognitive Science and Human Experience. MIT Press, Cambridge (2016)
9. Rosenboom, D.: Biofeedback and the Arts: Results of Early Experiments. Aesthetic Research Center, British Columbia (1976)
10. Blum, T.: Biofeedback and the arts - results of early experiments: a review. Computer Music **13**(4), 86–88 (1989)
11. Jones, C.: Sensorium: Embodied Experience, Technology, and Contemporary Art. MIT Press, Boston (2006)
12. Holmes, T.: Electronic and Experimental Music: Pioneers in Technology and Composition, 3rd edn. Routledge, New York (2008)
13. Lutters, B., Koehler, P.J.: Brainwaves in concert: the 20th century sonification of the electroencephalogram. Brain **139**(10), 2809–2814 (2016)
14. St. Louis, E., Frey, L.: Electroencephalography (EEG): An Introductory Text and Atlas of Normal and Abnormal Findings in Adults, Children, and Infants. American Epilepsy Society, Chicago (2016)
15. Teitelbaum, R.: Improvisation, computers and the unconscious mind. Contemp. Music Rev. **25**(5–6), 497–508 (2006)
16. Fajardo-Hill, C., Giunta, A.: Catálogo da Exposição Mulheres Radicais: Arte Latino-Americana 1965–1980: Curadoria e Textos. Pinacoteca de São Paulo, São Paulo (2018)

Author Index

A. L. Brooks (Ed.): ArtsIT 2023, LNICST 565, pp. 373–374, 2024.
https://doi.org/10.1007/978-3-031-55312-7

Printed in the United States
by Baker & Taylor Publisher Services